UNITEXT

La Matematica per il 3+2

Volume 184

The **UNITEXT - La Matematica per il 3+2** series is designed for undergraduate and graduate academic courses, and also includes books addressed to PhD students in mathematics, presented at a sufficiently general and advanced level so that the student or scholar interested in a more specific theme would get the necessary background to explore it.

Originally released in Italian, the series now publishes textbooks in English addressed to students in mathematics worldwide.

Some of the most successful books in the series have evolved through several editions, adapting to the evolution of teaching curricula.

Submissions must include at least 3 sample chapters, a table of contents, and a preface outlining the aims and scope of the book, how the book fits in with the current literature, and which courses the book is suitable for.

For any further information, please contact the Editor at Springer: Francesca Bonadei, francesca.bonadei@springer.com.

THE SERIES IS INDEXED IN SCOPUS.

Valter Moretti

Geometric Methods in Mathematical Physics I

Tensors, Special Relativity, Spinors

Springer

Valter Moretti
Department of Mathematics
University of Trento
Trento, Italy

ISSN 2038-5714 ISSN 2532-3318 (electronic)
UNITEXT
ISSN 2038-5722 ISSN 2038-5757 (electronic)
La Matematica per il 3+2
ISBN 978-3-032-31511-3 ISBN 978-3-032-31512-0 (eBook)
https://doi.org/10.1007/978-3-032-31512-0

This Springer imprint is published by the registered company Springer Nature Switzerland AG
The registered company address is: Gewerbestrasse 11, 6330 Cham, Switzerland

Contents

Chapter 1
Introduction

This introductory chapter explains the role of the volume within the broader course and fixes the basic algebraic conventions used later. Its purpose is not only to describe the contents of the book, but also to establish a common language for the subsequent chapters: vector bases, finite linear combinations, duality, tensor notation, and the relation between the algebraic formalism and its physical applications. In this way the reader can understand from the beginning why tensors will be treated both as intrinsic multilinear objects and as computational tools expressed through components.

The chapter also clarifies the connection with the following volume. The present text develops the algebraic and special-relativistic background; the second volume continues the same geometric viewpoint on smooth manifolds, tensor fields, connections, curvature, and General Relativity.

This book covers the first part of a graduate course in Modern Mathematical Physics at the University of Trento. The course has two versions, one geometric and the other analytic. This volume concerns only the geometric version of the course. The analytic version, regarding applications of linear functional analysis to quantum and relativistic quantum theories, is covered in my books [4, 5] and in the chapter [1].

The aim of this first part is to present a concise but rigorous introduction to the basic algebraic and geometric language of tensor calculus, with applications to mathematical and theoretical physics. In particular, this text aims to bridge the gap between two standard presentations of the subject. On the one hand, physicists usually employ tensors through the practical and efficient language of index notation and transformation laws. On the other hand, mathematicians often introduce tensors through multilinear maps, tensor products, universal properties, and coordinate-free constructions. Both viewpoints are essential. The former is indispensable in computations and in physical applications; the latter is necessary to understand the invariant meaning of the formalism. One of the main purposes of this book is to make these two languages communicate with each other.

V. Moretti, *Geometric Methods in Mathematical Physics I*, La Matematica per il 3+2 184, https://doi.org/10.1007/978-3-032-31512-0_1

The material starts with multilinear algebra, dual and conjugate spaces, tensor products, tensor algebra, abstract index notation, exterior algebra, scalar products, metric tensors, pseudo-tensors and tensor densities. These tools are then used in the geometric formulation of Special Relativity, in the study of the Lorentz and Poincaré groups, and in an introductory treatment of the relation between $SL(2, \mathbb{C})$ and $SO(1, 3)^{\uparrow}$, including Weyl and Dirac spinors. Several applications are presented as examples and exercises, in order to emphasize the role of tensors not only as algebraic objects, but also as the natural language of covariance, symmetry and physical invariance.

The exposition is logically organized in a sequential way, but the reader should not interpret this order as the only possible way of using the text. Readers mainly interested in the mathematical foundations may proceed linearly from the first chapters to the later applications. Readers mainly interested in physics may instead start directly from the chapters on Special Relativity, the Lorentz group, or spinors, and then return to the earlier mathematical chapters whenever a definition, a notation, or a structural result is needed. In particular, the chapters on tensor products, abstract index notation, scalar products and metric tools may be used as a reference for the subsequent physical sections. This non-linear use of the text is intentionally possible: the mathematical material is developed in sufficient detail so that it can support both a systematic study and a more application-driven reading.

This first volume is naturally connected with the second volume, where the geometric viewpoint is continued from tensor algebra and Special Relativity to smooth manifolds, tensor fields, connections, curvature, Lorentzian geometry and General Relativity. The present volume therefore provides the algebraic and relativistic background for the later differential-geometric formulation.

The reader is expected to be familiar with standard notions of linear algebra [3, 6], especially those concerning finite-dimensional vector spaces. From the end of Chap. 7 onward, some basic tools from Lie group theory and Lie group representation theory [2] are required. The definition of the Hilbert tensor product given at the end of Chap. 2 should be regarded as complementary material and requires that the reader be familiar with elementary notions of Hilbert spaces.

Some notions and results used several times throughout the text are recalled here. First of all, we stress that the notion of *linear independence* always refers to *finite* linear combinations. A possibly infinite subset $A \subset V$ of vectors of the vector space V over the field $\mathbb{K} = \mathbb{C}$ or $\mathbb{R}$ is said to consist of **linearly independent** elements if, for every *finite* subset $A' \subset A$,

$$\sum_{v \in A'} c_v v = 0 \quad \text{for some } c_v \in \mathbb{K},$$

entails

$$c_v = 0 \text{ for all } v \in A'.$$

Clearly, such an A cannot contain the zero vector.

Recall that, as a consequence of *Zorn's lemma* and as proved below, every vector space V admits a **vector basis**, that is, a *possibly infinite* set $B \subset V$ made of *linearly independent* elements such that every $v \in V$ can be written as

$$v = \sum_{i=1}^{n} c_i v_i$$

for some *finite* set of distinct elements $v_1, \ldots, v_n \in B$ depending on v and a corresponding set of not necessarily pairwise distinct scalars $c_1, \ldots, c_n \in \mathbb{K} \setminus \{0\}$ depending on v. The elements $v_1, \ldots, v_n$ and the coefficients $c_1, \ldots, c_n$ are uniquely determined by v itself in view of the linear independence of the elements of B. Indeed, if for some distinct elements $v'_1, \ldots, v'_m \in B$ and some not necessarily pairwise distinct scalars $c'_1, \ldots, c'_m \in \mathbb{K} \setminus \{0\}$ it again holds

$$v = \sum_{i=1}^{m} c'_i v'_i ,$$

then

$$0 = v - v = \sum_{i=1}^{n} c_i v_i - \sum_{j=1}^{m} c'_j v'_j .$$

In view of the linear independence of the finite set $\{v_1, \ldots, v_n, v'_1, \ldots, v'_m\}$, whose primed and non-primed elements are not necessarily distinct, and observing that $c_i \neq 0$ and $c'_j \neq 0$, the above identity implies that $n = m$, the set of the v_i's coincides with the set of the v'_j's, and the corresponding coefficients c_i and c'_j are equal.

Theorem 1.1 *Let V be a vector space over the field $\mathbb{K} = \mathbb{R}$ or $\mathbb{C}$. There is a vector basis $B \subset V$. Furthermore, if $v_1, \ldots, v_n \in V$ are linearly independent, there is a vector basis $B \subset V$ such that $B \ni v_1, \ldots, v_n$.*

Proof Consider the class $\mathscr{B}$ of all possible sets $L \subset V$ of linearly independent elements of V. Define on $\mathscr{B}$ the partial ordering relation given by set-theoretic inclusion $\subset$. If $\mathscr{T} \subset \mathscr{B}$ is totally ordered, then $\cup_{L \in \mathscr{T}} L \in \mathscr{B}$ is an upper bound for $\mathscr{T}$. Therefore Zorn's lemma entails that $\mathscr{B}$ admits maximal elements. Let $B \in \mathscr{B}$ be such a maximal element. Since B is maximal, if $u \in V \setminus \{0\}$ the set $\{u\} \cup B$ cannot consist of linearly independent elements. As B itself is made of linearly independent elements, this means that there are finitely many elements $u_1, \ldots, u_m \in B$ such that

$$cu + \sum_{k=1}^{m} c_k u_k = 0$$

for some $c \neq 0$ and some $c_k \in \mathbb{K}$. Hence every $u \in V$ can be written as a finite linear combination of elements of B, and thus B is a vector basis of V.

Regarding the second statement, it is sufficient to redefine $\mathscr{B}$ as the class of all possible sets $L \subset V$ of linearly independent elements of V which include $v_1, \ldots, v_n$. The same application of Zorn's lemma gives a maximal element B of this class, and such a B is necessarily a vector basis of V containing $v_1, \ldots, v_n$. $\square$

An initial draft of this text was generated with the assistance of an AI-based language tool. The author has subsequently revised, verified, and refined the content in its entirety.

References

1. Khavkine, I., Moretti, V.: Algebraic QFT in curved spacetime and quasifree hadamard states: an introduction. In: Brunetti, R., Dappiaggi, C., Fredenhagen, K., Yngvason, J. (eds.) Advances in Algebraic Quantum Field Theory, chap. 5. Springer, Berlin (2015)
2. Kirillov, A.: Éléments de la Théorie de Représentations. Mir, Mosca, 1974; M.A. Najmark and A.I. Stern, Theory of group representations, Mir, Moscow (1984)
3. Lang, S.: Linear Algebra. Undergraduate Texts in Mathematics, 3rd edn. Springer, Berlin (1987)
4. Moretti, V.: Spectral Theory and Quantum Mechanics: Mathematical Foundations of Quantum Theories, Symmetries and Introduction to the Algebraic Formulation, 2nd edn. Springer, Berlin (2017)
5. Moretti, V.: Fundamental Mathematical Structures of Quantum Theory Spectral Theory: Foundational Issues, Symmetries, Algebraic Formulation, Quantum Measurement Theory, 2nd revised and enlarged edn. Springer, Berlin (2026)
6. Sernesi, E.: Geometria I and II. Bollati Boringhieri, Torino (1990)

Chapter 2
Multilinear Maps and Tensors

This chapter contains the algebraic foundations on which the rest of the book is built. Starting from vector spaces and linear maps, it introduces dual spaces, conjugate spaces, pairings, adjoint maps, multilinear maps, and tensor products. The aim is to make precise the idea that tensors are not merely arrays of components, but intrinsic objects whose components depend on a choice of basis.

A central point of the chapter is the comparison between two complementary viewpoints. On the one hand, tensors can be represented as multilinear maps; on the other hand, they can be constructed abstractly by means of tensor products and their universal property. This double description is essential in mathematical physics, where one constantly moves between invariant statements and coordinate computations. The final part, devoted to the Hilbert tensor product, is included as supplementary material connecting the finite-dimensional theory with structures used in quantum theory.

2.1 Dual Space and Conjugate Space, Pairing, Adjoint Operator

As a first step, we introduce the dual space and the conjugate space of a given vector space.

Definition 2.1 (Dual Space, Conjugate Dual Space and Conjugate Space) Let V be a vector space over the field either $\mathbb{K} = \mathbb{R}$ or $\mathbb{C}$.

(a) The **dual space** of V, V^*, is the vector space of linear functionals on V, i.e., the linear maps $f : V \to \mathbb{K}$.

(b) If $\mathbb{K} = \mathbb{C}$, the **conjugate dual space** of V, $\overline{V^*}$, is the vector space of anti-linear functionals on V, i.e., the antilinear maps $g : V \to \mathbb{C}$. Finally, the **conjugate space** of V, $\overline{V}$, is the space $(\overline{V^*})^*$. ■

V. Moretti, *Geometric Methods in Mathematical Physics I*, La Matematica per il 3+2 184, https://doi.org/10.1007/978-3-032-31512-0_2

Remarks 2.2

(1) If V and V' are vector spaces over $\mathbb{C}$, a map $f : V \to V'$ is called *anti-linear* or *conjugate-linear* if it satisfies

$$f(\alpha u + \beta v) = \overline{\alpha} f(u) + \overline{\beta} f(v)$$

for all $u, v \in V$ and $\alpha, \beta \in \mathbb{C}$, $\overline{\lambda}$ denoting the complex conjugate of $\lambda \in \mathbb{C}$. If $V' = \mathbb{C}$ the given definition reduces to the definition of an *anti-linear functional.*

(2) V^*, $\overline{V^*}$ and $\overline{V}$ turn out to be vector spaces over the field $\mathbb{K}$ when the sum of vectors and the product by elements of the field are defined in the usual way. For instance, if $f, g \in V^*$ or $\overline{V^*}$, and $\alpha \in \mathbb{K}$ then $f + g$ and αf are functions such that:

$$(f + g)(u) := f(u) + g(u)$$

and

$$(\alpha f)(u) := \alpha f(u)$$

for all $u \in V$.

In Definition 2.3, we do not explicitly assume that V is finite-dimensional.

Definition 2.3 (Dual Basis, Conjugate Dual Basis and Conjugate Basis) Let V be a vector space over either $\mathbb{K} = \mathbb{R}$ or $\mathbb{C}$. Let $\{e_i\}_{i \in I}$ be a vector basis of V. The set $\{e^{*j}\}_{j \in I} \subset V^*$ whose elements are defined by

$$e^{*j}(e_i) := \delta_i^j$$

for all $i, j \in I$ is called the **dual basis** of $\{e_i\}_{i \in I}$.
Similarly, if $\mathbb{K} = \mathbb{C}$, the set of elements $\{\overline{e^{*j}}\}_{j \in I} \subset \overline{V^*}$ defined by:

$$\overline{e^{*j}}(e_i) := \delta_i^j$$

for all $i, j \in I$ is called the **conjugate dual basis** of $\{e_i\}_{i \in I}$.
Finally, the set of elements $\{\overline{e}_j\}_{j \in I} \subset \overline{V}$ defined by:

$$\overline{e}_p(\overline{e^{*q}}) := \delta_p^q$$

for all $p, q \in I$ is called the **conjugate basis** of $\{e_i\}_{i \in I}$. ■

Remarks 2.4

(1) One may wonder whether or not the *dual bases* (and the conjugate bases) are *proper* vector bases of the respective vector spaces. The following theorem gives a positive answer in the finite-dimensional case. This is not true, in general, for infinite-dimensional spaces V. We shall not try to improve the definition of dual basis in the general case, since we are interested only in algebraic features and the infinite-dimensional case should be approached by convenient topological tools, which are quite far from the goals of this introductory text.

(2) In *spinor* theory—employed, in particular, to describe the *spin* of quantum particles—there is a complex two-dimensional space V called the *space of Weyl spinors*. It is the representation space of the group $SL(2, \mathbb{C})$ viewed as a double-valued representation of the orthochronous proper Lorentz group.

Referring to a basis in V and the associated bases in the spaces V^*, $\overline{V}$ and $\overline{V}^*$, the following notation is often used in physics textbooks [4, 5]. The components of spinors, i.e. vectors in V, are denoted by ξ^A. The components of dual spinors, that is, vectors of V^*, are denoted by ξ_A. The components of conjugate spinors, that is, vectors of $\overline{V}$, are denoted either by $\xi^{A'}$ or, using the dotted-index notation, by $\xi^{\dot{A}}$. The components of dual conjugate spinors, that is, vectors of $\overline{V}^*$, are denoted either by $\xi_{A'}$ or by $\xi_{\dot{A}}$. ■

Theorem 2.5 *If $dim V < \infty$, then, concerning Definition 2.3, the dual basis, the conjugate dual basis and the conjugate basis of a basis $\{e_i\}_{i\in I} \subset V$ are proper vector bases of V^*, $\overline{V^*}$ and $\overline{V}$ respectively. As a consequence, $dim V = dim V^* = dim \overline{V^*} = dim \overline{V}$.*

Proof Consider the dual basis in V^*, $\{e^{*j}\}_{j\in I}$. We have to show that the functionals $e^{*j} : V \to \mathbb{K}$ are generators of V^* and are linearly independent.
(Generators.) Let us show that, if $f : V \mapsto \mathbb{K}$ is linear, then there are numbers $c_j \in \mathbb{K}$ such that $f = \sum_{j\in I} c_j e^{*j}$.
To this end define $f_j := f(e_j)$, $j \in I$, then we argue that $f = f'$ where $f' := \sum_{j\in I} f_j e^{*j}$. Indeed, any $v \in V$ may be decomposed as $v = \sum_{i\in I} v^i e_i$ and, by linearity, we have:

$$f'(v) = \sum_{j\in I} f_j e^{*j}(\sum_{i\in I} v^i e_i) = \sum_{i,j\in I} f_j v^i e^{*j}(e_i) = \sum_{i,j\in I} f_j v^i \delta_i^j$$

$$= \sum_{j\in I} v^j f_j = \sum_{j\in I} v^j f(e_j) = f(v)\,.$$

(Notice that above we have used the fact that one can extract the summation symbol from the argument of each e^{*j}. This is because the sum over the index i is *finite* by hypothesis, since $dim V < +\infty$.) Since $f'(v) = f(v)$ holds for all $v \in V$, we conclude that $f' = f$.

(Linear independence.) We have to show that if $\sum_{k\in I} c_k e^{*k} = 0$ then $c_k = 0$ for all $k \in I$.
To achieve that goal, notice that $\sum_{k\in I} c_k e^{*k} = 0$ means $\sum_{k\in I} c_k e^{*k}(v) = 0$ for all $v \in V$. Therefore, putting $v = e_i$ and using the definition of the dual basis, $\sum_{k\in I} c_k e^{*k}(e_i) = 0$ turns out to be equivalent to $c_k \delta_i^k = 0$, namely, $c_i = 0$. This result can be obtained for each $i \in I$ and thus $c_i = 0$ for all $i \in I$. The proof for the conjugate dual basis is completely analogous. The proof for the conjugate basis uses the fact that $\overline{V}$ is the dual space of $\overline{V^*}$ and thus the first part of the proof applies. The last statement of the thesis holds because, in the four considered cases, the set of indices I is the same. □

Remarks 2.6 Let $\{e_i\}_{i\in I}$ be a vector basis of the vector space V over the field $\mathbb{K}$ and consider the general case with I *infinite*. Each linear or anti-linear map $f : V \to \mathbb{K}$ is anyway completely determined by giving the values $f(e_i)$ for all $i \in I$. This is because, if $v \in V$ then $v = \sum_{i\in I_v} c^i e_i$ for some numbers $c^i \in \mathbb{K}$, $I_v \subset I$ being *finite*. Then the linearity of f yields $f(v) = \sum_{i\in I_v} c^i f(e_i)$. (This fact holds true no matter whether $dim V < +\infty$.) So, formally one may use the notation

$$f := \sum_{i\in I} f(e_i) e^{*i}$$

to expand any element $f \in V^*$ also if $\{e^{*i}\}_{i\in I}$ is not a vector basis of V^* in the proper sense (the linear combination in the right-hand side may be infinite). This can be done provided that one adopts the convention that, for $v = \sum_{k\in I} v^k e_k \in V$,

$$\sum_{i\in I} f(e_i) e^{*i} \sum_{j\in I} v^j e_j := \sum_{k\in I} f(e_k) v^k .$$

Notice that the last sum is always finite, so it makes sense. ■

Notation 2.7 From now on we take advantage of the following convention. Whenever an index appears twice, once as an upper index and once as a lower index, in whichever expression, the summation over the values of that index is understood in the notation. E.g.,

$$t_{ijkl} f^{rsil}$$

means

$$\sum_{i,l} t_{ijkl} f^{rsil} .$$

The range of the various indices is determined by the context or otherwise specified (Fig. 2.1). ■

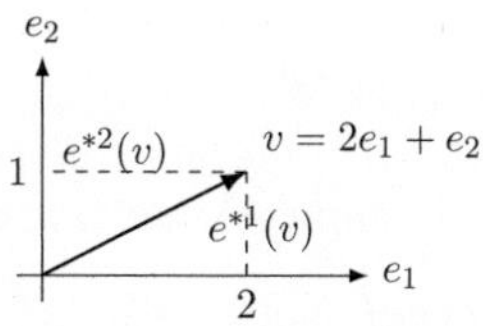

Fig. 2.1 The dual covectors e^{*1} and e^{*2} measure the components of v in the chosen basis

Examples 2.8 (Einstein Convention and Dual Basis in a Concrete Case) Let $V = \mathbb{R}^2$ and let $\{e_1, e_2\}$ be the standard basis, with dual basis $\{e^{*1}, e^{*2}\}$. If

$$v = 2e_1 + e_2 \quad \text{and} \quad f = 3e^{*1} - e^{*2} ,$$

then the pairing is

$$\langle v, f\rangle = f(v) = f_i v^i = 3 \cdot 2 + (-1) \cdot 1 = 5 .$$

Here the repeated index i means that the summation over $i = 1, 2$ is understood. Equivalently, e^{*1} and e^{*2} read off the two coordinates of v:

$$e^{*1}(v) = 2, \qquad e^{*2}(v) = 1 .$$

Notice that the numerical value of $\langle v, f\rangle$ does not depend on the chosen basis, even if the two lists of components v^i and f_i do. ■

We are naturally led to consider the following issue. It could seem that the definition of dual space, V^* (of a vector space V), may be implemented on V^* itself, producing the double dual space $(V^*)^*$ and so on, obtaining for instance $((V^*)^*)^*$ and, in that way, an infinite sequence of dual vector spaces. The theorem below shows that, *in the finite-dimensional case*, this is not the case because $(V^*)^*$ turns out to be *naturally isomorphic* to the initial space V and thus the apparently infinite sequence of dual spaces ends at the second step. Recall that a vector space isomorphism $F : V \to V'$ is a linear map which is also one-to-one, i.e., injective, and surjective. An isomorphism is called *natural* when it is built up using only the definition of the involved algebraic structures and it does not depend on "arbitrary choices". A more precise definition of *natural isomorphism* may be given by introducing the *theory of mathematical categories* and using the notion of *natural transformation* (between the identity functor in the category of finite-dimensional vector spaces over $\mathbb{K}$ and the functor induced by F in the same category).

Theorem 2.9 *Let V be a vector space over the field $\mathbb{K} = \mathbb{R}$ or $\mathbb{C}$. The following holds.*

(a) *There is an injective linear map $F : V \to (V^*)^*$ given by*

$$(F(v))\,(u) := u(v) ,$$

for all $u \in V^*$ and $v \in V$, so that V identifies naturally with a subspace of $(V^*)^*$.

(b) *If V has finite dimension, the map F is a natural isomorphism and V identifies naturally with the whole space $(V^*)^*$*

Proof Notice that $F(v) \in (V^*)^*$ because it is a linear functional on V^*:

$$(F(v))(\alpha u+\beta u') := (\alpha u+\beta u')(v) = \alpha u(v)+\beta u'(v) =: \alpha(F(v))(u)+\beta(F(v))(u').$$

Let us prove that $v \mapsto F(v)$ with $(F(v))\,(u) := u(v)$ is linear and injective in the general case, and surjective when V has finite dimension.
(Linearity.) We have to show that, for all $\alpha, \beta \in \mathbb{K}$, $v, v' \in V$,

$$F(\alpha v + \beta v') = \alpha F(v) + \beta F(v')\,.$$

This is equivalent, by the definition of F given above, to

$$u(\alpha v + \beta v') = \alpha u(v) + \beta u(v')\,,$$

for all $u \in V^*$. This follows immediately because u is a linear functional on V.
(Injectivity.) Due to linearity it is enough to show that $F(v) = 0$ implies $v = 0$. $(F(v))\,(u) = 0$ can be re-written as $u(v) = 0$. In our hypotheses, this holds true for all $u \in V^*$. Then define $e_1 := v$ and notice that, if $v \neq 0$, one can complete e_1 with other vectors to get an algebraic vector basis of V, $\{e_i\}_{i\in I}$ (I being infinite in general). Since u is arbitrary, we can pick out u as the unique element of V^* such that $u(e_1) = 1$ and $u(e_i) = 0$ if $i \neq 1$. (This is possible also if V—and thus V^*—is not finite-dimensional.) It contradicts the hypothesis $u(e_1) = u(v) = 0$. Consequently $v = 0$.
(Surjectivity.) Assume that V (and thus V^*) has finite dimension. Since $dim((V^*)^*) = dim(V^*)$ (V^* having finite dimension) and so $dim((V^*)^*) = dim V < +\infty$, injectivity implies surjectivity.
However it is interesting to give an explicit proof. We have to show that if $f \in (V^*)^*$, there is $v_f \in V$ such that $F(v_f) = f$.

Fix a basis $\{e_i\}_{i\in I}$ in V and the dual one in V^*. Since $\{e^{*i}\}_{i\in I}$ is a (proper) vector basis of V^* (it may be false if I is infinite!), $f \in (V^*)^*$ is completely defined by the coefficients $f(e^{*i})$. Then $v_f := f(e^{*i})e_i$ fulfills the requirement $F(v_f) = f$. □

Definition 2.10 (Pairing) Let V be a vector space over $\mathbb{K} = \mathbb{C}$ or $\mathbb{R}$ with dual space V^* and conjugate dual space $\overline{V^*}$ when $\mathbb{K} = \mathbb{C}$. The bi-linear map $\langle\, .,\, .\rangle : V \times V^* \to \mathbb{K}$ such that

$$\langle u, v\rangle := v(u)$$

for all $u \in V$, $v \in V^*$, is called the **pairing**.

If $\mathbb{K} = \mathbb{C}$, the map, linear in the right-hand entry and anti-linear in the left-hand entry, $\langle .,\ . \rangle : V \times \overline{V^*} \to \mathbb{C}$ such that

$$\langle u, v \rangle := v(u)$$

for all $u \in V$, $v \in \overline{V^*}$, is called the **(conjugate) pairing**. ■

Remarks 2.11 Because of the theorem proved above, we may indifferently think of $\langle u, v \rangle$ as representing either the action of $u \in (V^*)^*$ on $v \in V^*$ or the action of $v \in V^*$ on $u \in V$. The same happens for V and $\overline{V^*}$. ■

Notation 2.12 From now on

$$V \simeq W$$

indicates that the vector spaces V and W are isomorphic under some *natural* isomorphism. If the field of V and W is $\mathbb{C}$,

$$V \,\overline{\simeq}\, W$$

indicates that there is a *natural* anti-isomorphism, i.e. there is an injective, surjective, anti-linear map $G : V \to W$ built up using the abstract definition of vector space (including the abstract definitions of dual vector space and (dual) conjugate vector space). ■

We finally state a theorem, concerning conjugated spaces, that is analogous to Theorem 2.9 and has a strongly analogous proof, which is omitted.

Theorem 2.13 *If V is a finite-dimensional vector space over the field $\mathbb{C}$, one has:*

(i) $V^* \,\overline{\simeq}\, \overline{V^*}$*, where the anti-isomorphism* $G : V^* \to \overline{V^*}$ *is defined by*

$$(G(v))(u) := \overline{\langle u, v \rangle} \quad \textit{for all } v \in V^* \textit{ and } u \in V;$$

(ii) $V \,\overline{\simeq}\, \overline{V}$*, where the anti-isomorphism* $F : V \to \overline{V}$ *is defined by*

$$(F(v))(\omega) := \langle v, G^{-1}(\omega) \rangle \quad \textit{for all } v \in V \textit{ and } \omega \in \overline{V^*};$$

(iii) $\overline{V}^* \simeq \overline{V^*}$*, where the involved isomorphism* $H : \overline{V}^* \to \overline{V^*}$ *arises by applying Theorem 2.9 to the definition of* $\overline{V}$*.*

Finally, with respect to any fixed basis $\{e_i\}_{i \in I} \subset V$*, and the canonically associated bases* $\{\overline{e}_i\}_{i \in I} \subset \overline{V}$*,* $\{e^{*i}\}_{i \in I} \subset V^*$*,* $\{\overline{e^{*}}^i\}_{i \in I} \subset \overline{V^*}$*,* $\{\overline{e}^{*i}\}_{i \in I} \subset \overline{V}^*$*, one also has*

$$F : v^i e_i \mapsto \overline{v^i}\,\overline{e}_i\,, \quad G : v_i e^{*i} \mapsto \overline{v_i}\,\overline{e^*}^i\,, \quad H : v_i \overline{e^*}^i \mapsto v_i \overline{e}^{*i}$$

where the bar over the components denotes complex conjugation.

Exercises 2.14

1. Show that if $v \in V$ then $v = \langle v, e^{*j} \rangle e_j$, where $\{e_j\}_{j\in I}$ is any basis of the finite-dimensional vector space V.
(*Hint*. Decompose $v = c^i e_i$, compute $\langle v, e^{*k} \rangle$ taking the linearity of the left-hand entry into account. Alternatively, show that $v - \langle v, e^{*j} \rangle e_j = 0$ by proving that $f(v - \langle v, e^{*j} \rangle e_j) = 0$ for every $f \in V^*$.)
2. Show that $V^* \simeq \overline{V^*}$ if $dim V < \infty$ (and the field of V is $\mathbb{C}$). Similarly, show that $V \simeq \overline{V}$ under the same hypotheses, where the anti-isomorphism $G : V^* \to \overline{V^*}$ is defined by $(G(v))(u) := \overline{\langle u, v \rangle}$ and the anti-isomorphism $F : V \to \overline{V}$ is defined by $(F(v))(\omega) := \langle v, G^{-1}(\omega) \rangle$ for $\omega \in \overline{V^*}$.)
3. Show that if the finite-dimensional vector spaces V and V' are isomorphic or anti-isomorphic, then V^* and V'^* are isomorphic or anti-isomorphic respectively.

(*Hint*. If the initial (anti-)isomorphism is $F : V \to V'$ consider $G : V'^* \to V^*$ defined by $\langle F(u), v' \rangle = \langle u, G(v') \rangle$.) ■

The last notion we introduce is that of *adjoint operator*. We shall have very few occasions to employ this notion; however, it is an important mathematical tool and it deserves mention.

Consider a (linear) operator $T : V_1 \to V_2$, where V_1, V_2 are two vector spaces over the same field $\mathbb{K}$. One can define another operator: $T^* : V_2^* \to V_1^*$ completely determined by the requirement:

$$\left(T^* u^*\right)(v) := u^*\left(T v\right) , \quad \text{for all } v \in V_1 \text{ and all } u^* \in V_2^*.$$

The linearity of T^* follows immediately from the definition.

Definition 2.15 (Adjoint Operator) Let V_1, V_2 be a pair of vector spaces over the same field $\mathbb{K} = \mathbb{C}$ or $\mathbb{R}$ and of finite dimension. If $T : V_1 \to V_2$ is any linear operator, the operator $T^* : V_2^* \to V_1^*$ completely defined by the requirement

$$\langle v, T^* u^* \rangle_1 := \langle T v, u^* \rangle_2 , \quad \text{for all } v \in V_1 \text{ and all } u^* \in V_2^*,$$

is called the **adjoint operator** of T. ■

Remarks 2.16 Notice that we may drop the hypothesis of finite dimension of the involved spaces in the definition above without trouble: T^* would turn out to be well defined also in that case. However, this is not the full story, since such a definition, in the infinite-dimensional case, would not be very useful for applications. In fact, a generalized definition of the adjoint operator can be given when V_1, V_2 are infinite-dimensional and equipped with a suitable topology. In this case T is promoted to a continuous operator and thus one expects that T^* is continuous as well. This is not the case with the definition above as it stands: in general T^* would not be continuous even if one starts with T continuous, since the algebraic duals V_1^* and V_2^* are not equipped with any natural topology. However, if one replaces the algebraic duals with *topological duals*, the definition above gives rise to a continuous operator T^*

when T is continuous. This happens, in particular, whenever V_1 and V_2 are Banach spaces.

In this scenario, there is another definition of adjoint operator, in the context of vector spaces with scalar product (including Hilbert spaces). It is worth stressing that the adjoint operator in the sense of Hilbert space theory is different from the adjoint in the sense of Definition 2.15. However, the two notions enjoy a nice interplay that we shall discuss in Remark 6.2 later. ■

2.2 Multilinearity: Tensor Product, Tensors, Universality Theorem

2.2.1 Tensors as Multilinear Maps

Let us consider $n \geq 1$ vector spaces $V_1, \ldots, V_n$ over the common field $\mathbb{K} = \mathbb{R}$ or $\mathbb{C}$ and another vector space W over $\mathbb{K}$; all spaces are not necessarily finite-dimensional. In the following $\mathscr{L}(V_1, \ldots, V_n|W)$ denotes the vector space (over $\mathbb{K}$) of *multilinear maps* from $V_1 \times \ldots \times V_n$ to W. Recall that a map $f : V_1 \times \ldots \times V_n \to W$ is said to be *multilinear* if, arbitrarily fixing $n-1$ vectors, $v_i \in V_i$ for $i \neq k$, each map $v_k \mapsto f(v_1, \ldots, v_k, \ldots, v_n)$ is linear for every $k = 1, \ldots, n$. A direct verification shows that $\mathscr{L}(V_1, \ldots, V_n|W)$ is a *vector space* over $\mathbb{K}$, with respect to the usual sum of pairs of functions and product of an element of $\mathbb{K}$ and a function. If $W = \mathbb{K}$ we use the shorter notation $\mathscr{L}(V_1, \ldots, V_n) := \mathscr{L}(V_1, \ldots, V_n|\mathbb{K})$, $\mathbb{K}$ being the common field of $V_1, \ldots, V_n$.

Exercises 2.17

1. Suppose that $\{e_{k,i_k}\}_{i_k \in I_k}$ are bases of the vector spaces V_k, $k = 1, \ldots, n$ over the same field $\mathbb{K}$. Suppose also that $\{e_i\}_{i \in I}$ is a basis of the vector space W over $\mathbb{K}$. Show that each $f \in \mathscr{L}(V_1, \ldots, V_n|W)$ satisfies

$$f(v_1, \ldots, v_n) = v_1^{i_1} \cdots v_n^{i_n} \langle f(e_{1,i_1}, \ldots, e_{n,i_n}), e^{*i} \rangle e_i \ ,$$

for all $(v_1, \ldots, v_n) \in V_1 \times \cdots \times V_n$.

2. Endow $V_1 \times \cdots \times V_n$ with the structure of a vector space over the common field $\mathbb{K}$, viewed as the direct sum of $V_1, \ldots, V_n$. In other words, that vector space structure is the unique one compatible with the definition:

$$\alpha(v_1, \ldots, v_n) + \beta(v_1', \ldots, v_n') := (\alpha v_1 + \beta v_1', \ldots, \alpha v_n + \beta v_n')$$

for every pair of numbers $\alpha, \beta \in \mathbb{K}$ and vectors $v_i, v_i' \in V_i$, $i = 1, \ldots, n$.

If $f \in \mathscr{L}(V_1, \ldots, V_n|W)$, may we say that f is a linear map from $V_1 \times \cdots \times V_n$ to W? Generally speaking, is the range of f a subspace of W?

(*Hint*. No. Is the identity $f(\alpha(v_1, \ldots, v_n)) = \alpha f(v_1, \ldots, v_n)$ true? The answer to the second question is similarly no. This is already evident, for instance, for $W = \mathbb{R}^4$ and $V_1 = V_2 = \mathbb{R}^2$ considering the multilinear map

$$f((a, b), (c, d)) := (ac, ad, bc, bd) .$$

Its range is the set of 2×2 real matrices of rank at most one, identified with $\mathbb{R}^4$, and this set is not closed under addition.)
We have a first fundamental and remarkable theorem. To introduce it, we proceed in three steps.

(1) Take $f \in \mathscr{L}(V_1, ..., V_n|W)$. This means that f associates every string $(v_1, ..., v_n) \in V_1 \times \cdots \times V_n$ with a corresponding element $f(v_1, \ldots, v_n) \in W$, and the map

$$(v_1, \ldots, v_n) \mapsto f(v_1, \ldots, v_n)$$

is multilinear.
(2) Since, for *fixed* $(v_1, \ldots, v_n)$, the vector $f(v_1, \ldots, v_n)$ is an element of W, the action of $w^* \in W^*$ on $f(v_1, \ldots, v_n)$ makes sense, producing $\langle f(v_1, \ldots, v_n), w^* \rangle \in \mathbb{K}$.
(3) Allowing $v_1, \ldots v_n$ and w^* to range freely in the corresponding spaces, the map Ψ_f with

$$\Psi_f : (v_1, \ldots, v_n, w^*) \mapsto \langle f(v_1, \ldots, v_n), w^* \rangle ,$$

turns out to be multilinear by construction. Hence, by definition $\Psi_f \in \mathscr{L}(V_1, \ldots, V_n, W^*)$.

The theorem concerns the map F which associates f with Ψ_f. ■

Theorem 2.18 *Let $V_1, \ldots, V_n$ be (not necessarily finite-dimensional) vector spaces over the common field $\mathbb{K}$ (= $\mathbb{C}$ or $\mathbb{R}$), and let W be another finite-dimensional vector space over $\mathbb{K}$.*
The vector spaces $\mathscr{L}(V_1, \ldots, V_n|W)$ and $\mathscr{L}(V_1, \ldots, V_n, W^*)$ are naturally isomorphic by means of the map $F : \mathscr{L}(V_1, \ldots, V_n|W) \to \mathscr{L}(V_1, \ldots, V_n, W^*)$ with $F : f \mapsto \Psi_f$ defined by

$$\Psi_f(v_1, \ldots, v_n, w^*) := \langle f(v_1, \ldots, v_n), w^* \rangle,$$
$$\textit{for all } (v_1, \ldots, v_n) \in V_1 \times \ldots \times V_n \textit{ and } w^* \in W^*.$$

If $\dim W = +\infty$, the above-defined linear map F is injective in any case.

Proof Let us consider the map F defined above. We have only to establish that F is linear, injective and surjective. This ends the proof.

(Linearity.) We have to prove that $\Psi_{\alpha f+\beta g} = \alpha\Psi_f + \beta\Psi_g$ for all $\alpha, \beta \in \mathbb{K}$ and $f, g \in \mathscr{L}(V_1, \ldots, V_n|W)$.
In fact, making use of the left-hand linearity of the pairing, one has

$$\begin{aligned}\Psi_{\alpha f+\beta g}(v_1, \ldots, v_n, w^*) &= \langle(\alpha f + \beta g)(v_1, \ldots, v_n), w^*\rangle \\ &= \alpha\langle f(v_1, \ldots, v_n), w^*\rangle + \beta\langle g(v_1, \ldots, v_n), w^*\rangle\end{aligned}$$

and this is nothing but:

$$\Psi_{\alpha f+\beta g}(v_1, \ldots, v_n, w^*) = (\alpha\Psi_f + \beta\Psi_g)(v_1, \ldots, v_n, w^*)\,.$$

Since $(v_1, \ldots, v_n, w^*)$ is arbitrary, the thesis is proved.
(Injectivity.) Due to linearity we have only to show that if $\Psi_f = 0$ then $f = 0$.
In fact, if $\Psi_f(v_1, \ldots, v_n, w^*) = 0$ for all $(v_1, \ldots, v_n, w^*)$, using the definition of Ψ_f we have $\langle f(v_1, \ldots, v_n), w^*\rangle = 0$, for all $(v_1, \ldots, v_n)$ and w^*. Then define $e_1 := f(v_1, \ldots, v_n)$; if $e_1 \neq 0$ we can complete it to a basis of W. Fixing $w^* = e^{*1}$ we should have

$$0 = \Psi_f(v_1, \ldots, v_n, w^*) = \langle f(v_1, \ldots, v_n), w^*\rangle = 1$$

which is impossible. Therefore $f(v_1, \ldots, v_n) = 0$ for all $(v_1, \ldots, v_n) \in V_1 \times \ldots \times V_n$, in other words $f = 0$.
(Surjectivity.) We have to show that for each $\Phi \in \mathscr{L}(V_1, \ldots, V_n, W^*)$ there is an $f_\Phi \in \mathscr{L}(V_1, \ldots, V_n|W)$ with $\Psi_{f_\Phi} = \Phi$. To this end fix a basis $\{e_k\}_{k\in I} \subset W$ and the associated dual one $\{e^{*k}\}_{k\in I} \subset W^*$. Then, take $\Phi \in \mathscr{L}(V_1, \ldots, V_n, W^*)$ and define the map $f_\Phi \in \mathscr{L}(V_1, \ldots, V_n|W)$ by

$$f_\Phi(v_1, \ldots, v_n) := \Phi(v_1, \ldots, v_n, e^{*k})e_k\,.$$

By construction that map is multilinear. If $w^* = w_k e^{*k}$, then, using multilinearity and the duality relation $\langle e_k, e^{*i}\rangle = \delta^i_k$, we find

$$\begin{aligned}\Psi_{f_\Phi}(v_1, \ldots, v_n, w^*) &= \left\langle\Phi(v_1, \ldots, v_n, e^{*k})e_k, w_i e^{*i}\right\rangle \\ &= \Phi(v_1, \ldots, v_n, e^{*k})w_k = \Phi(v_1, \ldots, v_n, w^*)\,,\end{aligned}$$

for all $(v_1, \ldots, v_n, w^*)$, and this is equivalent to $\Psi_{f_\Phi} = \Phi$.
If $\dim W = +\infty$ the proof of surjectivity may fail since the sum over $k \in I$ used to define f_Φ may diverge. The proofs of linearity and injectivity are independent of the cardinality of I. □

Remarks 2.19 An overall difficult point in understanding and trusting the statement of the theorem above relies upon the fact that the function f has n entries, whereas the function Ψ_f has $n+1$ entries, but f is identified with Ψ_f by means of

F. This fact may seem quite weird at first glance and, consequently, the statement of the theorem may seem suspicious. The difficulty can be clarified from a practical point of view as follows.

If bases $\{e_{1,i_1}\}_{i_1\in I_1}$,...,$\{e_{n,i_n}\}_{i_n\in I_n}$ for V_1,...,V_n respectively, and $\{e_k\}_{k\in I}$ for W are fixed and $\Psi \in \mathscr{L}(V_1, \ldots, V_n, W^*)$, one has, for constant coefficients $P^k_{i_1\ldots i_n}$ depending on Ψ,

$$\Psi(v_1, \ldots, v_n, w^*) = P^k_{i_1\ldots i_n} v_1^{i_1} \ldots v_n^{i_n} w_k^* ,$$

where w_k^* are the components of w^* in the dual basis $\{e^{*k}\}_{k\in I}$ and $v_p^{i_p}$ are the components of v_p in the basis $\{e_{p,i_p}\}_{i_p\in I_p}$. Now consider the other map $f \in \mathscr{L}(V_1, \ldots, V_n|W)$ whose argument is the string $(v_1, \ldots, v_n)$, *no further vectors being necessary*:

$$f : (v_1, \ldots, v_n) \mapsto P^k_{i_1\ldots i_n} v_1^{i_1} \ldots v_n^{i_n} e_k$$

The former map Ψ, which deals with $n+1$ arguments, is associated with f, which deals with n arguments. The point is that we have taken advantage of the bases, $\{e_k\}_{k\in I}$ in particular.

Within this context, Theorem 2.18 just proves that *the correspondence which associates Ψ with f is linear, bijective and independent of the chosen bases.* ■

In the finite-dimensional case—and we are mainly interested in this case—Theorem 2.18 allows us to restrict our study to the spaces of multilinear *functionals* $\mathscr{L}(V_1, \ldots, V_k)$, since the spaces of multilinear *maps* are completely encompassed. Let us introduce the concept of *tensor product* of vector spaces. The following definitions can be extended to encompass the case of infinite-dimensional vector spaces by introducing suitable topological notions (e.g., Hilbert spaces).

Definition 2.20 (Tensor Product) Let $U_1, \ldots, U_n$ be $n \geq 1$ vector spaces (not necessarily finite-dimensional) over the common field $\mathbb{K} = \mathbb{R}$ or $\mathbb{C}$.

(1) if $(u_1, \ldots, u_n) \in U_1 \times \ldots \times U_n$, $u_1 \otimes \ldots \otimes u_n$ denotes the multilinear map in $\mathscr{L}(U_1^*, \ldots, U_n^*)$ defined by

$$(u_1 \otimes \ldots \otimes u_n)(v_1, \ldots, v_n) := \langle u_1, v_1\rangle \cdots \langle u_n, v_n\rangle ,$$

for all $(v_1, \ldots, v_n) \in U_1^* \times \ldots \times U_n^*$.
$u_1 \otimes \ldots \otimes u_n$ is called the **tensor product of vectors** $u_1, \ldots, u_n$.

(2) The map $\otimes : U_1 \times \ldots \times U_n \to \mathscr{L}(U_1^*, \ldots, U_n^*)$ given by: $\otimes : (u_1, \ldots, u_n) \mapsto u_1 \otimes \ldots \otimes u_n$, is called the **tensor product map**.

(3) The vector subspace of $\mathscr{L}(U_1^*, \ldots, U_n^*)$ *generated* by all $u_1 \otimes \ldots \otimes u_n$ for all $(u_1, \ldots, u_n) \in U_1 \times \ldots \times U_n$ is called the **tensor product of spaces** $U_1, \ldots, U_n$ and is denoted by $U_1 \otimes \ldots \otimes U_n$. The vectors in $U_1 \otimes \ldots \otimes U_n$ are called **tensors**.

■

Remarks

(1) $U_1 \otimes \ldots \otimes U_n$ is therefore made of all the *linear combinations* of the form

$$\sum_{j=1}^{N} \alpha_j u_{1,j} \otimes \ldots \otimes u_{n,j} ,$$

where $\alpha_j \in \mathbb{K}$, $u_{k,j} \in U_k$ and $N = 1, 2, \ldots$.

(2) A direct verification shows that the tensor product map

$$(u_1, \ldots, u_n) \mapsto u_1 \otimes \ldots \otimes u_n ,$$

is *multilinear.*

That is, for any $k \in \{1, 2, \ldots, n\}$, the following identity holds for all $u, v \in U_k$ and $\alpha, \beta \in \mathbb{K}$:

$$u_1 \otimes \ldots \otimes u_{k-1} \otimes (\alpha u + \beta v) \otimes u_{k+1} \otimes \ldots \otimes u_n$$

$$= \alpha(u_1 \otimes \ldots \otimes u_{k-1} \otimes u \otimes u_{k+1} \otimes \ldots \otimes u_n)$$

$$+ \beta(u_1 \otimes \ldots \otimes u_{k-1} \otimes v \otimes u_{k+1} \otimes \ldots \otimes u_n) .$$

As a consequence, it holds

$$(\alpha u) \otimes v = u \otimes (\alpha v) = \alpha(u \otimes v) ,$$

and similar identities hold when considering any number of factor spaces in a tensor product of vector spaces.

(3) From the given definition, if $U_1, \ldots, U_n$ are given vector spaces over the same field $\mathbb{K} = \mathbb{R}$ or $\mathbb{C}$, what we mean by $U_1 \otimes \ldots \otimes U_n$ or $U_1^* \otimes \ldots \otimes U_n^*$ but also, for instance, $U_1^* \otimes U_2 \otimes U_3$ or $U_1 \otimes U_2^* \otimes U_3^*$ is clear from the same convention. One simply has to take the action of U^* on $(U^*)^*$ into account.

By definition $U_1 \otimes \ldots \otimes U_n \subset \mathscr{L}(U_1^*, \ldots, U_n^*)$, more strongly, $U_1 \otimes \ldots \otimes U_n$ is, in general, a *proper* subspace of $\mathscr{L}(U_1^*, \ldots, U_n^*)$. The natural question which arises is whether there are hypotheses implying that $U_1 \otimes \ldots \otimes U_n$ coincides with the whole space $\mathscr{L}(U_1^*, \ldots, U_n^*)$. The following theorem gives an answer to that question. ■

Theorem 2.21 *If all the spaces U_i have finite dimension, one has:*

$$U_1 \otimes \ldots \otimes U_n = \mathscr{L}(U_1^*, \ldots, U_n^*) .$$

Proof It is sufficient to show that if $f \in \mathscr{L}(U_1^*, \ldots, U_n^*)$ then $f \in U_1 \otimes \ldots \otimes U_n$. To this end fix vector bases $\{e_{k,i}\}_{i \in I_k} \subset U_k$ for $k = 1, \ldots, n$ and consider also the associated dual bases $\{e_k^{*i}\}_{i \in I_k}$. f above is completely determined by the coefficients (their number is finite since $dimU_i^* = dimU_i < +\infty$ by hypothesis) $f_{i_1,\ldots,i_n} := f(e_1^{*i_1}, \ldots, e_n^{*i_n})$. Every $v_k \in U_k^*$ can be decomposed as $v_k = v_{k,i_k} e_k^{*i_k}$, where each sum is finite, and thus, by multilinearity:

$$f(v_1, \ldots, v_n) = v_{1,i_1} \cdots v_{n,i_n} f(e_1^{*i_1}, \ldots, e_n^{*i_n}) .$$

Then consider the tensor $t_f \in U_1 \otimes \ldots \otimes U_n$ defined by:

$$t_f := f(e_1^{*i_1}, \ldots, e_n^{*i_n}) e_{1,i_1} \otimes \ldots \otimes e_{n,i_n} .$$

Then, by Definition 2.20, one can directly prove that, by multilinearity,

$$t_f(v_1, \ldots, v_n) = v_{1,i_1} \cdots v_{n,i_n} f(e_1^{*i_1}, \ldots, e_n^{*i_n}) = f(v_1, \ldots, v_n) ,$$

for all $(v_1, \ldots, v_n) \in U_1^* \times \ldots \times U_n^*$. This is nothing but $t_f = f$. □

Another relevant result is stated by the theorem below.

Theorem 2.22 *Consider vector spaces U_i, $i = 1, \ldots, n$ over the same field $\mathbb{K}$. The following statements hold.*

(a) *If all spaces U_i are finite-dimensional, the dimension of $U_1 \otimes \ldots \otimes U_n$ is:*

$$dim(U_1 \otimes \ldots \otimes U_n) = \prod_{k=1}^{n} dim\, U_k .$$

(b) *If $\{e_{k,i_k}\}_{i_k \in I_k}$ is a basis of U_k, $k = 1, \ldots, n$, then $\{e_{1,i_1} \otimes \ldots \otimes e_{n,i_n}\}_{(i_1,\ldots,i_n) \in I_1 \times \ldots \times I_n}$ is a vector basis of $U_1 \otimes \ldots \otimes U_n$.*

(c) *If $t \in U_1 \otimes \ldots \otimes U_n$ and all spaces U_i are finite-dimensional, the components of t with respect to a basis $\{e_{1,i_1} \otimes \ldots \otimes e_{n,i_n}\}_{(i_1,\ldots,i_n) \in I_1 \times \ldots \times I_n}$ are given by:*

$$t^{i_1 \ldots i_n} = t(e_1^{*i_1}, \ldots, e_n^{*i_n})$$

and thus it holds

$$t = t(e_1^{*i_1}, \ldots, e_n^{*i_n}) e_{1,i_1} \otimes \ldots \otimes e_{n,i_n} .$$

Proof Notice that, in the finite-dimensional case, **(b)** trivially implies **(a)** because the elements $e_{1,i_1} \otimes \ldots \otimes e_{n,i_n}$ are exactly $\prod_{k=1}^{n} dimU_k$ in number. So it is sufficient to show that the second statement holds true. To this end, since the elements $e_{1,i_1} \otimes \ldots \otimes e_{n,i_n}$ are generators of $U_1 \otimes \ldots \otimes U_n$, it is sufficient to show that they are linearly independent. Consider the generic vanishing linear combination

$$C^{i_1 \ldots i_n} e_{1,i_1} \otimes \ldots \otimes e_{n,i_n} = 0 ,$$

We want to show that all the coefficients $C^{i_1 \dots i_n}$ vanish. The action of that linear combination of multilinear functionals on the generic element $(e_1^{*j_1}, \dots, e_n^{*j_n}) \in U_1^* \times \dots \times U_n^*$ produces the result

$$C^{j_1 \dots j_n} = 0 .$$

Since we can arbitrarily fix the indices $j_1, \dots j_n$ this proves the thesis.
The proof of **(c)** has already been given in the proof of Theorem 2.21. However, we repeat it for the sake of completeness. By uniqueness of the components of a vector with respect to a basis, it is sufficient to show that, defining (notice that the number of vectors $e_{1,i_1} \otimes \dots \otimes e_{n,i_n}$ is finite by hypothesis since $dim U_k$ is finite in the considered case)

$$t' := t(e_1^{*i_1}, \dots, e_n^{*i_n}) e_{1,i_1} \otimes \dots \otimes e_{n,i_n} ,$$

it holds

$$t'(v_1, \dots, v_n) = t(v_1, \dots, v_n) ,$$

for all $(v_1, \dots, v_n) \in U_1^* \times \dots \times U_n^*$. By multilinearity,

$$t(v_1, \dots, v_n) = v_{1\,i_1} \cdots v_{n\,i_n} t(e_1^{*i_1}, \dots, e_n^{*i_n}) = t'(v_1, \dots, v_n) .$$

□

We notice that, obviously, if some of the spaces U_i has infinite dimension, $U_1 \otimes \dots \otimes U_n$ cannot have finite dimension.

In the case of *finite dimension* of all involved spaces, there is an important result which, *together with the identification of* V *and* $(V^*)^*$, implies that all the spaces that one may build up using the symbols $\otimes$, V_k, * and () coherently are naturally isomorphic to spaces of the form $V_{i_1}^{(*)} \otimes \dots \otimes V_{i_n}^{(*)}$. The rule to produce spaces naturally isomorphic to a given initial space is that one has to (1) ignore parentheses—that is, $\otimes$ *is associative* –, (2) assume that * is *distributive with respect to* $\otimes$ and (3) assume that * is *involutive* (i.e. $(X^*)^* \simeq X$). For instance, one has:

$$((V_1^* \otimes V_2) \otimes (V_3 \otimes V_4^*))^* \simeq V_1 \otimes V_2^* \otimes V_3^* \otimes V_4 .$$

Let us state the theorems corresponding to rules (2) and (1) respectively (rule (3) being nothing but Theorem 2.9).

If not every V_i has finite dimension, the distributivity and involutivity properties of * may fail to hold, but associativity of $\otimes$ holds true anyway.

Theorem 2.23 *If* $V_1 \dots V_n$ *are finite-dimensional vector spaces over the same field* $\mathbb{K} = \mathbb{R}$ *or* $\mathbb{C}$, * *is* **distributive** *with respect to* $\otimes$ *by means of* **natural isomorphisms** $F : V_1^* \otimes \dots \otimes V_n^* \to (V_1 \otimes \dots \otimes V_n)^*$. *In other words it holds:*

$$(V_1 \otimes \dots \otimes V_n)^* \simeq V_1^* \otimes \dots \otimes V_n^* \,.$$

under F. *The isomorphism* F *is uniquely determined by the requirement*

$$\langle u_1 \otimes \dots \otimes u_n, F(v_1^* \otimes \dots \otimes v_n^*)\rangle = \langle u_1, v_1^*\rangle \cdots \langle u_n, v_n^*\rangle \,, \tag{2.1}$$

for every choice of $v_i^* \in V_i^*$, $u_i \in V_i$ *and* $i = 1, 2, \dots, n$.

Proof The natural isomorphism is determined by (2.1), thus employing only the general mathematical structures of dual space and tensor product. Consider, *if it exists* (see the discussion below), the linear map $F : V_1^* \otimes \dots \otimes V_n^* \to (V_1 \otimes \dots \otimes V_n)^*$ which satisfies, for all $u_1 \otimes \dots \otimes u_n \in V_1 \otimes \dots \otimes V_n$,

$$\big\langle u_1 \otimes \dots \otimes u_n, F(v_1^* \otimes \dots \otimes v_n^*)\big\rangle := \langle u_1, v_1^*\rangle \dots \langle u_n, v_n^*\rangle \,. \tag{2.2}$$

Using linearity and the involved definitions, one obtains that the map F is surjective. Indeed, equipping each space V_k with a basis $\{e_{k,i_k}\}_{i_k \in I_k}$ and the dual space V_k^* with the corresponding dual basis $\{e_k^{*i_k}\}_{i_k \in I_k}$, if $f \in (V_1 \otimes \cdots \otimes V_n)^*$, defining the numbers $f_{i_1 \cdots i_n} := f(e_{1,i_1} \otimes \cdots \otimes e_{n,i_n})$, by direct inspection one finds that:

$$\langle u_1 \otimes \dots \otimes u_n, F(f_{i_1 \cdots i_n} e_1^{*i_1} \otimes \cdots \otimes e_n^{*i_n})\rangle = \langle u_1 \otimes \dots \otimes u_n, f\rangle \,,$$

for all $u_1 \otimes \dots \otimes u_n$. By linearity this result extends to any element $u \in V_1 \otimes \dots \otimes V_n$ in place of $u_1 \otimes \dots \otimes u_n$ so that $f(u) = \left(F(f_{i_1 \cdots i_n} e_1^{*i_1} \otimes \cdots \otimes e_n^{*i_n})\right)(u)$ for every $u \in V_1 \otimes \cdots \otimes V_n$. Therefore the functionals $f, F(f_{i_1 \cdots i_n} e_1^{*i_1} \otimes \cdots \otimes e_n^{*i_n}) \in (V_1 \otimes \dots \otimes V_n)^*$ coincide:

$$f = F(f_{i_1 \cdots i_n} e_1^{*i_1} \otimes \cdots \otimes e_n^{*i_n}) \,.$$

Finally, by the theorems proved previously,

$$dim(V_1 \otimes \cdots \otimes V_n)^* = dim(V_1 \otimes \cdots \otimes V_n) = dim(V_1^* \otimes \cdots \otimes V_n^*)$$

As a consequence, F must also be injective and thus it is an isomorphism. □

Theorem 2.24 *The tensor product is* **associative** *by means of* **natural isomorphisms**. *In other words, considering a space which is made of tensor products of vector spaces over the same field* $\mathbb{R}$ *or* $\mathbb{C}$, *one may omit parentheses everywhere, obtaining a space which is naturally isomorphic to the initial one. So, for instance,*

$$V_1 \otimes V_2 \otimes V_3 \simeq (V_1 \otimes V_2) \otimes V_3 \,.$$

Where the natural isomorphism $F_1 : V_1 \otimes V_2 \otimes V_3 \to (V_1 \otimes V_2) \otimes V_3$ *satisfies*

$$F_1 : v_1 \otimes v_2 \otimes v_3 \to (v_1 \otimes v_2) \otimes v_3$$

for every choice of $v_i \in V_i$, $i = 1, 2, 3$.
Consequently, if the space $\{e_{i,j_i}\}_{j_i \in J_i} \subset V_i$ *is a basis for* $i = 1, 2, 3$, *the action of* F *can be written as:*

$$F : t^{j_1 j_2 j_3} e_{1,j_1} \otimes e_{2,j_2} \otimes e_{3,j_3} \mapsto t^{j_1 j_2 j_3} (e_{1,j_1} \otimes e_{2,j_2}) \otimes e_{3,j_3} \,.$$

Proof The natural isomorphism is constructed as a linear mapping, *if it exists* (see discussion below), which linearly extends the action $F_1 : V_1 \otimes V_2 \otimes V_3 \to (V_1 \otimes V_2) \otimes V_3$ such that

$$F_1 : v_1 \otimes v_2 \otimes v_3 \mapsto (v_1 \otimes v_2) \otimes v_3 \,,$$

for all $v_1 \otimes v_2 \otimes v_3 \in V_1 \otimes V_2 \otimes V_3$.
Injectivity and surjectivity can be proved by considering the linear map *if it exists* (see discussion below), $F'_1 : (V_1 \otimes V_2) \otimes V_3 \to V_1 \otimes V_2 \otimes V_3$ such that

$$F'_1 : (v_1 \otimes v_2) \otimes v_3 \mapsto v_1 \otimes v_2 \otimes v_3 \,,$$

for all $(v_1 \otimes v_2) \otimes v_3 \in (V_1 \otimes V_2) \otimes V_3$. By construction $F_1 \circ F'_1 = id_{(V_1 \otimes V_2) \otimes V_3}$ and $F'_1 \circ F_1 = id_{V_1 \otimes V_2 \otimes V_3}$ are the identity maps so that F_1 is linear and bijective. The last statement immediately follows from the definition of F. □

Remarks 2.25 The central point is that the maps F and F_1 (and F'_1) above have been given by specifying their action on tensor products of elements (e.g., $F(v_1 \otimes \ldots \otimes v_n)$) and not on *linear combinations* of these tensor products of elements. Recall that, for instance, $V_1^* \otimes \ldots \otimes V_n^*$ is not the set of products $u_1^* \otimes \ldots \otimes u_n^*$ but it is the set of *linear combinations* of those products. Hence, in order to completely define F and F_1, one must require that F and F_1 admit *uniquely determined linear extensions* on their initial domains in order to encompass the whole tensor spaces generated by linear combinations of simple tensor products. In other words one has to complete the given definition, in the former case, by adding the further requirement

$$F(\alpha\, u_1^* \otimes \ldots \otimes u_n^* + \beta\, v_1^* \otimes \ldots \otimes v_n^*) = \alpha F(u_1^* \otimes \ldots \otimes u_n^*) + \beta F(v_1^* \otimes \ldots \otimes v_n^*) \,,$$

Actually, one has to assume in addition that, for any fixed $v_1^* \otimes \cdots \otimes v_n^*$, there is a unique linear extension of the application $F(v_1^* \otimes \cdots \otimes v_n^*)$ defined in (2.2), when the argument is a linear combination of vectors of type $u_1 \otimes \cdots \otimes u_n$.

Despite the fact that these could seem trivial requirements (and they are, in the present case), they are not trivial in general. The point is that the class of all possible products $v_1^* \otimes \ldots \otimes v_n^*$ is a system of generators of $V_1^* \otimes \ldots \otimes V_n^*$, but it is by no means a basis *because these elements are linearly dependent*. So, in general, any attempt to assign a linear map with domain $V_1^* \otimes \ldots \otimes V_n^*$ by assigning its values on all the elements $v_1^* \otimes \ldots \otimes v_n^*$ may give rise to contradictions. For instance, $v_1^* \otimes \ldots \otimes v_n^*$ can be re-written using linear combinations:

$$v_1^* \otimes \ldots \otimes v_n^* = [(v_1^* + u_1^*) \otimes \ldots \otimes v_n^*] - [u_1^* \otimes \ldots \otimes v_n^*] .$$

Now consider the identities, which have to hold as a consequence of the assumed linearity of F, such as:

$$F(v_1^* \otimes \ldots \otimes v_n^*) = F((v_1^* + u_1^*) \otimes \ldots \otimes v_n^*) - F(u_1^* \otimes \ldots \otimes v_n^*)$$

Above $F(v_1^* \otimes \ldots \otimes v_n^*)$, $F(u_1^* \otimes \ldots \otimes v_n^*)$, $F((v_1^* + u_1^*) \otimes \ldots \otimes v_n^*)$ are *independently* defined as we said at the beginning, and there is no reason, in principle, for the validity of the constraint:

$$F(v_1^* \otimes \ldots \otimes v_n^*) = F((v_1^* + u_1^*) \otimes \ldots \otimes v_n^*) - F(u_1^* \otimes \ldots \otimes v_n^*) .$$

Similar problems may arise concerning F_1. ■

The general problem which arises in the two considered cases can be stated as follows. Suppose we are given a tensor product of vector spaces $V_1 \otimes \ldots \otimes V_n$ and we are interested in the possible *linear extensions* of a map f on $V_1 \otimes \ldots \otimes V_n$, with values in some vector space W, when f is initially defined only on the whole class of the simple products $v_1 \otimes \ldots \otimes v_n \in V_1 \otimes \ldots \otimes V_n$. *Is there any general prescription on the specification of values $f(v_1 \otimes \ldots \otimes v_n)$ which assures that f can be extended, uniquely, to a linear map from $V_1 \otimes \ldots \otimes V_n$ to W?*
The general obstruction is evident: if the class of all possible tensors $v_1 \otimes \ldots \otimes v_n$ were a basis of $V_1 \otimes \ldots \otimes V_n$, the class of values $f(v_1 \otimes \ldots \otimes v_n)$ would determine a (unique) linear map from $V_1 \otimes \ldots \otimes V_n$ to W. *Instead we have here a set which is much larger than a basis, so assigning independent values thereon for a linear function is impossible, in view of the linearity constraint.*
An answer is given by the following very important **universality theorem** discussed in the following subsection.

2.2.2 *Universality Theorem and its Applications*

Theorem 2.26 (Universality Theorem) *Given $n \geq 1$ vector spaces $U_1, \ldots, U_n$ (not necessarily finite-dimensional) over the same field $\mathbb{K} = \mathbb{R}$ or $\mathbb{C}$, the following so-called* **universal property** *holds for the pair $(\otimes, U_1 \otimes \ldots \otimes U_n)$. For any vector space W (not necessarily finite-dimensional) and any multilinear map $f : U_1 \times \ldots \times U_n \to W$, there is a unique linear map $f^{\otimes} : U_1 \otimes \ldots \otimes U_n \to W$ such that the diagram below* **commutes** *(in other words, $f^{\otimes} \circ \otimes = f$).*

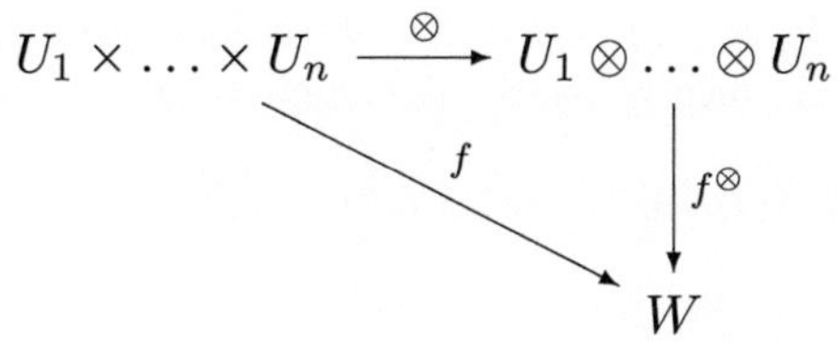

Proof Fix bases $\{e_{k,i_k}\}_{i_k \in I_k} \subset U_k$, $k = 1, \ldots, n$. By linearity, a linear map $g : U_1 \otimes \ldots \otimes U_n \to W$ is uniquely determined by fixing the set of vectors

$$\{g(e_{1,i_1} \otimes \cdots \otimes e_{n,i_n})\}_{i_1 \in I_1, \ldots, i_n \in I_n} \subset W \,.$$

This remains true even if some of the spaces U_k, W have infinite dimension. Define the linear function $f^{\otimes} : U_1 \otimes \ldots \otimes U_n \to W$ as the unique linear map with

$$f^{\otimes}(e_{1,i_1} \otimes \cdots \otimes e_{n,i_n}) := f(e_{1,i_1}, \ldots, e_{n,i_n}) \,.$$

By construction, using linearity of $f^{\otimes}$ and multilinearity of f, one gets immediately:

$$f^{\otimes}(v_1 \otimes \ldots \otimes v_n) = f(v_1, \ldots, v_n)$$

for all $(v_1, \ldots, v_n) \in U_1 \times \ldots \times U_n$. In other words $f^{\otimes} \circ \otimes = f$.
The uniqueness of the map $f^{\otimes}$ follows directly: suppose there is another map $g^{\otimes}$ with $g^{\otimes} \circ \otimes = f$ then $(f^{\otimes} - g^{\otimes}) \circ \otimes = 0$. This means in particular that

$$(f^{\otimes} - g^{\otimes})(e_{1,i_1} \otimes \ldots \otimes e_{n,i_n}) = 0 \,,$$

for all $i_k \in I_k$, $k = 1, \ldots, n$. Since the coefficients above completely determine a map, the considered map must be the null map and thus: $f^{\otimes} = g^{\otimes}$. □

Let us now explain how the universality theorem gives a precise answer to the question formulated before the universality theorem. The theorem says that a *linear extension* of any function f with values in W, initially defined only on simple tensor products, $f(v_1 \otimes \ldots \otimes v_n)$ with $v_1 \otimes \ldots \otimes v_n \in U_1 \otimes \ldots \otimes U_n$, does *exist* on the whole domain space $U_1 \otimes \ldots \otimes U_n$ and it is *uniquely* determined provided $f(v_1 \otimes \ldots \otimes v_n) = g(v_1, \ldots, v_n)$ where $g : U_1 \times \ldots \times U_n \to W$ is some *multilinear function*.
Concerning the maps F and F_1 introduced above, we may profitably use the universality theorem to show that they are well defined on the whole domain made of linear combinations of simple tensor products. In fact, consider F for example. We can define the *multilinear* map $G : V_1^* \times \ldots \times V_n^* \to (V_1 \otimes \ldots \otimes V_n)^*$ such

that $G(v_1^*, \ldots, v_n^*) : V_1 \otimes \ldots \otimes V_n \to \mathbb{C}$ is the unique linear function (assumed to exist, see below) which linearly extends the requirement

$$\left(G(v_1^*, \ldots, v_n^*)\right)(u_1 \otimes \cdots \otimes u_n) := \langle u_1, v_1^* \rangle \cdots \qquad \langle u_n, v_n^* \rangle,$$
$$\text{for all } (u_1, \ldots, u_n) \in U_1 \times \cdots \times U_n \tag{2.3}$$

to the whole space $V_1 \otimes \ldots \otimes V_n$. Then the universality theorem with $U_k = V_k^*$ and $W = (V_1 \otimes \ldots \otimes V_n)^*$ assures the existence of a linear map $F : V_1^* \otimes \ldots \otimes V_n^* \to (V_1 \otimes \ldots \otimes V_n)^*$ with the required properties because $F := G^{\otimes}$ is such that

$$F(v_1^* \otimes \ldots \otimes v_n^*) = (G^{\otimes} \circ \otimes)(v_1^*, \ldots, v_n^*) = G(v_1^*, \ldots, v_n^*)\,,$$

so that (2.2) is fulfilled due to (2.3). Finally, the existence of the linear function $G(v_1^*, \ldots, v_n^*) : V_1 \otimes \ldots \otimes V_n \to \mathbb{K}$ fulfilling (2.3) can be proved, once again, employing the universality theorem for $U_k := V_k$ and $W := \mathbb{K}$. Starting from the multilinear function $G'(v_1^*, \ldots, v_n^*) : V_1 \times \ldots \times V_n \to \mathbb{C}$ such that

$$\left(G'(v_1^*, \ldots, v_n^*)\right)(u_1, \cdots, u_n) := \langle u_1, v_1^* \rangle \cdots \langle u_n, v_n^* \rangle$$

it is enough to define $G(v_1^*, \ldots, v_n^*) := G'(v_1^*, \ldots, v_n^*)^{\otimes}$.
A similar multilinear map G_1 can be found for F_1:

$$G_1 : (v_1, v_2, v_3) \mapsto (v_1 \otimes v_2) \otimes v_3\,.$$

Then: $F_1 := G_1^{\otimes}$ can be used in the proof of Theorem 2.24. If V_1, V_2, V_3 have finite dimension, the proof of Theorem 2.24 ends because the map F_1 so obtained is trivially surjective and it is also injective since

$$\begin{aligned} dim((V_1 \otimes V_2) \otimes V_3) &= dim(V_1 \otimes V_2) \times dim V_3 \\ &= dim(V_1) \times dim(V_2) \times dim(V_3) = dim(V_1 \otimes V_2 \otimes V_3) \end{aligned}$$

by Theorem 2.22.

If not every V_1, V_2, V_3 has finite dimension, the proof of Theorem 2.24 is more complicated and the existence of the function F_1' used in its proof has to be explicitly checked. The construction of the function F_1' used in the proof of Theorem 2.24 can be obtained with a little change in the first statement of the universality theorem. With trivial changes one gets that:

Proposition 2.27 *Given $n \geq 1$ vector spaces $U_1, \ldots, U_n$ over the same field $\mathbb{K} = \mathbb{R}$ or $\mathbb{C}$, the following statements hold.*
If $h < k = 2, \ldots, n$ are fixed, for any vector space W and any multilinear map

$$f : U_1 \times \ldots \times U_n \to W\,,$$

there is a unique linear map

$$f^{\dot{\otimes}} : U_1 \otimes \cdots \otimes U_{h-1} \otimes (U_h \otimes \cdots \otimes U_k) \otimes U_{k+1} \otimes \cdots \otimes U_n \to W$$

such that the diagram below commutes.

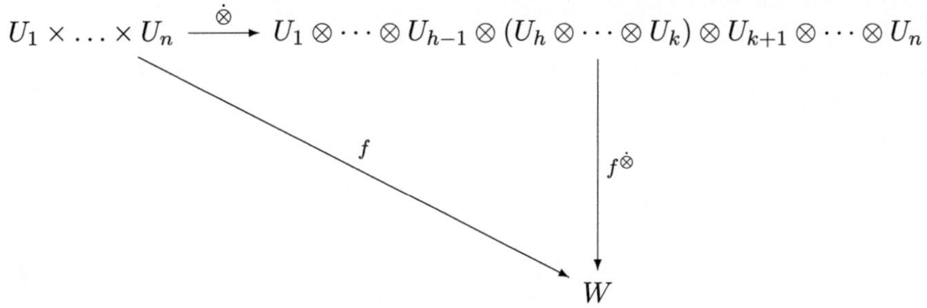

Above

$$\dot{\otimes} : U_1 \times \ldots \times U_n \to U_1 \otimes \cdots \otimes U_{h-1} \otimes (U_h \otimes \cdots \otimes U_k) \otimes U_{k+1} \otimes \cdots \otimes U_n \,.$$

is defined as the multilinear function such that:

$$\dot{\otimes} : (u_1, \ldots, u_n) \mapsto u_1 \otimes \cdots \otimes u_{h-1} \otimes (u_h \otimes \cdots \otimes u_k) \otimes u_{k+1} \otimes \cdots \otimes u_n \,.$$

Now the linear map $F_1' : (V_1 \otimes V_2) \otimes V_3 \to V_1 \otimes V_2 \otimes V_3$ which linearly extends

$$(v_1 \otimes v_2) \otimes v_3 \to v_1 \otimes v_2 \otimes v_3$$

is nothing but the function $H_1^{\dot{\otimes}}$

$$H : (v_1, v_2, v_3) \to v_1 \otimes v_2 \otimes v_3$$

when $n = 3, h = 1, k = 2$ and $W := V_1 \otimes V_2 \otimes V_3$.

Exercises 2.28

1. Consider a finite-dimensional vector space V and its dual V^*. Show, by the universality theorem, that there is a natural isomorphism such that

$$V \otimes V^* \simeq V^* \otimes V \,.$$

(*Hint*. Consider the bilinear map $f : V \times V^* \to V^* \otimes V$ with $f : (v_1, v_2^*) \mapsto v_2^* \otimes v_1$. Show that $f^{\otimes}$ is injective and thus surjective because $dim(V^* \otimes V) = dim(V \otimes V^*)$.)

2. Extend the result in Exercise 1 to the infinite-dimensional case.

3. If $\psi_i : V_i \to U_i$ are $\mathbb{K}$-vector-space isomorphisms for $i = 1, \ldots, n$, prove that

$$\Psi^{\otimes} : \bigotimes_{i=1}^{n} V_i \to \bigotimes_{i=1}^{n} U_i$$

is a $\mathbb{K}$-vector-space isomorphism if

$$\Psi : V_1 \times \cdots \times V_n \ni (v_1, \ldots, v_n) \mapsto \psi_1(v_1) \otimes \cdots \otimes \psi_n(v_n) \in U_1 \otimes \cdots \otimes U_n \,.$$

(*Hint*. Prove that $(\Psi^{\otimes})^{-1} = \Phi^{\otimes}$ for

$$\Phi : U_1 \times \cdots \times U_n \ni (u_1, \ldots, u_n) \mapsto \psi_1^{-1}(u_1) \otimes \cdots \otimes \psi_n^{-1}(u_n) \in V_1 \otimes \cdots \otimes V_n \,,$$

using the universal property of the tensor product.) ■

2.2.3 *Abstract Definition of the Tensor Product of Vector Spaces*

The property of the pair $(\otimes, U_1 \otimes \ldots \otimes U_n)$ stated in Theorem 2.26, the *universality property (of the tensor product)*, is very important from a theoretical point of view. As a matter of fact, it can be used, adopting a more advanced theoretical point of view, to *define* the tensor product of given vector spaces. This is due to the following important result.

Theorem 2.29 *Given $n \geq 1$ vector spaces $U_1, \ldots, U_n$ (not necessarily finite-dimensional) over the same field $\mathbb{K}$ (not necessarily $\mathbb{R}$ or $\mathbb{C}$)*[1] *suppose there is a pair (T, U_T) where U_T is a vector space over $\mathbb{K}$ and $T : U_1 \times \ldots \times U_n \to U_T$ is a multilinear map, fulfilling the universal property. That is, for any vector space W over $\mathbb{K}$ and any multilinear map $f : U_1 \times \ldots \times U_n \to W$, there is a unique linear map $f^T : U_T \to W$ such that the diagram below commutes,*

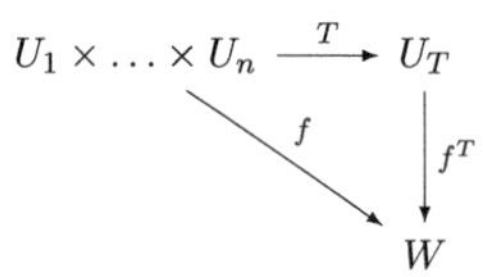

Under these hypotheses the pair (T, U_T) is determined up to vector-space isomorphisms.

[1] The theorem can be proved assuming that $U_1, \ldots, U_n, U_T$ are *modules* over a common *commutative ring* R. A module is defined as a vector space with the exception that the field of scalars is here replaced by a *ring*, which may also be non-commutative.

In other words, for any other pair (S, V_S) fulfilling the universality property with respect to $U_1, \ldots, U_n$, there is a unique isomorphism $\phi : V_S \to U_T$ such that $f^T \circ \phi = f^S$ for every multilinear map $f : U_1 \times \ldots \times U_n \to W$ and every vector space W.

Proof Suppose that there is another pair (S, V_S) fulfilling the universality property with respect to $U_1, \ldots, U_n$. Then, using the universal property with $f = T$ and $W = U_T$, we have the diagram below, with the former diagram commutation relation:

$$T^S \circ S = T\,. \tag{2.4}$$

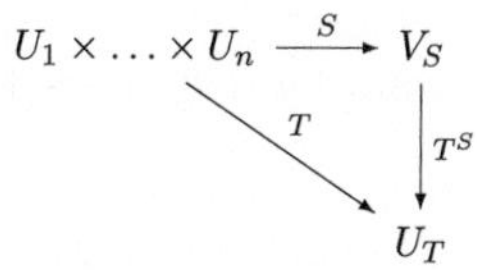

On the other hand, using the analogous property of the pair (T, U_T) with $f = S$ and $W = V_S$, we also have the commutative diagram

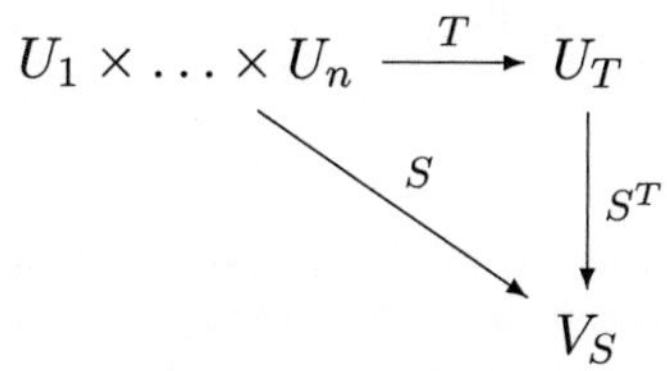

which involves a second diagram commutation relation:

$$S^T \circ T = S\,.$$

The two obtained relations imply

$$(T^S \circ S^T) \circ T = T$$

and

$$(S^T \circ T^S) \circ S = S\,,$$

Now observe that the following corresponding identities are also trivially valid:

$$id_{U_T} \circ T = T$$

and

$$id_{V_S} \circ S = S\,.$$

The former can be seen as a special case of $f^T \circ T = f$ for $f = T$ and the latter as a special case of $f^S \circ S = f$ for $f = S$. Since these identities uniquely define the respective function f due to the universal property, we conclude that $T^S \circ S^T = Id_{U_T}$ and $S^T \circ T^S = Id_{V_S}$. Summarizing, we have found that the linear map $\phi := T^S : V_S \to U_T$ is a vector-space isomorphism whose inverse is S^T. As far as the property $f^T \circ \phi = f^S$ is concerned, we observe that, by definition of f^S and f^T, the maps f^S and $f^T \circ \phi$ satisfy the universal property for f with respect to the pair (S, V_S):

$$f^S \circ S = f \quad \text{and} \quad (f^T \circ \phi) \circ S = f \,, \tag{2.5}$$

and thus f^S and $f^T \circ \phi$ have to coincide.
Indeed, the former in (2.5) is valid by definition. Furthermore, using (2.4), one achieves the latter:

$$(f^T \circ \phi) \circ S = f^T \circ T^S \circ S = f^T \circ (T^S \circ S) = f^T \circ T = f \,.$$

To conclude we prove that $\phi : V_S \to U_T$ verifying $f^T \circ \phi = f^S$ is unique. Indeed if $\phi' : V_S \to U_T$ is another isomorphism verifying $f^T \circ \phi' = f^S$, we have $f^T \circ (\phi' - \phi) = 0$ for every multilinear $f : U_1 \times \cdots \times U_n \to W$. Choose $W = U_T$ and $f = T$ so that $f^T = id_{U_T}$ and $f^T \circ (\phi' - \phi) = 0$ implies $\phi' - \phi = 0$, that is $\phi' = \phi$. □

Obviously, first of all this theorem applies to the pair $(\otimes, U_1 \otimes \cdots \otimes U_n)$ over the field $\mathbb{R}$ or $\mathbb{C}$. In fact, we wish to stress that, taking advantage of such a result, given generic vector spaces $U_1, \ldots, U_n$ over a generic field $\mathbb{K}$, one may say that a vector space U_T over $\mathbb{K}$, equipped with a multilinear map $T : U_1 \times \ldots \times U_n \to U_T$, is the **tensor product** of the spaces and T is the **tensor product map**, if the pair (T, U_T) fulfills the universality property. In this way (T, U_T) turns out to be defined *up to vector-space isomorphisms*. Obviously, if one aims to follow this very general way of defining tensor products, one still has to show that a pair (T, U_T) exists for the considered set of vector spaces $U_1, \ldots, U_n$. This is exactly what we have done within our constructive approach in the case when $\mathbb{K} = \mathbb{R}$ or $\mathbb{C}$. However, the existence of a pair (T, U_T) satisfying the universality property can be given similarly also for a larger class of vector spaces than that of finite-dimensional ones over the field $\mathbb{R}$ or $\mathbb{C}$ and also for *modules* over *commutative* rings. Moreover, the explicit construction of a pair (T, U_T) fulfilling the universality property can be produced, within a more abstract approach, by taking the quotient of a suitably freely generated module with respect to suitable submodules [1].

2.2.4 Hilbert Tensor Product of Hilbert Spaces

This subsection is devoted to defining the Hilbert tensor product of Hilbert spaces, which plays a key role in applications to Quantum Mechanics in reference to composite systems (see, e.g., [2]). The issue is quite far from the main line of this text and it may be omitted by the reader who is not strictly interested, also because it involves some advanced mathematical tools [3] which will not be treated in detail here (however see Chap. 5 for some elementary pieces of information about Hilbert spaces). We assume here that the reader is familiar with the basic notions of complex (generally nonseparable) Hilbert space and the notion of H bases (also called orthonormal systems).

Consider a set of complex Hilbert spaces $(H_i, (\cdot|\cdot)_i)$, where $i = 1, \ldots, n < +\infty$ and the spaces are not necessarily separable. Here $(\cdot|\cdot)_i$ denotes the Hermitian scalar product on H_i. (In this text, concerning Hermitian scalar products, we always assume the convention that each $(\cdot|\cdot)_i$ is antilinear in the left-hand entry.) Since the H_i are vector spaces over $\mathbb{C}$, their tensor product $H_1 \otimes \cdots \otimes H_n$ is defined as a purely algebraic object, but it does not support any preferred Hermitian scalar product. We wish to specialize the definition of tensor product in order to endow $H_1 \otimes \cdots \otimes H_n$ with the structure of a Hilbert space, naturally induced by the structures of the $(H_i, (\cdot|\cdot)_i)$. Actually, this specialization requires an extension of $H_1 \otimes \cdots \otimes H_n$ to assure the completeness of the space. The key tool is the following theorem whose nature is very close to that of the universality theorem (Theorem 2.26). This result allows us to define a preferred Hermitian product $(\cdot|\cdot)$ on $H_1 \otimes \cdots \otimes H_n$ induced by the $(\cdot|\cdot)_i$. The Hilbert structure generated by that scalar product on $H_1 \otimes \cdots \otimes H_n$ will be the desired one.

Theorem 2.30 *Given $n \geq 1$ complex Hilbert spaces $(H_i, (\cdot|\cdot)_i)$, $i = 1, \ldots, n < +\infty$ the following holds.*

(a) *For any complex vector space W and a map $f : (H_1 \times \cdots \times H_n) \times (H_1 \times \cdots \times H_n) \to W$ that is anti-linear in each one of the first n entries and linear in each one of the remaining n ones, there is a unique map $f^{\boxtimes} : (H_1 \otimes \ldots \otimes H_n) \times (H_1 \otimes \ldots \otimes H_n) \to W$ which is anti-linear in the left-hand argument and linear in the right-hand one and such that the diagram below commutes (in other words, $f^{\boxtimes} \circ \boxtimes = f$), where we have defined:*

$$\boxtimes : (v_1, \ldots, v_n, u_1, \ldots, u_n) \mapsto (v_1 \otimes \cdots \otimes v_n, u_1 \otimes \cdots \otimes u_n) \quad \text{for all } u_i, v_i \in H_i,\ i = 1, \ldots, n.$$

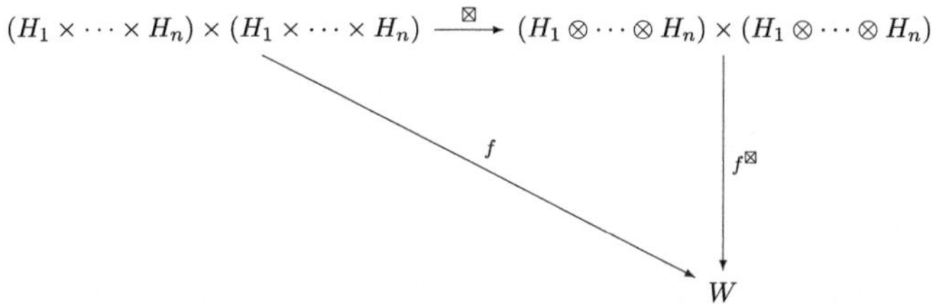

(b) *If $W := \mathbb{C}$ and it holds*

$$\begin{aligned} &f((v_1, \ldots, v_n), (u_1, \ldots, u_n)) \\ &\quad = \overline{f((u_1, \ldots, u_n), (v_1, \ldots, v_n))} \quad \text{for all } u_i, v_i \in H_i, i = 1, \ldots, n, \end{aligned}$$

then the map $f^{\boxtimes}$ as in (a) fulfills

$$f^{\boxtimes}(t, s) = \overline{f^{\boxtimes}(s, t)} \quad \text{for all } s, t \in H_1 \otimes \cdots \otimes H_n.$$

Proof (a) Fix (algebraic) bases $\{e_{k,i_k}\}_{i_k \in I_k} \subset H_k$, $k = 1, \ldots, n$. By (anti-)linearity, a map $g : (H_1 \otimes \ldots \otimes H_n) \times (H_1 \otimes \ldots \otimes H_n) \to W$, which is anti-linear in the left-hand entry and linear in the right-hand one, is uniquely determined by fixing the set of vectors

$$\{g(e_{1,i_1} \otimes \cdots \otimes e_{n,i_n}, e_{1,j_1} \otimes \cdots \otimes e_{n,j_n})\}_{i_1 \in I_1, \ldots, i_n \in I_n, j_1 \in I_1, \ldots, j_n \in I_n} \subset W\,.$$

Define the function $f^{\boxtimes} : (H_1 \otimes \ldots \otimes H_n) \times (H_1 \otimes \ldots \otimes H_n) \to W$ as the unique map, anti-linear in the left-hand argument and linear in the remaining one, with

$$f^{\boxtimes}(e_{i_1} \otimes \cdots \otimes e_{i_n}, e_{j_1} \otimes \cdots \otimes e_{j_n}) := f(e_{i_1}, \ldots, e_{i_n}, e_{j_1}, \ldots, e_{j_n})\,.$$

By construction, using (anti-)linearity of $f^{\boxtimes}$ and multi-(anti-)linearity of f, one gets immediately:

$$f^{\boxtimes}((v_1 \otimes \ldots \otimes v_n), (u_1 \otimes \ldots \otimes u_n)) = f(v_1, \ldots, v_n, u_1, \ldots, u_n)$$

for all $(v_1, \ldots, v_n), (u_1, \ldots, u_n) \in H_1 \times \ldots \times H_n$. In other words $f^{\boxtimes} \circ \boxtimes = f$. The uniqueness of the map $f^{\boxtimes}$ follows directly: suppose there is another map $g^{\boxtimes}$ with $g^{\boxtimes} \circ \boxtimes = f$ then $(f^{\boxtimes} - g^{\boxtimes}) \circ \boxtimes = 0$. This means in particular that

$$(f^{\boxtimes} - g^{\boxtimes})(e_{i_1} \otimes \ldots \otimes e_{i_n}, e_{j_1} \otimes \ldots \otimes e_{j_n}) = 0\,,$$

for all $i_k \in I_k$, $j_k \in I_k$, $k = 1, \ldots, n$. Since the coefficients above completely determine a map, the considered map must be the null map and thus: $f^{\boxtimes} = g^{\boxtimes}$. The proof of (b) follows straightforwardly from the construction of $f^{\boxtimes}$. □

Equipped with the established result, consider the map $f : (H_1 \times \cdots \times H_n) \times (H_1 \times \cdots \times H_n) \to \mathbb{C}$ that is anti-linear in the first n entries and linear in the remaining n entries and is defined by the requirement:

$$\begin{aligned} &f((v_1, \ldots, v_n), (u_1, \ldots, u_n)) := (v_1|u_1)_1 (v_2|u_2)_2 \cdots (v_n|u_n)_n, \\ &\quad \text{for all } v_i, u_i \in H_i, i = 1, \ldots, n. \end{aligned}$$

As $(v_k|u_k)_k = \overline{(u_k|v_k)_k}$ by definition of Hermitian scalar product, the hypothesis in (b) of the theorem above is verified and thus the application

$$f^{\boxtimes} : (H_1 \otimes \cdots \otimes H_n) \times (H_1 \otimes \cdots \otimes H_n) \to \mathbb{C}$$

defined in (a) of the mentioned theorem, is linear in the right-hand argument, anti-linear in the left-hand one and fulfills

$$f^{\boxtimes}(t, s) = \overline{f^{\boxtimes}(s, t)}\,, \quad \text{for all } s, t \in H_1 \otimes \cdots \otimes H_n.$$

If we were able to prove that $f^{\boxtimes}(\cdot, \cdot)$ is positive-definite, i.e. $f^{\boxtimes}(s, s) \geq 0$ for every $s \in H_1 \otimes \cdots \otimes H_n$ as well as $f^{\boxtimes}(s, s) = 0$ implies $s = 0$, then $f^{\boxtimes}(\cdot, \cdot)$ would define a Hermitian scalar product on $H_1 \otimes \cdots \otimes H_n$. This is just the last ingredient we need to give the desired definition of H tensor product. Let us prove it in the form of a proposition.

Proposition 2.31 *Given $n \geq 1$ complex Hilbert spaces $(H_i, (\cdot|\cdot)_i)$, $i = 1, \ldots, n < +\infty$, the unique map $(\cdot|\cdot) : (H_1 \otimes \cdots \otimes H_n) \times (H_1 \otimes \cdots \otimes H_n) \to \mathbb{C}$ which is linear in the right-hand argument and anti-linear in the left-hand one and verifies*

$$(v_1 \otimes \cdots \otimes v_n | u_1 \otimes \cdots \otimes u_n) := (v_1|u_1)_1 (v_2|u_2)_2 \cdots (v_n|u_n)_n,$$
$$\textit{for all } v_i, u_i \in H_i, i = 1, \ldots, n,$$

is a Hermitian scalar product on $H_1 \otimes \cdots \otimes H_n$.

Proof The only point we have to demonstrate is that $(\cdot|\cdot)$ is positive-definite. To this end, fix an algebraic basis $\{e_{k,i_k}\}_{i_k \in I_k}$ for every H_k. If $S = S^{i_1 \cdots i_n} e_{1,i_1} \otimes \cdots \otimes e_{n,i_n} \in H_1 \otimes \cdots \otimes H_n$, the sum over repeated indices is always finite in view of the algebraic nature of the bases, even if each I_k may be infinite and uncountable. Similarly, fix a Hilbert basis $\{h_{k,\alpha_k}\}_{\alpha_k \in A_k}$ for every H_k. Notice that the set of indices A_k may also be uncountable; however, in the standard Hilbert-space decomposition, $(u|v)_k = \sum_{\alpha_k \in A_k} (u|h_{k,\alpha_k})_k (h_{k,\alpha_k}|v)_k$ the set of non-vanishing terms $(u|h_{k,\alpha_k})_k$ and $(h_{k,\alpha_k}|v)_k$ is countable and the sum reduces to an (absolutely convergent) standard series. In the following we do *not* omit the summation symbol over the indices varying in every A_k, but we do follow that convention concerning algebraic indices $i_k, j_k \in I_k$. Notice also that the order of infinite sums cannot be interchanged in general. Under our hypotheses we have for every $S \in H_1 \otimes \cdots \otimes H_n$:

$$(S|S) = \overline{S^{i_1 \cdots i_n}} S^{j_1 \cdots j_n} (e_{1,i_1}|e_{1,j_1})_1 \cdots (e_{n,i_n}|e_{n,j_n})_n$$

$$= \sum_{\alpha_1 \in A_1} \cdots \sum_{\alpha_n \in A_n} \overline{S^{i_1 \cdots i_n}} S^{j_1 \cdots j_n} (e_{1,i_1}|h_{1,\alpha_1})_1 (h_{1,\alpha_1}|e_{1,j_1})_1$$
$$\cdots (e_{n,i_n}|h_{n,\alpha_n})_n (h_{n,\alpha_n}|e_{n,j_n})_n$$

$$= \sum_{\alpha_1 \in A_1} \cdots \sum_{\alpha_n \in A_n} \overline{S^{i_1 \cdots i_n}} S^{j_1 \cdots j_n} \overline{(h_{1,\alpha_1}|e_{1,i_1})_1} (h_{1,\alpha_1}|e_{1,j_1})_1$$

$$\cdots \overline{(h_{n,\alpha_n}|e_{n,i_n})_n} \cdots (h_{n,\alpha_n}|e_{n,j_n})_n$$

$$= \sum_{\alpha_1 \in A_1} \cdots \sum_{\alpha_n \in A_n} \overline{S^{i_1 \cdots i_n} (h_{1,\alpha_1}|e_{1,i_1})_1 \cdots (h_{n,\alpha_n}|e_{n,i_n})_n} \; S^{j_1 \cdots j_n} (h_{1,\alpha_1}|e_{1,j_1})_1$$

$$\cdots (h_{n,\alpha_n}|e_{n,j_n})_n$$

$$= \sum_{\alpha_1 \in A_1} \cdots \sum_{\alpha_n \in A_n} \left| S^{j_1 \cdots j_n} (h_{1,\alpha_1}|e_{1,j_1})_1 \cdots (h_{n,\alpha_n}|e_{n,j_n})_n \right|^2 \geq 0 .$$

Semi-positivity is thus established. We have to prove that $(S|S) = 0$ entails $S = 0$. From the last line of the expansion of $(S|S)$ obtained above, we conclude that $(S|S) = 0$ implies

$$S^{j_1 \cdots j_n} (h_{1,\alpha_1}|e_{1,j_1})_1 \cdots (h_{n,\alpha_n}|e_{n,j_n})_n = 0$$

for every given set of elements $h_{1,\alpha_1}, \ldots, h_{n,\alpha_n}$. Now fix $\alpha_2, \ldots, \alpha_n$ to arbitrarily assigned values and define

$$T_{\alpha_2 \ldots \alpha_n} := S^{j_1 j_2 \cdots j_n} (h_{2,\alpha_2}|e_{2,j_2})_2 \cdots (h_{n,\alpha_n}|e_{n,j_n})_n e_{1,j_1} \in H_1 .$$

The preceding identity says precisely that

$$(h_{1,\alpha_1}|T_{\alpha_2 \ldots \alpha_n})_1 = 0$$

for every $\alpha_1 \in A_1$. Since $\{h_{1,\alpha_1}\}_{\alpha_1 \in A_1}$ is a Hilbert basis, $T_{\alpha_2 \ldots \alpha_n} = 0$. Since $\{e_{1,i_1}\}_{i_1 \in I_1}$ is an algebraic basis, all the corresponding coefficients vanish, namely

$$S^{i_1 j_2 \cdots j_n} (h_{2,\alpha_2}|e_{2,j_2})_2 \cdots (h_{n,\alpha_n}|e_{n,j_n})_n = 0 .$$

Iterating the same argument with $H_2, \ldots, H_n$ we achieve the final identity:

$$S^{i_1 i_2 \cdots i_n} = 0 , \quad \text{for every } i_1 \in I_1, \ldots, i_n \in I_n,$$

that is $S = 0$. □

We have just established that the algebraic tensor product $H_1 \otimes \cdots \otimes H_n$ can be equipped with a natural Hermitian scalar product $(\cdot|\cdot)$ naturally induced by the Hermitian scalar products in the factors H_i. At this point, recall that, given a complex vector space V with a Hermitian scalar product $(\cdot, \cdot)$, there is a unique (up to Hilbert-space isomorphisms) Hilbert space which admits V as a dense subspace and whose scalar product is obtained as a continuous extension of $(\cdot, \cdot)_V$. This distinguished Hilbert space is the **Hilbert completion** of $(V, (\cdot, \cdot)_V)$ [3]. We have all the ingredients to state the definition of the H tensor product.

Definition 2.32 (Hilbert Tensor Product) Given $n \geq 1$ complex Hilbert spaces $(H_i, (\cdot|\cdot)_i)$, $i = 1, \ldots, n < +\infty$, the **Hilbert tensor product** $H_1 \otimes_H \cdots \otimes_H H_n$ of those spaces is the Hilbert completion of the space $H_1 \otimes \cdots \otimes H_n$ with respect to the unique scalar product $(\cdot|\cdot) : H_1 \otimes \cdots \otimes H_n \times H_1 \otimes \cdots \otimes H_n \to \mathbb{C}$ verifying

$$(v_1 \otimes \cdots \otimes v_n | u_1 \otimes \cdots \otimes u_n) := (v_1|u_1)_1 (v_2|u_2)_2 \cdots (v_n|u_n)_n ,$$

$$\text{for all } v_i, u_i \in H_i, i = 1, \ldots, n .$$

■

To conclude, we prove the following useful result.

Proposition 2.33 *If $\{h_{k,\alpha_k}\}_{\alpha_k \in A_k}$ is a Hilbert basis of the complex Hilbert space H_k, for every $k = 1, \ldots, n$, then $\{h_{1,\alpha_1} \otimes \cdots \otimes h_{n,\alpha_n}\}_{\alpha_k \in A_k, k=1,\ldots,n}$ is a Hilbert basis of $H_1 \otimes_H \cdots \otimes_H H_n$.*
As a consequence, $H_1 \otimes_H \cdots \otimes_H H_n$ is separable whenever all the factors H_k are separable.

Proof In the following $B_k := \{h_{k,\alpha_k}\}_{\alpha_k \in A_k}$, $< B_k >$ denotes the dense subspace of H_k generated by B_k, $B := \{h_{1,\alpha_1} \otimes \cdots \otimes h_{n,\alpha_n}\}_{\alpha_k \in A_k, k=1,\ldots,n}$, and $< B >$ is the subspace of $H_1 \otimes_H \cdots \otimes_H H_n$ generated by B. Finally, in the rest of the proof, the isometric linear map with dense range $j : H_1 \otimes \cdots \otimes H_n \to H_1 \otimes_H \cdots \otimes_H H_n$ defining the Hilbert completion of $H_1 \otimes \cdots \otimes H_n$ will be omitted in order to simplify the notation (one can always reduce to this case). Using the definition of the scalar product $(\cdot|\cdot)$ in $H_1 \otimes_H \cdots \otimes_H H_n$ given in Definition 2.32 one sees that, trivially, B is an orthonormal set. Therefore, to conclude the proof, it is sufficient to prove that $\overline{< B >} = H_1 \otimes_H \cdots \otimes_H H_n$, where the bar indicates the topological closure. To this end consider also algebraic bases $D_k := \{e_{k,i_k}\}_{i_k \in I_k} \subset H_k$ made of *normalized vectors*. Since $\overline{< B_k >} = H_k$, given $m = 1, 2, \ldots$, there is $u_{k,i_k} \in < B_k >$ such that $||e_{k,i_k} - u_{k,i_k}||_k < 1/m$. Thus

$$||u_{k,i_k}||_k \leq ||e_{k,i_k}||_k + ||e_{k,i_k} - u_{k,i_k}||_k < 1 + 1/m .$$

Therefore, using $||v_1 \otimes \cdots \otimes v_n|| = ||v_1||_1 \cdots ||v_n||_n$ and the triangular inequality (see the remark after Definition 5.3), we achieve: $||e_{1,i_1} \otimes \cdots \otimes e_{n,i_n} - u_{1,i_1} \otimes \cdots \otimes u_{n,i_n}||$

$$= ||(e_{1,i_1} - u_{1,i_1}) \otimes e_{2,i_2} \otimes \cdots \otimes e_{n,i_n} + u_{1,i_1} \otimes (e_{2,i_2} - u_{2,i_2}) \otimes \cdots \otimes e_{n,i_n}$$
$$+ \cdots + u_{1,i_1} \otimes u_{2,i_2} \otimes \cdots \otimes (e_{n,i_n} - u_{n,i_n})||$$

$$\leq ||(e_{1,i_1} - u_{1,i_1}) \otimes e_{2,i_2} \otimes \cdots \otimes e_{n,i_n}||$$

$$+ ||u_{1,i_1} \otimes (e_{2,i_2} - u_{2,i_2}) \otimes \cdots \otimes e_{n,i_n}|| + \cdots$$

$$+ ||u_{1,i_1} \otimes u_{2,i_2} \otimes \cdots \otimes (e_{n,i_n} - u_{n,i_n})|| \leq n\frac{(1+1/m)^{n-1}}{m} \leq n\frac{2^{n-1}}{m} .$$

We conclude that, fixing $\epsilon > 0$, choosing m large enough, there is $y_{i_1 \ldots i_n} \in < B >$ such that

$$||e_{1,i_1} \otimes \cdots \otimes e_{n,i_n} - y_{i_1 \ldots i_n}|| < \epsilon . \tag{2.6}$$

The vectors $e_{1,i_1} \otimes \cdots \otimes e_{n,i_n}$ generate $H_1 \otimes \cdots \otimes H_n$ and linear combinations of elements $y_{i_1 \ldots i_n}$ as above belong to $< B >$. Hence, (2.6) together with the homogeneity property of the norm as well as the triangular inequality (see the remarks after Definition 5.3), imply that, for every $v \in H_1 \otimes \cdots \otimes H_n$ and every $\epsilon > 0$, there is $u \in < B >$ such that:

$$||u - v|| \leq \epsilon/2 . \tag{2.7}$$

Now notice that, for every $v_0 \in H_1 \otimes_H \cdots \otimes_H H_n$ and any fixed $\epsilon > 0$, there is $v \in H_1 \otimes \cdots \otimes H_n$ with $||v_0 - v|| \leq \epsilon/2$, because $\overline{H_1 \otimes \cdots \otimes H_n} = H_1 \otimes_H \cdots \otimes_H H_n$. Therefore, fixing $u \in < B >$ as in (2.7), we have that, for every $v_0 \in H_1 \otimes_H \cdots \otimes_H H_n$ and any fixed $\epsilon > 0$, there is $u \in < B >$ with $||v_0 - u|| \leq \epsilon$. In other words $\overline{< B >} = H_1 \otimes_H \cdots \otimes_H H_n$ as wanted. □

Remarks 2.34 Consider the case, quite usual in quantum mechanics, where the relevant Hilbert spaces are spaces of *square-integrable functions* [3], $H_i := L^2(X_i, d\mu_i)$, where every X_i is a σ-finite space with measure μ_i, $i = 1, \ldots, n$. For instance $X_i = \mathbb{R}^3$ and $\mu_i = \mu$ is the standard Lebesgue measure (in this case every H_i is the state space of a spinless quantum particle). In the considered case, on the one hand one can define the tensor product $H_1 \otimes_H \cdots \otimes_H H_n = L^2(X_1, d\mu_1) \otimes_H \cdots \otimes_H L^2(X_n, d\mu_n)$; on the other hand another natural object, relying on the given mathematical structures, is the Hilbert space $H := L^2(X_1 \times \cdots \times X_n, d\mu_1 \otimes \cdots \otimes d\mu_n)$, where the product measure $\mu_1 \otimes \cdots \otimes \mu_n$ is well defined since the spaces X_i are σ-finite. It turns out that these two spaces are naturally isomorphic and the isomorphism is the unique linear and continuous extension of the map:

$$L^2(X_1, d\mu_1) \otimes_H \cdots \otimes_H L^2(X_n, d\mu_n) \ni \psi_1 \otimes \cdots \otimes \psi_n \mapsto \psi_1 \cdots \psi_n \in L^2(X_1 \times \cdots \times X_n, d\mu_1 \otimes \cdots \otimes d\mu_n)$$

where

$$\psi_1 \cdots \psi_n(x_1, \ldots, x_n) := \psi_1(x_1) \cdots \psi_n(x_n) \quad \text{for all } x_i \in X_i, i = 1, \ldots, n..$$

Some textbooks on quantum mechanics define the Hilbert space of a composite system of n spinless particles as $L^2(X_1 \times \cdots \times X_n, d\mu \otimes \cdots \otimes d\mu)$ rather than $L^2(\mathbb{R}^3, d\mu)\otimes_H \cdots \otimes_H L^2(\mathbb{R}^3, d\mu)$. Actually, the two definitions are mathematically equivalent. ■

References

1. Lang, S.: Algebra. Graduate Texts in Mathematics, rev. 3rd edn. Springer, Berlin (1995)
2. Moretti, V.: Spectral Theory and Quantum Mechanics: Mathematical Foundations of Quantum Theories, Symmetries and Introduction to the Algebraic Formulation, 2nd edn. Springer, Berlin (2017)
3. Rudin, W.: Real and Complex Analysis. McGraw-Hill, New York (1987)
4. Streater, R.F., Wightman, A.S.: PCT, Spin and Statistics, and All That with Revised Preface and Corrections. Princeton University Press, Princeton (2000)
5. Wald, R.M.: General Relativity. University of Chicago Press, Chicago (1984)

Chapter 3
Tensor Algebra, Abstract Index Notation and Some Applications

This chapter turns the tensor product construction into a systematic calculus. Once a vector space is fixed, the various tensor spaces generated by the space and its dual are organized into a tensor algebra, and the usual operations on tensors are described in a unified way. The abstract index notation is then introduced as a bridge between the intrinsic language of multilinear algebra and the component notation familiar in physics.

The guiding idea is that indices should not be regarded only as labels for components: they also encode the tensorial type of an object, the admissible contractions, and the invariant meaning of a formula. This is why the chapter discusses canonical bases, tensor transformations, contractions, decompositions, and the construction of new tensors from old ones. The final sections apply this language to affine and Euclidean spaces, preparing the formalism later used for spacetime, reference frames, and relativistic quantities.

3.1 Tensor Algebra Generated by a Vector Space

Let V be a vector space over the field $\mathbb{K} = \mathbb{R}$ or $\mathbb{C}$. Equipped with this algebraic structure, in principle, we may build up several different tensor spaces. Notice that we have to consider also $\mathbb{K}$, $\mathbb{K}^*$ and V^* as admissible tensor factors. (In the following we shall not be interested in conjugate spaces.) Obviously, we are interested in tensor products which are not identifiable by some natural isomorphism.

First consider the dual $\mathbb{K}^*$ when we regard $\mathbb{K}$ as a vector space over the field $\mathbb{K}$ itself. $\mathbb{K}^*$ is made of linear functionals from $\mathbb{K}$ to $\mathbb{K}$. Each $c^* \in \mathbb{K}^*$ has the form

V. Moretti, *Geometric Methods in Mathematical Physics I*, La Matematica per il 3+2 184, https://doi.org/10.1007/978-3-032-31512-0_3

$c^*(k) := c \cdot k$, for all $k \in \mathbb{K}$, where $c \in \mathbb{K}$ is a fixed field element which completely determines c^*. The map $c \mapsto c^*$ is a natural vector space isomorphism. Therefore

$$\mathbb{K} \simeq \mathbb{K}^* .$$

The natural isomorphism $\mathbb{K} \simeq \mathbb{K}^*$, taking advantage of the result in (3) of Exercises 2.28, immediately induces an analogous isomorphism $\mathbb{K} \otimes \ldots \otimes \mathbb{K} \simeq \mathbb{K}^* \otimes \ldots \otimes \mathbb{K}^*$. We therefore pass to consider products $\mathbb{K} \otimes \ldots \otimes \mathbb{K} \simeq \mathbb{K}^* \otimes \ldots \otimes \mathbb{K}^* = \mathscr{L}(\mathbb{K}, \ldots, \mathbb{K})$. Each multilinear map $f \in \mathscr{L}(\mathbb{K}, \ldots, \mathbb{K})$ is completely determined by the number $f(1, \ldots, 1)$, since $f(k_1, \ldots, k_n) = k_1 \cdots k_n f(1, \ldots, 1)$. A direct verification shows that the map $f \mapsto f(1, \ldots, 1)$ is a natural vector space isomorphism between $\mathbb{K} \otimes \ldots \otimes \mathbb{K}$ and $\mathbb{K}$ itself. Therefore

$$\mathbb{K} \otimes \ldots \otimes \mathbb{K} \simeq \mathbb{K} .$$

The found isomorphism satisfies $c_1 \otimes \ldots \otimes c_n \mapsto c_1 \cdots c_n$, and thus the tensor product map reduces to the ordinary product in the field.

We now pass to consider the product $\mathbb{K} \otimes V \simeq \mathbb{K}^* \otimes V = \mathscr{L}(\mathbb{K}, V^*)$. Each multilinear functional f in $\mathscr{L}(\mathbb{K}, V^*)$ is completely determined by the functional on V^*, $f(1, \cdot) : V^* \to \mathbb{K}$, which maps each $v^* \in V^*$ to $f(1, v^*)$. Again, a direct verification shows that $f \mapsto f(1, \cdot)$ is a natural vector space isomorphism between $\mathbb{K} \otimes V$ and V itself:

$$\mathbb{K} \otimes V \simeq V .$$

Notice that the found isomorphism satisfies $k \otimes v \mapsto kv$ and thus the tensor product map reduces to the ordinary product of a field element and a vector.

Recalling that $\otimes$ is associative and * is involutive and, in the finite-dimensional case, it is also distributive, as established in Chap. 2, we conclude that only the spaces $\mathbb{K}$, V, V^* and all the tensor products of V and V^* (in whatever order and number) may be significantly different. This result leads us to the following definition, which we extend to the infinite-dimensional case (it is however worth stressing that, in the infinite-dimensional case, some of the natural isomorphisms encountered in the finite-dimensional case are not valid, for instance $(V \otimes V)^*$ may be larger than $V^* \otimes V^*$).

Definition 3.1 (Tensor Algebra Generated by V) Let V be a vector space over the field $\mathbb{K}$.

(1) The **tensor algebra** $\mathscr{A}_{\mathbb{K}}(V)$ generated by V over the field $\mathbb{K}$ is the disjoint union of the vector spaces $\mathbb{K}$, V, V^* and all tensor products of factors V and V^* in whatever order and number.

(2) (i) The tensors of $\mathbb{K}$ are called **scalars**,
(ii) the tensors of V are called **contravariant vectors**,
(iii) the tensors of V^* are called **covariant vectors**,

(iv) the tensors of spaces $V^{n\otimes} := \underbrace{V \otimes \ldots \otimes V}_{n\ times}$ are called **contravariant tensors of order** n or **tensors of order** $(n, 0)$,

(v) the tensors of spaces $V^{*n\otimes} := \underbrace{V^* \otimes \ldots \otimes V^*}_{n\ times}$ are called **covariant tensors of order** n or **tensors of order** $(0, n)$,

(vi) the remaining tensors, belonging to spaces containing n factors V and m factors V^* in a prescribed order, are called **tensors of order** (n, m).

■

Remarks 3.2 Obviously, pairs of tensor spaces made of the same number of factors V and V^* in different order are naturally isomorphic (see Exercise 2.28 1). However, for practical reasons it is convenient to consider these spaces as different spaces and use the identifications when and if necessary. ■

Definition 3.3 (Associative Algebra) An *associative algebra* over $\mathbb{K} = \mathbb{R}$ or $\mathbb{C}$ is a vector space $\mathscr{A}$ over $\mathbb{K}$ equipped with a map $\circ : \mathscr{A} \times \mathscr{A} \to \mathscr{A}$ which is an *associative product*, i.e.,

$$A \circ (B \circ C) = (A \circ B) \circ C$$

if $A, B, C \in \mathscr{A}$, such that it is *two-sided distributive* with respect to the vector space sum:

$$A \circ (B + C) = A \circ B + A \circ C\,, \quad (B + C) \circ A = B \circ A + C \circ A$$

if $A, B, C \in \mathscr{A}$ and finally satisfies

$$a(A \circ B) = (aA) \circ B = A \circ (aB)$$

for all $A, B \in \mathscr{A}$ and $a \in \mathbb{K}$.
The algebra $\mathscr{A}$ is said to be **commutative** or **Abelian** if $A \circ B = B \circ A$ for every $A, B \in \mathscr{A}$.

The algebra $\mathscr{A}$ is said to be **unital** if there is $I\!I \in \mathscr{A}$, called the **unit element**, such that $A \circ I\!I = I\!I \circ A = A$ for all $A \in \mathscr{A}$. ■

A simple example of a (non-commutative for $n > 1$) unital algebra over $\mathbb{K}$ is the set of the $n \times n$ matrices with coefficients in $\mathbb{K}$ and equipped with the standard structure of vector space over $\mathbb{K}$, the associative product being the standard product of matrices and the unit element being the identity matrix I.

According to our previous definition $\mathscr{A}_{\mathbb{K}}(V)$ is *not* an *associative algebra.* Indeed, even if it has a notion of product, it does not enjoy a structure of vector space over $\mathbb{K}$, since linear combinations of tensors of different type are not defined. For this reason $\mathscr{A}_{\mathbb{K}}(V)$ is sometimes called the "weak tensor algebra".

It is however possible to strengthen this structure, obtaining a proper associative algebra in the sense of the stated definition, if one suitably interprets the tensor product. In this way, the various spaces of tensors turn out to be proper subspaces of the constructed structure. However, linear combinations of tensors of different type have no natural interpretation, especially in physical applications.

As a matter of fact, let us denote by S_α the generic tensor space in $\mathscr{A}_{\mathbb{K}}(V)$. More precisely, if

$$\mathscr{A}_{\mathbb{K}}(V) = \bigsqcup_{\alpha \in A} S_\alpha \,,$$

where A is a set of indices with $S_\alpha \neq S_\beta$ if $\alpha \neq \beta$, then we can define a natural $\mathbb{K}$-vector space structure on

$$\bigoplus_{\alpha \in A} S_\alpha := \left\{ \{v_\alpha\}_{\alpha \in A} \;\middle|\; v_\alpha \in S_\alpha \text{ and } v_\alpha = 0 \text{ for all but finitely many } \alpha \right\} ,$$

the notation stressing that we are now considering the direct sum. The notion of linear combination is given by, for $a, b \in \mathbb{K}$,

$$a\{v_\alpha\}_{\alpha \in A} + b\{u_\alpha\}_{\alpha \in A} := \{a v_\alpha + b u_\alpha\}_{\alpha \in A} \,.$$

Finally the associative algebra structure arises by defining the associative product on $\bigoplus_{\alpha \in A} S_\alpha$:

$$\{v_\alpha\}_{\alpha \in A} \bigotimes \{u_\beta\}_{\beta \in A} := \{w_\gamma\}_{\gamma \in A} \,,$$

where

$$w_\gamma := \sum_{S_\alpha \otimes S_\beta \simeq S_\gamma} v_\alpha \otimes u_\beta \,.$$

Here we have identified the elements $v_\alpha \otimes u_\beta$ with the corresponding elements of tensor spaces S_γ in $\mathscr{A}_{\mathbb{K}}(V)$ making use of the vector space isomorphisms in Theorem 2.24. Notice that products $v_\alpha \otimes u_\beta$ and $v_{\alpha'} \otimes u_{\beta'}$ with $\alpha \neq \alpha'$ and $\beta \neq \beta'$ may belong to the same tensor space S_γ in $\mathscr{A}_{\mathbb{K}}(V)$.

3.2 The Abstract Index Notation and Rules to Handle Tensors

3.2.1 Canonical Bases and Abstract Index Notation

Let us introduce the **abstract index notation**. Consider a finite-dimensional vector space V over the field $\mathbb{K} = \mathbb{R}$ or $\mathbb{C}$. After specifying a basis $\{e_i\}_{i \in I} \subset V$ and the corresponding basis in V^*, each tensor is completely determined by giving its components with respect to the induced basis in the corresponding tensor space in $\mathscr{A}_{\mathbb{K}}(V)$. We are interested in the transformation rule of these components under a change of basis in V. Suppose we fix another basis $\{e'_j\}_{j\in I} \subset V$ with $e_i = A^j{}_i e'_j$. The coefficients $A^j{}_i$ determine a matrix $A := [A^j{}_i]$ in the matrix group $GL(dimV, \mathbb{K})$, i.e., the group (see Sect. 4.1) of $dimV \times dimV$ matrices with coefficients in $\mathbb{K}$ and non-vanishing determinant. First consider a contravariant vector $t = t^i e_i$. Passing to the other basis, we have $t = t^i e_i = t'^j e'_j$, and thus $t'^j e'_j = t^i A^j{}_i e'_j$. This is equivalent to $(t'^j - A^j{}_i t^i) e'_j = 0$, which implies

$$t'^j = A^j{}_i t^i \ ,$$

because of the linear independence of the vectors e'_j. Similarly, if we specify a set of components in $\mathbb{K}$, $\{t^i\}_{i\in I}$, for each basis $\{e_i\}_{i\in I} \subset V$ and these components, changing the basis to $\{e'_j\}_{j\in I}$, transform as

$$t'^j = A^j{}_i t^i \ ,$$

where the coefficients $A^j{}_i$ are defined by

$$e_i = A^j{}_i e'_j \ ,$$

then a contravariant tensor t is defined. It is determined by $t := t^i e_i$ in each basis. The proof is self-evident.

Concerning covariant vectors, a similar result holds. Indeed a covariant vector $u \in V^*$ is completely determined by the specification of a set of components $\{u_i\}_{i\in I}$ for each basis $\{e^{*i}\}_{i\in I} \subset V^*$ (dual basis of $\{e_i\}_{i\in I}$ above) when these components, changing the basis to $\{e'^{*j}\}_{j\in I}$ (dual basis of $\{e'_j\}_{j\in I}$ above), transform as

$$u'_j = B_j{}^i u_i \ ,$$

where the coefficients $B_l{}^r$ are defined by

$$e^{*i} = B_j{}^i e'^{*j} \ .$$

The matrices $A = [A^j{}_i]$ and $B := [B_k{}^h]$ are related by

$$\delta^i_j = \langle e_j, e^{*i} \rangle = A^l{}_j B_k{}^i \langle e'_l, e'^{*k} \rangle = A^l{}_j B_k{}^i \delta^k_l = A^k{}_j B_k{}^i \ .$$

In other words it has to hold $I = A^t B$, which is equivalent to

$$B = A^{-1t} .$$

(Notice that ^{t} and $^{-1}$ commute.)
Proper tensors have components which transform similarly to the vector case. For instance, consider $t \in V \otimes V^*$. Fix a basis $\{e_i\}_{i\in I} \subset V$, the dual one $\{e^{*i}\}_{i\in I} \subset V^*$ and consider the induced basis in $V \otimes V^*$, $\{e_i \otimes e^{*j}\}_{(i,j)\in I\times I}$. Then $t = t^i{}_j e_i \otimes e^{*j}$. By bi-linearity of the tensor product map, if we pass to consider another basis $\{e'_i\}_{i\in I} \subset V$ and those associated in the relevant spaces as above, concerning the components $t'^k{}_l$ of t in the new tensor-space basis, one obtains

$$t'^k{}_l = A^k{}_i B_l{}^j t^i{}_j ,$$

where the matrices $A = [A^j{}_i]$ and $B := [B_k{}^h]$ are those considered above. The specification of a tensor of $V \otimes V^*$ is completely equivalent to the specification of a set of components for each basis of $V \otimes V^*$, $\{e_i \otimes e^{*j}\}_{(i,j)\in I\times I}$, provided these components transform as specified above under change of basis.
We can generalize the obtained results after a definition.

Definition 3.4 (Canonical Bases) Let $\mathscr{A}_{\mathbb{K}}(V)$ be the tensor algebra generated by the finite-dimensional vector space V over the field $\mathbb{K}$ ($= \mathbb{C}$ or $\mathbb{R}$). If $B = \{e_i\}_{i\in I}$ is a basis in V with dual basis $B^* = \{e^{*i}\}_{i\in I} \subset V^*$, the **canonical bases** associated with the former are the bases in the tensor spaces of $\mathscr{A}_{\mathbb{K}}(V)$ obtained by tensor products of elements of B and B^*. ■

Remarks 3.5 Notice that also $\{e_i \otimes e'_j\}_{i,j\in I}$ is a basis of $V \otimes V$ if $\{e_i\}_{i\in I}$ and $\{e'_j\}_{j\in I}$ are bases of V. However, $\{e_i \otimes e'_j\}_{i,j\in I}$ is *not* canonical unless $e_i = e'_i$ for all $i \in I$. ■

Theorem 3.6 *Consider the tensor algebra $\mathscr{A}_{\mathbb{K}}(V)$ generated by a finite-dimensional vector space V over the field $\mathbb{K} = \mathbb{R}$ or $\mathbb{C}$, and take a tensor space $V^{n\otimes} \otimes V^{*m\otimes} \in \mathscr{A}_{\mathbb{K}}(V)$. Defining a tensor $t \in V^{n\otimes} \otimes V^{*m\otimes}$ is completely equivalent to assigning a set of components*

$$\{t^{i_1\ldots i_n}{}_{j_1\ldots j_m}\}_{i_1,\ldots,i_n,j_1,\ldots,j_m\in I}$$

with respect to each canonical basis of $V^{n\otimes} \otimes V^{*m\otimes}$,

$$\{e_{i_1} \otimes \ldots \otimes e_{i_n} \otimes e^{*j_1} \otimes \ldots \otimes e^{*j_m}\}_{i_1,\ldots,i_n,j_1,\ldots,j_m\in I}$$

which, under change of basis to

$$\{e'_{i_1} \otimes \ldots \otimes e'_{i_n} \otimes e'^{*j_1} \otimes \ldots \otimes e'^{*j_m}\}_{i_1,\ldots,i_n,j_1,\ldots,j_m\in I} ,$$

transform as

$$t'^{i_1\ldots i_n}{}_{j_1\ldots j_m} = A^{i_1}{}_{k_1}\ldots A^{i_n}{}_{k_n} B_{j_1}{}^{l_1}\ldots B_{j_m}{}^{l_m} t^{k_1\ldots k_n}{}_{l_1\ldots l_m}\,,$$

where

$$e_i = A^j{}_i e'_j\,,$$

and the coefficients $B_j{}^l$ are those of the matrix

$$B = A^{-1t}\,,$$

with $A := [A^j{}_i]$. The associated tensor t is represented by

$$t = t^{i_1\ldots i_n}{}_{j_1\ldots j_m} e_{i_1}\otimes\ldots\otimes e_{i_n}\otimes e^{*j_1}\otimes\ldots\otimes e^{*j_m}$$

for each considered canonical basis. Analogous results hold for tensor spaces whose factors V and V^* take different positions.

Notation 3.7 In the **abstract index notation** a tensor is indicated by writing its generic component in a non-specified basis. E.g. $t \in V^*\otimes V$ is indicated by $t_i{}^j$. ■

We shall occasionally adopt a *cumulative index notation*, i.e., letters $A, B, C, \ldots$ denote sets of covariant, contravariant or mixed indices. For instance $t^{ijk}{}_{lm}$ can be written as t^A with ${}^A = {}^{ijk}{}_{lm}$. Similarly e_A denotes the element of a canonical basis $e_i\otimes e_j\otimes e_k\otimes e^{*l}\otimes e^{*m}$. Moreover, if ^{A} and ^{B} are cumulative indices, the indices of the cumulative index ^{AB} are those of ^{A} immediately followed by those of ^{B}, e.g., if ${}^A = {}^{ijk}{}_{lm}$ and ${}^B = {}^{pq}{}_u{}^n$, ${}^{AB} = {}^{ijk}{}_{lm}{}^{pq}{}_u{}^n$.

3.2.2 Rules to Compose, Decompose, Produce Tensors from Tensors

Let us specify the allowed mathematical rules to produce tensors from given tensors. To this end we shall use both the synthetic and the index notation.

Linear Combinations of Tensors of a Fixed Tensor Space Take a tensor space S in $\mathscr{A}_{\mathbb{K}}(V)$. This is a vector space by definition, and thus picking out $s, t \in S$, and $\alpha, \beta \in \mathbb{K}$, linear combinations can be formed which still belong to S. In other words we may define the tensor of S

$$u := \alpha s + \beta t \quad \text{or, in the abstract index notation} \quad u^A = \alpha s^A + \beta t^A\,.$$

The definition of u above given by the abstract index notation means that the components of u are related to the components of s and t by a linear combination which has *the same form in every canonical basis of the space* S and the coefficients α, β *do not depend on the basis.*

Products of Tensors of Generally Different Tensor Spaces Take two tensor spaces S, S' in $\mathscr{A}_{\mathbb{K}}(V)$ and pick out $t \in S$, $t' \in S'$. Then the tensor $t \otimes t' \in S \otimes S'$ is well defined. Using the associativity of the tensor product by natural isomorphisms, we find a unique tensor space S'' in $\mathscr{A}_{\mathbb{K}}(V)$ which is isomorphic to $S \otimes S'$, and thus we can identify $t \otimes t'$ with a tensor in S'' which we shall indicate by $t \otimes t'$ once again, with a slight abuse of notation. $t \otimes t'$ is called the **product** of tensors t and t'. Therefore, the product of tensors coincides with the usual tensor product of tensors up to a natural isomorphism. What about the abstract index notation? By Theorem 2.24 one sees that, fixing a basis in V, the natural isomorphism which identifies $S \otimes S'$ with S'' transforms the products of elements of canonical bases in S and S' into the corresponding elements of the canonical basis of S'' obtained by cancelling every parenthesis; e.g. for $S = V \otimes V^*$ and $S' = V^*$, the natural isomorphism from $(V \otimes V^*) \otimes V^*$ to $V \otimes V^* \otimes V^*$ transforms each $(e_k \otimes e^{*h}) \otimes e^{*r}$ into $e_k \otimes e^{*h} \otimes e^{*r}$. As a consequence

$$(t \otimes t')^{AB} = t^A t'^B \, .$$

Therefore, for instance, if $S = V \otimes V^*$ and $S' = V^*$, the tensors t and t' are respectively indicated by $t^i{}_j$ and s_k, and thus $(t \otimes s)^i{}_{jk} = t^i{}_j s_k$.

Contractions Consider a tensor space of $\mathscr{A}_{\mathbb{K}}(V)$ of the form

$$U_1 \otimes \ldots \otimes U_k \otimes V \otimes U_{k+1} \otimes \ldots \otimes U_l \otimes V^* \otimes U_{l+1} \otimes \ldots \otimes U_n \, ,$$

where U_i denotes either V or V^*. Everything we are going to say can be re-stated for the analogous space

$$U_1 \otimes \ldots \otimes U_k \otimes V^* \otimes U_{k+1} \otimes \ldots \otimes U_l \otimes V \otimes U_{l+1} \otimes \ldots \otimes U_n \, .$$

Then consider the *multilinear* map C with domain

$$U_1 \times \ldots \times U_k \times V \times U_{k+1} \times \ldots \times U_l \times V^* \times U_{l+1} \times \ldots \times U_n \, ,$$

and values in

$$U_1 \otimes \ldots \otimes U_k \otimes U_{k+1} \otimes \ldots \otimes U_l \otimes U_{l+1} \otimes \ldots \otimes U_n \, ,$$

defined by

$$(u_1, \ldots, u_k, v, u_{k+1}, \ldots, u_l, v^*, u_{l+1}, \ldots, u_n)$$
$$\mapsto \langle v, v^* \rangle u_1 \otimes \ldots \otimes u_k \otimes u_{k+1} \otimes \ldots \otimes u_l \otimes u_{l+1} \otimes \ldots \otimes u_n \,.$$

By the universality theorem there is a *linear* map $C^{\otimes}$, called the **contraction** of V and V^*, defined on the whole tensor space

$$U_1 \otimes \ldots \otimes U_k \otimes V \otimes U_{k+1} \otimes \ldots \otimes U_l \otimes V^* \otimes U_{l+1} \otimes \ldots \otimes U_n \,,$$

taking values in

$$U_1 \otimes \ldots \otimes U_k \otimes U_{k+1} \otimes \ldots \otimes U_l \otimes U_{l+1} \otimes \ldots \otimes U_n \,,$$

such that, on simple tensor products of vectors, it reduces to

$$u_1 \otimes \ldots \otimes u_k \otimes v \otimes u_{k+1} \otimes \ldots \otimes u_l \otimes v^* \otimes u_{l+1} \otimes \ldots \otimes u_n$$
$$\mapsto \langle v, v^* \rangle u_1 \otimes \ldots \otimes u_k \otimes u_{k+1} \otimes \ldots \otimes u_l \otimes u_{l+1} \otimes \ldots \otimes u_n \,.$$

This linear map takes tensors in a tensor-product space with $n + 2$ factors and produces tensors in a space with n factors. The simplest case arises for $n = 0$. In that case $C : V \times V^* \to \mathbb{K}$ is nothing but the bilinear pairing $C : (v, v^*) \mapsto \langle v, v^* \rangle$, and $C^{\otimes}$ is the linear associated map by the universality theorem.

Finally, let us represent the contraction map within the abstract-index picture. It is quite simple to show that $C^{\otimes}$ takes a tensor with components

$$t^{AiB\ C}_{\ \ \ \ j}$$

where ^{A}, ^{B} and ^{C} are arbitrary cumulative indices, and produces the tensor with components

$$(C^{\otimes} t)^{ABC} := t^{AkB\ C}_{\ \ \ \ k} \,,$$

where we stress the convention of summation over the repeated index k. To show that the abstract-index representation of contractions is that above notice that contractions are linear and thus

$$C^{\otimes}\left(t^{AiB\ C}_{\ \ \ \ j} e_A \otimes e_i \otimes e_B \otimes e^{*j} \otimes e_C\right)$$
$$= t^{AiB\ C}_{\ \ \ \ j} C^{\otimes}\left(e_A \otimes e_i \otimes e_B \otimes e^{*j} \otimes e_C\right) = t^{AiB\ C}_{\ \ \ \ j} \delta^j_i e_A \otimes e_B \otimes e_C \,,$$

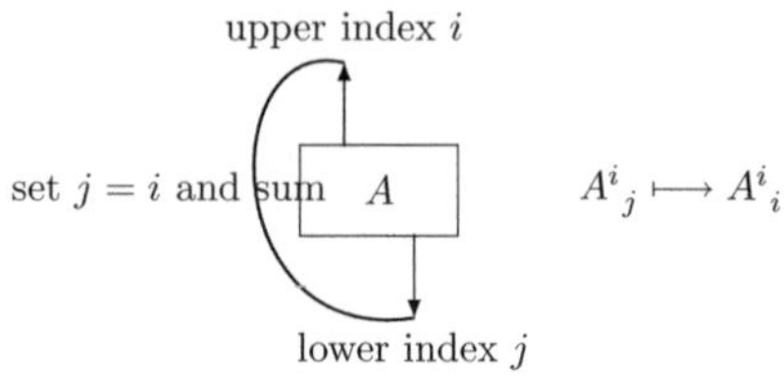

Fig. 3.1 Contraction joins one upper index with one lower index and sums over the repeated label

and thus

$$C^{\otimes}\left(t^{AiB}{}_j{}^C e_A \otimes e_i \otimes e_B \otimes e^{*j} \otimes e_C\right) = t^{AkB}{}_k{}^C e_A \otimes e_B \otimes e_C\,.$$

This is nothing but

$$(C^{\otimes}t)^{ABC} := t^{AkB}{}_k{}^C\,.$$

Examples 3.8 (The Trace of an Endomorphism as a Contraction) Let V be finite-dimensional and let

$$A = A^i{}_j e_i \otimes e^{*j} \in V \otimes V^*\,.$$

The contraction of the contravariant index with the covariant index gives the scalar

$$C^{\otimes}(A) = A^i{}_i\,.$$

If $V = \mathbb{R}^2$ and, in the chosen basis,

$$[A^i{}_j] = \begin{pmatrix} 1 & 2 \\ 3 & 4 \end{pmatrix},$$

then

$$C^{\otimes}(A) = A^1{}_1 + A^2{}_2 = 1 + 4 = 5\,.$$

Thus the ordinary trace of a matrix is the simplest concrete manifestation of tensor contraction (Fig. 3.1).

This is also why contractions are independent of the basis: although the components of A change under a change of basis, the scalar $A^i{}_i$ is the same in every canonical basis. ■

3.2.3 Linear Transformations of Tensors Are Tensors Too

To conclude we pass to consider a final theorem which shows that there is a one-to-one correspondence between linear maps on tensors and tensors themselves.

Theorem 3.9 (Linear Maps and Tensors) *Let S, S' be a pair of tensor spaces of a tensor algebra $\mathscr{A}_{\mathbb{K}}(V)$ where V is finite dimensional. The vector space of linear maps from S to S' is naturally isomorphic to $S^* \otimes S'$ (which is naturally isomorphic to the corresponding tensor space of $\mathscr{A}_{\mathbb{K}}(V)$). The isomorphism $F : S^* \otimes S' \to \mathscr{L}(S|S')$ is such that $F : t \mapsto f_t$, where the linear map $f_t : S \to S'$ is defined by*

$$f_t(s) := C^{\otimes}(s \otimes t)\,, \quad \textit{for all } s \in S,$$

where $C^{\otimes}$ is the contraction of all the indices of s with the corresponding indices in S^* of t. Moreover, fixing a basis $\{e_i\}_{i\in I}$ in V, let $\{e_A\}$ denote the canonical basis induced in S, $\{e'_B\}$ that induced in S' and $\{e'^{*C}\}$ that induced in S'^*. With those definitions if $s = s^A e_A \in S$,

$$f_t(s)^C = s^A t_A{}^C$$

and

$$t_A{}^C = \langle f_t(e_A), e'^{*C}\rangle\,.$$

The isomorphism F^{-1} is the composition of the isomorphisms in Theorems 2.18 and 2.9: $\mathscr{L}(S|S') \simeq \mathscr{L}(S, S'^*) = S^* \otimes (S'^*)^* \simeq S^* \otimes S'$.

Proof Defining f_t as said above it is evident that the map F is linear. The given definition implies that, in components, $f_t(s)^C = s^A t_A{}^C$. Since $S^* \otimes S'$ and $\mathscr{L}(S|S') \simeq \mathscr{L}(S, S'^*) \simeq S^* \otimes S'$ have the same dimension, injectivity of F implies surjectivity. Injectivity has a straightforward proof: $f_t = 0$ is equivalent to, fixing canonical bases as indicated in the hypotheses, $s^A t_A{}^C = 0$ for every choice of coefficients s^A. This is equivalent, in turn, to $t_A{}^C = 0$ for every component of t and thus $t = 0$. The formula $t_A{}^C = \langle f_t(e_A), e'^{*C}\rangle$ can be proved as follows: if $s = e_A$, the identity $f_t(s)^C = s^A t_A{}^C$ reduces to $f_t(e_A)^C = t_A{}^C$, therefore $f_t(e_A) = f_t(e_A)^C e'_C = t_A{}^C e'_C$ and so $\langle f_t(e_A), e'^{*C}\rangle = t_A{}^C$. The last statement in the thesis can be obtained by direct inspection. □

Let us illustrate how one may use that theorem. For instance consider a linear map $f : V \to V$ where V is finite-dimensional with field $\mathbb{K} = \mathbb{C}$ or $\mathbb{R}$. Then f defines a tensor $t \in V^* \otimes V$. In fact, fixing a basis $\{e_i\}_{i\in I}$ in V and considering those canonically associated, by the linearity of f and the pairing:

$$f(v) = \langle f(v), e^{*k}\rangle e_k = v^i \langle f(e_i), e^{*k}\rangle e_k\,.$$

The tensor $t \in V^* \otimes V$ associated with f in view of the proved theorem has components

$$t_i{}^k := \langle f(e_i), e^{*k} \rangle .$$

Moreover, using the abstract index notation, one has

$$(f(v))^k = v^i t_i{}^k .$$

In other words the action of f on v reduces to (1) a product of the involved tensors:

$$v^i t_j{}^k ,$$

(2) followed by a convenient contraction:

$$(f(v))^k = v^i t_i{}^k .$$

More complicated cases can be treated similarly. For example, linear maps f from $V^* \otimes V$ to $V^* \otimes V \otimes V$ are one-to-one with tensors $t^i{}_{jk}{}^{lm}$ of $V \otimes V^* \otimes V^* \otimes V \otimes V$, and their action on tensors $u_p{}^q$ of $V^* \otimes V$ is

$$f(u)_k{}^{lm} = u_i{}^j t^i{}_{jk}{}^{lm}$$

i.e., a product of tensors and two contractions. Obviously

$$t^i{}_{jk}{}^{lm} = \langle f(e^{*i} \otimes e_j), e_k \otimes e^{*l} \otimes e^{*m} \rangle .$$

Remarks 3.10 In applications it is often convenient to define the tensor t in $S' \otimes S^*$ instead of $S^* \otimes S'$. In this case the isomorphism between $S' \otimes S^*$ and $\mathscr{L}(S|S')$ associates $t \in S' \otimes S^*$ with $g_t \in \mathscr{L}(S|S')$ where

$$g_t(s) := C^{\otimes}(t \otimes s) , \quad \text{for all } s \in S,$$

where $C^{\otimes}$ is the contraction of all the indices of s with the corresponding indices in S^* of t. ■

3.3 Physical Invariance of the Form of Laws and Tensors

A physically relevant result is that the rules given above to produce a new tensor from given tensors have the same form independently of the basis one uses to handle tensors. For that reason the abstract index notation makes sense. In physics the choice of a basis is associated with the choice of a reference frame. Relativity principles (*Galileian Principle*, *Special Relativity Principle* and *General Relativity Principle*) assume that

the laws of Physics can be written in the same form in every reference frame.

The allowed reference frames range in a class singled out by the considered relativity principle; for instance, in Special Relativity the relevant class is that of inertial reference frames.

Tensorial formulations and the associated rules for composing and decomposing tensors are especially useful in implementing relativity principles. In fact the theories of Relativity can be profitably formulated in terms of tensors and operations on tensors just to assure the invariance of physical laws under change of reference frame. When physical laws are given in terms of tensorial relations one says that those laws are *covariant*. It is worthwhile stressing that covariance is *not* the only way to state physical laws which preserve their form under changes of reference frames. For instance the Hamiltonian formulation of mechanics, in relativistic contexts, is invariant under change of reference frame but it is not formulated in terms of tensor relations (in spacetime).

3.4 Tensors on Affine and Euclidean Spaces

We recall the definitions of *affine space* and **Euclidean space**. We assume that the reader is familiar with these notions and in particular with the notion of scalar product and orthonormal basis [1]. A general theory of scalar product from the point of view of tensor theory will be developed in Chap. 5.

Definition 3.11 (Affine Space) A **real** n**-dimensional affine space** [1] is a triple $(\mathbb{A}^n, V, \vec{\cdot})$ where $\mathbb{A}^n$ is a set whose elements are called **points**, V is a real n-dimensional vector space called **space of translations** and $\vec{\cdot} : \mathbb{A}^n \times \mathbb{A}^n \to V$ is a map such that the two following requirements are fulfilled.

(i) For each pair $p \in \mathbb{A}^n$, $v \in V$ there is a *unique* point $q \in \mathbb{A}^n$ such that $\overrightarrow{pq} = v$.
(ii) $\overrightarrow{pq} + \overrightarrow{qr} = \overrightarrow{pr}$ for all $p, q, r \in \mathbb{A}^n$.

$\overrightarrow{pq}$ is called the vector with **initial point** p and **final point** q. ■

Definition 3.12 (Euclidean Space) A (real) **Euclidean** space is a pair $(\mathbb{E}^n, (\cdot|\cdot))$, where $\mathbb{E}^n$ is an affine space with space of translations V and $(\cdot|\cdot) : V \times V \to \mathbb{R}$ is a scalar product on V. ■

A **Cartesian coordinate system** on the affine space $\mathbb{A}^n$ is a bijective map $f : \mathbb{A}^n \to \mathbb{R}^n$ defined as follows. Fix a point $O \in \mathbb{A}^n$ called **origin** of the coordinates and fix a vector basis $\{e_i\}_{i=1,\dots,n} \subset V$ whose elements are called **axes** of the coordinates. Varying $p \in \mathbb{A}^n$, the components $(x^1(p), \dots, x^n(p)) \in \mathbb{R}^n$ of the vector $\overrightarrow{Op}$ with respect to the chosen basis define a map $f : \mathbb{A}^n \to \mathbb{R}^n$. Property (i) above implies that f is bijective.

It is simply proved that, if f' is another Cartesian coordinate system with origin O' and axes $\{e'_j\}_{j=1,\ldots,n}$, the following coordinate transformation law holds:

$$x'^j = \sum_{i=1}^{n} A^j{}_i x^i + b^j \,, \tag{3.1}$$

where $e_i = \sum_j A^j{}_i e'_j$ and $\overrightarrow{O'O} = \sum_j b^j e'_j$.

An **orthonormal Cartesian coordinate system** on the Euclidean space $(\mathbb{E}^n, (\cdot|\cdot))$ is a Cartesian coordinate system whose axes form an orthonormal basis of the vector space V associated with $\mathbb{E}^n$ with respect to $(\cdot|\cdot)$, namely $(e_i|e_j) = \delta_{ij}$.

In that case the matrices A of elements $A^j{}_i$ employed to connect different orthonormal Cartesian coordinate systems are all elements of the **orthogonal group** of order n:

$$O(n) = \{R \in M(n, \mathbb{R}) \mid R^t R = I \,.\}$$

This is evident from the requirement that both the bases appearing in the identities $e_i = \sum_j A^j{}_i e'_j$ are orthonormal.

Remarks 3.13

(1) The simplest example of affine space is $\mathbb{R}^n$ itself with $\mathbb{A}^n = \mathbb{R}^n$ and $V = \mathbb{R}^n$. The map $\vec{\cdot} : \mathbb{A}^n \times \mathbb{A}^n \to V$ in this case is trivial: $\overrightarrow{(x^1, \ldots, x^n)(y^1, \ldots, y^n)} = (y^1 - x^1, \ldots, y^n - x^n)$.
From a physical point of view affine and Euclidean spaces are more interesting than $\mathbb{R}^n$ to describe physical space (of inertial reference frames) because the affine structure can naturally be used to represent homogeneity and isotropy of space. Conversely $\mathbb{R}^n$ has an unphysical structure breaking those symmetries: it admits a preferred coordinate system!

(2) Three-dimensional Euclidean spaces are the spaces of usual Euclidean geometry.

(3) The whole class of Cartesian coordinate systems allows the introduction of a Hausdorff second-countable topology on $\mathbb{A}^n$. This is done by fixing a Cartesian coordinate system $f : \mathbb{A}^n \to \mathbb{R}^n$ and defining the open sets on $\mathbb{A}^n$ as the sets of the form $f^{-1}(B)$, $B \subset \mathbb{R}^n$ being any open set of $\mathbb{R}^n$. Since affine transformations (3.1) from $\mathbb{R}^n$ to $\mathbb{R}^n$ are homeomorphisms, the given definition does not depend on the Cartesian coordinate system used. In this way every Cartesian coordinate system becomes a homeomorphism and thus Hausdorff and second-countability properties are inherited from the topology on $\mathbb{R}^n$.

(4) It is possible to give the notion of a *differentiable function* $h : A \to B$, where $A \subset \mathbb{A}^r$, $B \subset \mathbb{A}^s$ are open sets. In particular it could be either $\mathbb{A}^r = \mathbb{R}^r$ or $\mathbb{A}^s = \mathbb{R}^s$. ■

Definition 3.14 If $A \subset \mathbb{A}^r$, $B \subset \mathbb{A}^s$ are open sets in the corresponding affine spaces, we say that $h : A \to B$ is of **class** $C^k(A)$ if $f' \circ h \circ f^{-1} : f(A) \to \mathbb{R}^s$ is of class $C^k(f(A))$ for a pair of Cartesian coordinate systems $f : \mathbb{A}^r \to \mathbb{R}^r$ and $f' : \mathbb{A}^s \to \mathbb{R}^s$. ■

The definition is well posed, since affine transformations (3.1) are C^∞, and thus the given definition does not depend on the choice of f and f'.

Remarks 3.15 As a final remark we notice that, from a more abstract point of view, the previous comments prove that every n-dimensional affine (or Euclidean) space has a natural structure of *n-dimensional C^∞ differentiable manifold* induced by the class of all Cartesian coordinate systems. ■

3.4.1 Tensor Spaces, Cartesian Tensors

Consider an affine space $\mathbb{A}^n$ and a Cartesian coordinate system $f : \mathbb{A}^n \ni p \mapsto (x^1(p), \ldots, x^n(p)) \in \mathbb{R}^n$ with origin O and axes $e_1, \ldots, e_n$. A standard notation to denote the elements of that basis and those of the associated dual basis is:

$$\frac{\partial}{\partial x^i} := e_i \,, \quad \text{and} \quad dx^i := e^{*i} \,, \quad \text{for } i = 1, \ldots, n. \tag{3.2}$$

This notation is standard in differential geometry; however we do not want to address here the general definition of vectors tangent and cotangent to a manifold. The simplest explanation of such a notation is that, under changes of coordinates (3.1), formal computations based on the fact that

$$\frac{\partial x'^j}{\partial x^i} = A^j{}_i \,,$$

show that (writing the summation over repeated indices explicitly)

$$\frac{\partial}{\partial x^i} = \sum_{j=1}^n A^j{}_i \frac{\partial}{\partial x'^j} \,, \quad \text{and} \quad dx'^j = \sum_{i=1}^n A^j{}_i dx^i \,. \tag{3.3}$$

And, under the employed notation, those correspond to the *correct* relations

$$e_i = \sum_{j=1}^n A^j{}_i e'_j \,, \quad \text{and} \quad e'^{*j} = \sum_{i=1}^n A^j{}_i e^{*i} \,.$$

Very often, the notation for $\frac{\partial}{\partial x^i}$ is shortened to ∂_{x^i}. We make use of the shortened notation from now on.

Let us pass to consider tensors in the algebra $\mathscr{A}_{\mathbb{R}}(V)$. A change of Cartesian coordinates

$$x'^j = \sum_{i=1}^{n} A^j{}_i x^i + b^j \, ,$$

involves a corresponding change in the associated bases of V and V^*. Under these changes of bases, the transformation of components of contravariant tensors uses the same coefficients $A^j{}_i$ appearing in the relation connecting different Cartesian coordinate systems: if $t \in V \otimes V$ one has

$$t = t^{ij} \partial_{x^i} \otimes \partial_{x^j} = t'^{kl} \partial_{x'^k} \otimes \partial_{x'^l} \, , \quad \text{where } t'^{kl} = A^k{}_i A^l{}_j t^{ij} \, .$$

Covariant tensors use the coefficients of the matrix A^{-1t} where A is the matrix of coefficients $A^j{}_i$. Therefore, if we restrict ourselves to employing only bases in V and V^* associated with Cartesian coordinate systems, the transformation law of tensor components uses the same coefficients as those appearing in the transformation law of Cartesian coordinates. For this reason tensors in affine spaces are said to be *Cartesian tensors*. A straightforward application of the given notation is the tangent vector to a differentiable curve in $\mathbb{A}^n$.

Definition 3.16 (Tangent Vector) Let $\mathbb{A}^n$ be an affine space with space of translations V. If $\gamma : (a, b) \to \mathbb{A}^n$ is $C^1((a, b))$, one defines the **tangent vector** $\dot{\gamma}(t)$ at γ in $\gamma(t)$ as

$$\dot{\gamma}(t) := \frac{dx^i}{dt} \partial_{x^i} \, ,$$

where $f : \mathbb{A}^n \ni p \mapsto (x^1(p), \ldots, x^n(p)) \in \mathbb{R}^n$ is any Cartesian coordinate system on $\mathbb{A}^n$. ■

The definition is well posed because $\dot{\gamma}(t)$ does not depend on the chosen Cartesian coordinate system. Indeed, under a change of coordinates (3.1) one finds

$$\frac{dx'^j}{dt} = A^j{}_i \frac{dx^i}{dt} \, ,$$

and thus

$$\frac{dx'^j}{dt} \partial_{x'^j} = A^j{}_i \frac{dx^i}{dt} \partial_{x'^j} \, ,$$

so that, by (3.3),

$$\frac{dx'^j}{dt}\partial_{x'^j} = \frac{dx^i}{dt}\partial_{x^i} .$$

■

3.4.2 Applied Tensors

In several physical applications it is convenient to view a tensor as applied at a point $p \in \mathbb{A}^n$. This notion is based on the following straightforward definition, which is equivalent to that given using the structure of differentiable manifold of an affine space.

Definition 3.17 (Tangent and Cotangent Space) If $\mathbb{A}^n$ is an affine space with space of translations V and $p \in \mathbb{A}^n$, the **tangent space** at p is the vector space

$$T_p\mathbb{A}^n := \{(p, v) \mid v \in V\}$$

with vector-space structure naturally induced by that of V, namely

$$a(p, v) + b(p, u) := (p, au + bv) , \quad \text{for all } a, b \in \mathbb{R} \text{ and } u, v \in V.$$

The **cotangent space** at p is the vector space

$$T_p^*\mathbb{A}^n := \{(p, v^*) \mid v^* \in V^*\}$$

with vector-space structure naturally induced by that of V^*.

The bases of $T_p\mathbb{A}^n$ and $T_p^*\mathbb{A}^n$ associated with a Cartesian coordinate system $f : \mathbb{A}^n \ni p \mapsto (x^1(p), \ldots, x^n(p)) \in \mathbb{R}^n$ are respectively denoted by $\{\partial_{x^i}|_p\}_{i=1,\ldots,n}$ and $\{dx^i|_p\}_{i=1,\ldots,n}$ (but the indication of the application point p is very often omitted).

■

It follows immediately that $T_p^*\mathbb{A}^n$ is canonically isomorphic to the dual space $(T_p\mathbb{A}^n)^*$ and every tensor space $S_p \in \mathscr{A}_{\mathbb{R}}(T_p\mathbb{A}^n)$ is canonically isomorphic to a vector space

$$\{(p, t) \mid t \in S\} ,$$

where $S \in \mathscr{A}_{\mathbb{R}}(V)$.

Definition 3.18 (Tensor Fields) Referring to Definition 3.17, the **tensor algebra of applied tensors** at p is the algebra $\mathscr{A}(T_p\mathbb{A}^n)$.

An assignment $\mathbb{A}^n \ni p \mapsto t_p \in S_p$ such that every space $S_p \in \mathscr{A}(T_p\mathbb{A}^n)$ is of the same order, and the components of t_p define C^k functions with respect to the bases of tensors applied at every $p \in \mathbb{A}^n$ associated with a fixed Cartesian coordinate system, is a C^k **tensor field** on $\mathbb{A}^n$. The order of the tensor field is defined as the order of each tensor t_p. ■

Another remarkable aspect of Cartesian tensor fields is the fact that it is possible to coherently define a notion of derivative of each of them with respect to a vector field.

Definition 3.19 (Derivative of Tensor Fields) Referring to Definitions 3.17 and 3.18, given a C^k tensor field $\mathbb{A}^n \ni p \mapsto t_p \in S_p$, with $k \geq 1$, and a contravariant vector field $\mathbb{A}^n \ni p \mapsto v_p$, the **derivative of** t **with respect to** v is the tensor field $\mathbb{A}^n \ni p \mapsto \nabla_v t_p \in S_p$ (thus of the same type as t) that in every Cartesian coordinate system has the components

$$(\nabla_v t_p)^A = v^a(p) \left.\frac{\partial t^A}{\partial x^a}\right|_p .$$

The **derivative** of t (without fixing any vector field) is the tensor field $\mathbb{A}^n \ni p \mapsto \nabla t_p \in T_p^*\mathbb{A}^n \otimes S_p$ such that, in every Cartesian coordinate system, it has the components

$$\nabla_a t^A(p) := (\nabla t_p)_a{}^A := t(p)^A{}_{,a} := \left.\frac{\partial t^A}{\partial x^a}\right|_p ,$$

so that

$$(\nabla_v t)^A = v^a (\nabla_a t)^A = v^a t^A{}_{,a} .$$

■

The given definitions are well posed in the sense that they do not depend on the choice of the Cartesian coordinate system used. This is due to the fact that the components of a fixed tensor field transform, under changes of *Cartesian coordinates*, by means of a linear transformation with constant coefficients. The fact that these coefficients are *constant* implies that the definition of $(\nabla t)_a{}^A$ is well posed: If $B = A^{-1}$, the component transformation above gives

$$t^{ij} = B^i{}_p B^j{}_q t'^{pq} ,$$

and the chain rule, applied to $x'^l = A^l{}_k x^k + b^l$, gives

$$\frac{\partial}{\partial x^k} = A^l{}_k \frac{\partial}{\partial x'^l} .$$

Therefore

$$\frac{\partial t^{ij}}{\partial x^k} = \frac{\partial}{\partial x^k}\left(B^i{}_p B^j{}_q t'^{pq}\right) = A^l{}_k \frac{\partial}{\partial x'^l}\left(B^i{}_p B^j{}_q t'^{pq}\right) = A^l{}_k B^i{}_p B^j{}_q \frac{\partial}{\partial x'^l} t'^{pq} \,,$$

so that

$$\frac{\partial t^{ij}}{\partial x^k} = A^l{}_k B^i{}_p B^j{}_q \frac{\partial t'^{pq}}{\partial x'^l} \,.$$

3.4.3 The Natural Isomorphism Between V and V* for Cartesian Vectors in Euclidean Spaces

Let us consider a Euclidean space $\mathbb{E}^n$ and, from now on, *we restrict ourselves to using only orthonormal Cartesian coordinate frames*. We are going to discuss the existence of a natural isomorphism between V and V^* that we shall consider much more extensively later from a wider viewpoint.

If $\{e_i\}_{i=1,\ldots,n}$ and $\{e^{*j}\}_{j=1,\ldots,n}$ are an orthonormal basis in V and its associated basis in V^* respectively, we can consider the isomorphism

$$H : V \ni v = v^i e_i \mapsto v_j e^{*j} \in V^* \quad \text{where } v_j := v^j \text{ for } j = 1, 2, \ldots, n.$$

In principle that isomorphism may seem to depend on the chosen basis $\{e_i\}_{i=1,\ldots,n}$, actually it does not. Indeed, if we choose another *orthonormal* basis and its dual one $\{e'_k\}_{k=1,\ldots,n}$ and $\{e'^{*h}\}_{h=1,\ldots,n}$, so that $e_i = A^j{}_i e'_j$ and $e^{*j} = B_h{}^j e'^{*h}$, the isomorphism above turns out to be written:

$$H : V \ni v = v'^j e'_j \mapsto v'_h e'^{*h} \in V^*$$

$$\text{where } v'_h := B_h{}^k (A^{-1})^k{}_j v'^j \text{ for } j = 1, 2, \ldots, n.$$

In fact, since we are simply writing down the same isomorphism H defined above using another basis, v'_h are the components of $v_j e^{*j}$ but computed with respect to the basis $\{e'^{*h}\}_{h=1,\ldots,n}$ and so $v'_h = B_h{}^k v_k = B_h{}^k v^k = B_h{}^k (A^{-1})^k{}_j v'^j$.

We know that $B = A^{-1t}$, however, in our case $A \in O(n)$, so that $A^t = A^{-1}$ and thus $B = A$. Consequently

$$B_h{}^k (A^{-1})^k{}_j = (AA^{-1})_{hj} = \delta_{hj} \,.$$

We conclude that the isomorphism H always takes the form

$$H : V \ni v = v^i f_i \mapsto v_j f^{*j} \in V^* \quad \text{where } v_j := v^j \text{ for } j = 1, 2, \ldots, n,$$

for every choice of the *orthonormal* basis $\{f_i\}_{i=1,\ldots,n} \subset V$. So H is actually independent of the initially chosen orthonormal basis and, in that sense, is natural. In other words, in the presence of a scalar product there is a natural identification of V and V^* that can be written down as, in the abstract index notation,

$$v^i \mapsto v_j \quad \text{where } v_j := v^j \text{ for } j = 1, 2, \ldots, n,$$

provided we confine ourselves to handling vectors by employing orthonormal bases only. Exploiting the universality property, this isomorphism easily propagates to the whole tensor algebra generated by V. So, for instance there exist natural isomorphisms between $V \otimes V^*$ and $V \otimes V$ or $V \otimes V^*$ and $V^* \otimes V$ acting as follows in the abstract index notation and referring to canonical bases associated with orthogonal bases in V:

$$t^i{}_j \mapsto t^{ij} := t_i{}^j \ ,$$

or, respectively,

$$t_i{}^j \mapsto t^i{}_j := t_i{}^j \ .$$

In practice the height of indices does not matter in vector spaces equipped with a scalar product, provided one uses orthonormal bases only.

All that has been said immediately applies to Cartesian tensors in Euclidean spaces when V is the space of translations or is the tangent space $T_p\mathbb{E}^n$ at some point $p \in \mathbb{E}^n$.

Examples 3.20 In classical continuum mechanics, consider an internal portion $\mathscr{C}$ of a continuous body $\mathscr{B}$. Suppose that $\mathscr{C}$ is represented by a regular set in the Euclidean space $\mathbb{E}^3$, the physical space where the body is supposed to be at rest. Let $\partial\mathscr{C}$ be the boundary of $\mathscr{C}$, assumed to be a regular surface. If $p \in \partial\mathscr{C}$, the part of $\mathscr{B}$ external to $\mathscr{C}$ acts on $\mathscr{C}$ through the element of surface $dS \subset \partial\mathscr{C}$ about p, by a density of force called the *stress vector* $\mathbf{s}(p, \mathbf{n})$. $\mathbf{n}$ denotes the versor orthogonal to dS at p. In this context, it is convenient to assume that

$$\mathbf{s}(p, \mathbf{n}), \mathbf{n}_p \in T_p\mathbb{E}^3 \ .$$

In general one would expect a complicated dependence of the stress vector on $\mathbf{n}$. Actually a celebrated theorem due to Cauchy proves that (under standard hypotheses concerning the dynamics of the continuous body) this dependence must be linear.

More precisely, if one defines $\mathbf{s}(p, \mathbf{0}) := \mathbf{0}$ and

$$\mathbf{s}(p, \mathbf{v}) := |\mathbf{v}|\mathbf{s}(p, \mathbf{v}/|\mathbf{v}|) \ ,$$

it turns out that

$$T_p\mathbb{E}^3 \ni \mathbf{v} \mapsto \mathbf{s}(p, \mathbf{v})$$

is linear. As a consequence, there is a tensor $\sigma_p \in T_p^*\mathbb{E}^3 \otimes T_p\mathbb{E}^3$ (depending on the nature of $\mathscr{B}$ and on p) called the *Cauchy stress tensor*, such that

$$\mathbf{s}(p, \mathbf{n}) = \sigma_p(\mathbf{n}) \,. \tag{3.4}$$

In order to go on with continuum mechanics it is assumed that, varying $p \in \mathbb{E}^3$, the map $p \mapsto \sigma(p) := \sigma_p$ gives rise to a sufficiently regular (usually C^2) tensor field of order $(1, 1)$.

Exploiting the abstract index notation:

$$s^i(p, \mathbf{n}) = \sigma(p)^i{}_j n^j \,.$$

It is possible to prove, taking the equation for the angular momentum into account, that the stress tensor is symmetric. In other words, passing to the completely covariant expression for it (taking advantage of the existence of the natural isomorphism between $V \otimes V^*$ and $V^* \otimes V^*$—for $V = T_p\mathbb{E}^3$—discussed above, one has

$$\sigma(p)_{ij} = \sigma(p)_{ji} \,.$$

Reference

1. Sernesi, E.: Geometria I and II. Bollati Boringhieri, Torino (1990)

Chapter 4
Some Applications to General Group Theory

This chapter collects a number of algebraic applications of the tensor formalism to group theory and representation theory. The discussion begins with basic notions about groups, matrix groups, homomorphisms, direct products, and semidirect products, because these structures appear naturally in the study of symmetries in geometry and physics.

The main point is to show how tensor products interact with group representations. If a group acts on vector spaces, then it also acts naturally on tensor products, on symmetric tensors, and on antisymmetric tensors. This mechanism is the algebraic origin of many constructions used later for rotations, Lorentz transformations, spinors, and physical covariance. The chapter also introduces the permutation group, tensor symmetries, exterior algebra, and the interior product, all of which will reappear in geometric and physical contexts.

4.1 Groups

4.1.1 Basic Notions on Groups and Matrix Groups

A **group** is an algebraic structure $(G, \circ)$, where G is a set and $\circ : G \times G \to G$ is a map called the **composition rule** of the group or **group product**. Moreover, the following three conditions have to be satisfied.

(1) $\circ$ is *associative*, i.e.,

$$g_1 \circ (g_2 \circ g_3) = (g_1 \circ g_2) \circ g_3 \,, \quad \text{for all} \quad g_1, g_2, g_3 \in G \,.$$

(2) There is a **group unit** or **unit element**, i.e., there is $e \in G$ such that

$$e \circ g = g \circ e = g \,, \quad \text{for all} \quad g \in G \,.$$

V. Moretti, *Geometric Methods in Mathematical Physics I*, La Matematica per il 3+2 184, https://doi.org/10.1007/978-3-032-31512-0_4

(3) Each element $g \in G$ admits an **inverse element**, i.e.,

$$\text{for each} \quad g \in G \quad \text{there is} \quad g^{-1} \in G \quad \text{with} \quad g \circ g^{-1} = g^{-1} \circ g = e\,.$$

The unit element is unique, and so is the inverse element of each element of the group; these uniqueness properties are immediate consequences of the axioms. A group $(G, \circ)$ is said to be **commutative** or **Abelian** if $g \circ g' = g' \circ g$ for each pair of elements $g, g' \in G$; otherwise it is said to be **non-commutative** or **non-Abelian**. A subset $G' \subset G$ of a group is called a **subgroup** if it is a group with respect to the restriction to $G' \times G'$ of the composition rule of G. A subgroup N of a group G is called a **normal subgroup** if it is invariant under **conjugation**, that is,. for each element $n \in N$ and each $g \in G$, the element $g \circ n \circ g^{-1}$ still belongs to N.
If $(G_1, \circ_1)$ and $(G_2, \circ_2)$ are groups, a group **homomorphism** from G_1 to G_2 is a map $h : G_1 \to G_2$ which *preserves the group structure*, i.e., the following requirement has to be fulfilled:

$$h(g \circ_1 g') = h(g) \circ_2 h(g') \quad \text{for all} \quad g, g' \in G_1\,.$$

As a consequence, one also has, with obvious notations,

$$h(e_1) = e_2\,,$$

and

$$h(g^{-1}) = (h(g))^{-1} \quad \text{for each} \quad g \in G_1\,.$$

Indeed, if $g \in G_1$ one has $h(g) \circ e_2 = h(g) = h(g \circ e_1) = h(g) \circ h(e_1)$. Applying $h(g)^{-1}$ on the left, one finds $e_2 = h(e_1)$. On the other hand $h(g)^{-1} \circ h(g) = e_2 = h(e_1) = h(g^{-1} \circ g) = h(g^{-1}) \circ h(g)$ implies $h(g)^{-1} = h(g^{-1})$.
The **kernel** of a group homomorphism $h : G \to G'$ is the subgroup $K \subset G$ whose elements g satisfy $h(g) = e'$, e' being the unit element of G'. h is injective if and only if its kernel contains only the unit element. A group **isomorphism** is a *bijective* group homomorphism. A group isomorphism $h : G \to G$, so that the domain and the codomain are the same group, is called a group **automorphism** of G. The set of group automorphisms of a given group G is denoted by $Aut(G)$ and it is a group in its own right when the group product is the standard composition of maps.

Examples 4.1 The algebra $M(n, \mathbb{K})$, that is the vector space of the $n \times n$ matrices with coefficients in $\mathbb{K}$ equipped with matrix multiplication, contains several interesting *matrix groups*. If $n > 1$, most interesting groups in $M(n, \mathbb{K})$ are non-commutative.

(1) The first example is $GL(n, \mathbb{K})$, which is the set of the $n \times n$ matrices A with components in the field $\mathbb{K} = \mathbb{C}$ or $\mathbb{R}$ and $\det A \neq 0$. It is a group, with group composition rule given by the usual product of matrices.

(2) An important subgroup of $GL(n, \mathbb{K})$ is the *special linear group* $SL(n, \mathbb{K})$, i.e., the set of matrices A in $GL(n, \mathbb{K})$ with $\det A = 1$. (One checks that $SL(n, \mathbb{K})$ is a subgroup of $GL(n, \mathbb{K})$.)

(3) If $\mathbb{K} = \mathbb{R}$, another interesting subgroup is the *orthogonal group* $O(n)$ containing all matrices satisfying $R^t R = I$.

(4) If $\mathbb{K} = \mathbb{C}$, an interesting subgroup of $GL(n, \mathbb{C})$ is the *unitary group* $U(n)$ (and the associated subgroup $SU(n) := SL(n, \mathbb{C}) \cap U(n)$). It contains the matrices A satisfying $\overline{A^t} A = I$ (the bar denoting complex conjugation) and is of great relevance in quantum physics.

(5) In Sect. 5.2.1 we shall encounter the pseudo-orthogonal groups $O(m, p)$ and, one of them, the Lorentz group $O(1, 3)$ will be studied in detail in the last two chapters.

Notice that there are groups which *are not defined* as groups of matrices, e.g., $(\mathbb{Z}, +)$. Some of them can be represented in terms of matrices anyway (for example $(\mathbb{Z}, +)$ can be represented as a matrix subgroup of $GL(2, \mathbb{R})$; the proof is a useful exercise). There are, however, groups which cannot be represented in terms of matrices, such as the so-called *universal covering of* $SL(2, \mathbb{R})$. An example of a group which is not defined as a matrix group (but admits such representations) is given by the *group of permutations of* n *elements*, which we shall consider in the next subsection.

Exercises 4.2

1. Prove the uniqueness of the unit element and of the inverse element in any group.
2. Show that in any group G the unique element e such that $e^2 = e$ is the unit element.
3. Show that if G' is a subgroup of G, the unit element of G' must coincide with the unit element of G, and, if $g \in G'$, the inverse element g^{-1} in G' coincides with the inverse element in G.
4. Show that if $h : G_1 \to G_2$ is a group homomorphism, then $h(G_1)$ is a subgroup of G_2.

4.1.2 *Direct Product and Semi-direct Product of Groups*

We recall two important elementary notions of group theory, the direct product and the semi-direct product of groups.

If G_1 and G_2 are groups, their **direct product**, $G_1 \times G_2$, is another group defined as follows. The elements of $G_1 \times G_2$ are, as the notation suggests, the elements (g_1, g_2) of the *Cartesian product* of the *sets* G_1 and G_2. Moreover, the composition rule is

$$(g_1, g_2) \circ (f_1, f_2) := (g_1 \circ_1 f_1, g_2 \circ_2 f_2)$$

for all $(g_1, g_2), (f_1, f_2) \in G_1 \times G_2$. Obviously the unit of $G_1 \times G_2$ is (e_1, e_2), where e_1 and e_2 are the unit elements of G_1 and G_2 respectively. A direct verification shows that these definitions determine a correct group structure on the set $G_1 \times G_2$. Notice that, with the given definition of direct product, G_1 and G_2 can be identified with normal subgroups of the direct product $G_1 \times G_2$.

Sometimes there are groups with a structure close to that of a direct product but a bit more complicated. These structures are relevant, especially in physics, and are captured by the following definition.

Suppose that $(G_1, \circ_1)$ and $(G_2, \circ_2)$ are groups and that for every $g_1 \in G_1$ there is a group isomorphism $\psi_{g_1} : G_2 \to G_2$ satisfying the property

$$\psi_{g_1} \circ \psi_{g_1'} = \psi_{g_1 \circ_1 g_1'} ,$$

which easily implies $\psi_{e_1} = id_{G_2}$ as previously discussed, where $\circ$ is the usual composition of functions and e_1 the unit element of G_1. In other words, $\psi_g \in Aut(G_2)$ for every $g \in G_1$ and the map $\psi : G_1 \ni g \mapsto \psi_g$ is a group homomorphism from G_1 to $Aut(G_2)$ viewed as a group with respect to composition of maps. In this case a natural group structure can be assigned to the Cartesian product $G_1 \times G_2$. This is done by defining the composition rule between a pair of elements $(g_1, g_2), (f_1, f_2) \in G_1 \times G_2$ as

$$(g_1, g_2) \circ_\psi (f_1, f_2) := (g_1 \circ_1 f_1,\ g_2 \circ_2 \psi_{g_1}(f_2)) .$$

The well-posedness of the given group composition rule follows by a straightforward verification. Again, the unit element turns out to be (e_1, e_2). The obtained group structure $(G_1 \times_\psi G_2, \circ_\psi)$ is called the **semi-direct product** of G_1 and G_2.

Examples 4.3

(1) Consider the group $O(3)$ of orthogonal matrices in $\mathbb{R}^3$, and $\mathbb{R}^3$ itself viewed as an additive Abelian group. Define, for every $R \in O(3)$, the automorphism $\psi_R \in Aut(\mathbb{R}^3)$ as

$$\psi_R : \mathbb{R}^3 \ni \mathbf{v} \mapsto R\mathbf{v} \in \mathbb{R}^3 .$$

We can endow $O(3) \times \mathbb{R}^3$ with the structure of a semi-direct product group $O(3) \times_\psi \mathbb{R}^3$ by defining the composition rule

$$(R, \mathbf{v}) \circ_\psi (R', \mathbf{v}') := (RR', \mathbf{v} + R\mathbf{v}') .$$

The meaning of the structure so defined is the following: a pair $(R, \mathbf{v})$ is a roto-translation, i.e. a generic isometry of the space $\mathbb{R}^3$ (equipped with the standard metric structure). In other words $(R, \mathbf{v})$ acts on a point $\mathbf{u} \in \mathbb{R}^3$ by a rotation followed by a translation,

$$(R, \mathbf{v}) : \mathbf{u} \mapsto \mathbf{v} + R\mathbf{u} .$$

The composition rule $\circ_\psi$ is nothing but the composition of two such transformations. It is possible to prove that every isometry of $\mathbb{R}^3$ equipped with the standard Euclidean structure is necessarily an element of $O(3) \times_\psi \mathbb{R}^3$.

(2) As further examples, we may mention the Poincaré group as will be presented in Definition 7.27, the group of isometries of a Euclidean affine space, the Galileo group in classical mechanics, the BMS group in general relativity, and the inhomogeneous $SL(2, \mathbb{C})$ group used in relativistic spinor theory. Similar structures appear very frequently in physics.

Exercises 4.4

1. Consider a semi-direct product of groups $(G \times_\psi N, \circ_\psi)$ as defined above (notice that the roles of G and N are not interchangeable). Prove that N is a normal subgroup of $G \times_\psi N$ and that

$$\psi_g(n) = g \circ_\psi n \circ_\psi g^{-1} \quad \text{for every } g \in G \text{ and } n \in N.$$

2. Consider a group $(H, \circ)$, let N be a normal subgroup of H and let G be a subgroup of H. Suppose that $N \cap G = \{e\}$, where e is the unit element of H, and that $H = GN$, in the sense that, for every $h \in H$ there are $g \in G$ and $n \in N$ such that $h = gn$. Prove that (g, n) is uniquely determined by h and that H is isomorphic to $G \times_\psi N$, where

$$\psi_g(n) := g \circ n \circ g^{-1} \quad \text{for every } g \in G \text{ and } n \in N.$$

4.2 Tensor Products of Group Representations

4.2.1 Linear Representation of Groups and Tensor Representations

We are interested in the notion of a (linear) *representation of a group on a vector space*. In order to state the corresponding definition, notice that, if V is a (not necessarily finite-dimensional) vector space, $\mathscr{L}(V|V)$ contains an important group. This is $GL(V)$, which is the set of both injective and surjective elements of $\mathscr{L}(V|V)$ equipped with the usual composition rule of maps. We can give the following definition.

Definition 4.5 (Linear Group on a Vector Space) If V is a vector space, $GL(V)$ denotes the group of linear maps $f : V \to V$ such that f is injective and surjective, with group composition rule given by the usual composition of maps. $GL(V)$ is called the **linear group on** V. ■

Remarks 4.6

(1) If $V := \mathbb{K}^n$ then $GL(V) = GL(n, \mathbb{K})$.
(2) If $V \neq V'$, it is not possible to define an analogue of the group $GL(V)$ by considering some subset of $\mathscr{L}(V|V')$. (There is no composition law with fixed domain and codomain in this case.)
(3) Due to Theorem 3.9, if $dim V < +\infty$, we have that $GL(V)$ can be viewed as a subset of $V \otimes V^*$. ■

Definition 4.7 (Linear Group Representation on a Vector Space) Let $(G, \circ)$ be a group and V a vector space. A (linear group) **representation of** G **on** V is a homomorphism $\rho : G \to GL(V)$. Moreover a representation $\rho : G \to GL(V)$ is called:

(1) **faithful** if it is injective,
(2) **free** if, for any $v \in V \setminus \{0\}$, the subgroup of G made of the elements h_v such that $\rho(h_v)v = v$ contains only the unit element of G,
(3) **transitive** if for each pair $v, v' \in V \setminus \{0\}$ there is $g \in G$ with $v' = \rho(g)v$,
(4) **irreducible** if there is no *proper* vector subspace $S \subset V$ which is **invariant** under the action of $\rho(G)$, i.e., which satisfies $\rho(g)S \subset S$ for all $g \in G$.
Finally, the representation $\rho^d : G \to GL(V^*)$ defined by

$$\rho^d(g) := \rho(g^{-1})^* \quad \text{for all } g \in G,$$

where $\rho(g^{-1})^* : V^* \to V^*$ is the adjoint operator of $\rho(g^{-1}) : V \to V$, as defined in Definition 2.15, is called the **dual representation** associated with ρ. ■

Remarks 4.8

(1) The definitions given above for representations can be extended to the case of general (nonlinear) representations of a group G by replacing $GL(V)$ with a group of bijective transformations $f : V \to V$. Assuming that V is a differential manifold, $GL(V)$ can be replaced by the group $Diff(V)$ of *diffeomorphisms* of V. In that case the homomorphism $\rho : G \to Diff(V)$ is a representation of G in terms of diffeomorphisms. When V is a Riemannian manifold, representations in terms of *isometries* can be defined similarly. In these extended contexts the definitions of faithful, free (dropping the constraint $v \neq 0$), and transitive (dropping the constraint $v, v' \neq 0$) representations can be stated as well.
(2) The presence of the inverse element in the definition of $\rho^d(g) := \rho(g^{-1})^*$ accounts for the fact that adjoint operators compose in reverse order, i.e. $(T_1T_2)^* = T_2^*T_1^*$. Thanks to the inverse element g^{-1} in the definition of $\rho^d(g)$, one gets $\rho^d(g_1)\rho^d(g_2) = \rho^d(g_1g_2)$ as required for representations, instead of $\rho^d(g_1)\rho^d(g_2) = \rho^d(g_2g_1)$ as it would be if one simply defined $\rho^d(g) := \rho(g)^*$. ■

Equipped with the above-given definitions, we are able to study the simplest interplay between tensors and group representations. We want to show that the notion of tensor product allows the definition of *tensor products of representations*. That mathematical object is of fundamental importance in applications to Quantum Mechanics, in particular as far as systems with many components are concerned.

Consider a group G (from now on we omit the symbol of the composition rule whenever it does not produce misunderstandings) and several representations of the group $\rho_i : G \to GL(V_i)$, where $V_1, \ldots, V_n$ are finite-dimensional vector spaces over the same field $\mathbb{K} = \mathbb{C}$ or $\mathbb{R}$. For each $g \in G$, we may define a multilinear map

$$[\rho_1(g), \ldots, \rho_n(g)] \in \mathscr{L}(V_1, \ldots, V_n | V_1 \otimes \ldots \otimes V_n)$$

given by, for all $(v_1, \ldots, v_n) \in V_1 \times \ldots \times V_n$,

$$[\rho_1(g), \ldots, \rho_n(g)] : (v_1, \ldots, v_n) \mapsto (\rho_1(g)v_1) \otimes \ldots \otimes (\rho_n(g)v_n) .$$

That map is multilinear because of the multilinearity of the tensor-product map and the linearity of the operators $\rho_k(g)$. Using the universality theorem, we *uniquely* find a linear map which we indicate by

$$\rho_1(g) \otimes \ldots \otimes \rho_n(g) : V_1 \otimes \ldots \otimes V_n \to V_1 \otimes \ldots \otimes V_n$$

such that

$$\rho_1(g) \otimes \ldots \otimes \rho_n(g)(v_1 \otimes \ldots \otimes v_n) = (\rho_1(g)v_1) \otimes \ldots \otimes (\rho_n(g)v_n) .$$

Definition 4.9 (Tensor Product of Representations) Let $V_1, \ldots, V_n$ be finite-dimensional vector spaces over the same field $\mathbb{K} = \mathbb{C}$ or $\mathbb{R}$ and suppose there are n representations $\rho_k : G \to GL(V_k)$ of the same group G on the given vector spaces. The family of linear maps

$$\{\rho_1(g) \otimes \ldots \otimes \rho_n(g) : V_1 \otimes \ldots \otimes V_n \to V_1 \otimes \ldots \otimes V_n \mid g \in G\}$$

defined above is called the **tensor product of the representations** $\rho_1, \ldots, \rho_n$. ■

The relevance of the definition above is evident because of the following theorem.

Theorem 4.10 *Referring to the definition above, the map*

$$G \ni g \mapsto \rho_1(g) \otimes \ldots \otimes \rho_n(g)$$

defines a ***linear group representation of*** *G* ***on the tensor-product space*** $V_1 \otimes \ldots \otimes V_n$.

Proof We have to show that the map

$$g \mapsto \rho_1(g) \otimes \ldots \otimes \rho_n(g)$$

is a group homomorphism from G to $GL(V_1 \otimes \ldots \otimes V_n)$. Using the fact that each ρ_i is a group homomorphism, if $g, g' \in G$, one has

$$\rho_1(g') \otimes \ldots \otimes \rho_n(g')\Big(\rho_1(g) \otimes \ldots \otimes \rho_n(g)(v_1 \otimes \ldots \otimes v_n)\Big)$$

$$= \rho_1(g') \otimes \ldots \otimes \rho_n(g')\Big((\rho_1(g)v_1) \otimes \ldots \otimes (\rho_n(g)v_n)\Big)$$

and this is

$$(\rho_1(g'g)v_1) \otimes \ldots \otimes (\rho_n(g'g)v_n) .$$

The obtained result clearly holds also for a canonical basis of $V_1 \otimes \ldots \otimes V_n$ made of elements $e_{1,i_1} \otimes \ldots \otimes e_{n,i_n}$ in place of $v_1 \otimes \ldots \otimes v_n$. By linearity, it follows that the found identity is valid if one replaces $v_1 \otimes \cdots \otimes v_n$ by any tensor in $V_1 \otimes \cdots \otimes V_n$. Therefore,

$$(\rho_1(g') \otimes \ldots \otimes \rho_n(g'))(\rho_1(g) \otimes \ldots \otimes \rho_n(g)) = \rho_1(g'g) \otimes \ldots \otimes \rho_n(g'g) .$$

To conclude, notice that $\rho_1(g) \otimes \ldots \otimes \rho_n(g) \in GL(V_1 \otimes \ldots \otimes V_n)$ because $\rho_1(g) \otimes \ldots \otimes \rho_n(g)$ is linear and bijective. The latter can be proved as follows:

$$\Big(\rho_1(g^{-1}) \otimes \ldots \otimes \rho_n(g^{-1})\Big) \circ \Big(\rho_1(g) \otimes \ldots \otimes \rho_n(g)\Big)$$

$$= \Big(\rho_1(g) \otimes \ldots \otimes \rho_n(g)\Big) \circ \Big(\rho_1(g^{-1}) \otimes \ldots \otimes \rho_n(g^{-1})\Big)$$

$$= \rho_1(e) \otimes \ldots \otimes \rho_n(e) = I .$$

The last identity follows by linearity from

$$(\rho_1(e) \otimes \ldots \otimes \rho_n(e))(v_1 \otimes \cdots \otimes v_n) = v_1 \otimes \cdots \otimes v_n .$$

□

More generally, if $A_k : V_k \to U_k$ are n linear maps (operators), and all involved vector spaces are finite dimensional and over the same field $\mathbb{K} = \mathbb{C}$ or $\mathbb{R}$, one defines the *tensor product of operators*.

Definition 4.11 (Tensor Product of Operators) If $A_k : V_k \to U_k$, $k = 1, \ldots, n$, are n linear maps (operators), and all the vector spaces U_i, V_j are finite dimensional over the same field $\mathbb{K} = \mathbb{C}$ or $\mathbb{R}$, the **tensor product of** $A_1, \ldots, A_n$ is the linear map

$$A_1 \otimes \ldots \otimes A_n : V_1 \otimes \ldots \otimes V_n \to U_1 \otimes \ldots \otimes U_n$$

uniquely determined by the universality theorem and the requirement

$$(A_1 \otimes \ldots \otimes A_n) \circ \otimes = A_1 \times \ldots \times A_n \,,$$

where

$$A_1 \times \ldots \times A_n : V_1 \times \ldots \times V_n \to U_1 \otimes \ldots \otimes U_n$$

is the multilinear map such that, for all $(v_1, \ldots, v_n) \in V_1 \times \ldots \times V_n$,

$$A_1 \times \ldots \times A_n : (v_1, \ldots, v_n) \mapsto (A_1 v_1) \otimes \ldots \otimes (A_n v_n) \,.$$

■

Remarks 4.12

(1) Employing the given definition and the same proof used for the relevant part of the proof of Theorem 4.10, it is simply shown that if $A_i : V_i \to U_i$ and $B_i : U_i \to W_i$, $i = 1, \ldots, n$, are $2n$ linear maps and all involved spaces V_i, U_j, W_k are finite dimensional over the same field $\mathbb{K} = \mathbb{C}$ or $\mathbb{R}$, then

$$(B_1 \otimes \ldots \otimes B_n)\,(A_1 \otimes \ldots \otimes A_n) = (B_1 A_1) \otimes \ldots \otimes (B_n A_n) \,.$$

(2) Suppose that $A : V \to V$ and $B : U \to U$ are linear operators. How does $A \otimes B$ work in components? Or, what is the same, what is the explicit expression of $A \otimes B$ in the abstract index notation?

To answer this question we will make use of Theorem 3.9. Fix a basis $\{e_i\}_{i \in I} \subset V$ and another $\{f_j\}_{j \in J} \subset U$, and consider the associated canonical bases in tensor spaces constructed out of V, U and their dual spaces. In view of the mentioned theorem we know that if $v \in V$ and $u \in U$, then, indicating again by $A \in V \otimes V^*$ and $B \in U \otimes U^*$ the tensors associated with the linear maps A and B, respectively, one has

$$(Av)^i = A^i{}_j v^j \quad \text{and} \quad (Bu)^r = B^r{}_s u^s \,.$$

Passing to tensor products and making use of the definition of tensor product of linear operators,

$$((A \otimes B)(v \otimes u))^{ir} = ((Av) \otimes (Bu))^{ir} = A^i{}_j v^j B^r{}_s u^s = A^i{}_j B^r{}_s v^j u^s \,.$$

Now, since $A \otimes B : V \otimes U \to V \otimes U$ is linear, we can write in general, for $t = t^{js} e_j \otimes f_s$,

$$((A \otimes B)t)^{ir} = A^i{}_j B^r{}_s t^{js} . \tag{4.1}$$

We conclude that the tensor in $V \otimes V^* \otimes U \otimes U^*$ associated with $A \otimes B$ has components

$$A^i{}_j B^r{}_s ,$$

and, as usual, the action of $A \otimes B : V \otimes U \to V \otimes U$ on $t \in V \otimes U$ is obtained by the tensor product of t and the tensor associated with $A \otimes B$, followed by a suitable contraction, as written in (4.1). ■

4.2.2 *An Example from Continuum Mechanics*

Referring to Example 3.20, consider once more an internal portion $\mathscr{C}$ of a continuous body $\mathscr{B}$. Suppose that $\mathscr{C}$ is represented by a regular set in the Euclidean space $\mathbb{E}^3$. As we said, there is a tensor $\sigma_p \in T_p^*\mathbb{E}^3 \otimes T_p\mathbb{E}^3$ (depending on the nature of $\mathscr{B}$ and on p) called the Cauchy stress tensor, such that

$$\mathbf{s}(p, \mathbf{n}) = \sigma_p(\mathbf{n}) . \tag{4.2}$$

We henceforth write σ in place of σ_p for shortness, as p is supposed to be fixed.

Suppose now to rotate $\mathscr{B}$ about p by a rotation $R \in SO(3)$. If the body is isolated from any other external system, and the rest frame is inertial, assuming *isotropy of space*, we expect that the relationship between **s**, **n** and σ "remains fixed" under the rotation R, provided one replaces the vectors **s**, **n** and the tensor σ with the corresponding objects transformed under R. How does the rotation R act on the tensor σ?

We start from the action of a rotation on vectors. We stress that we are assuming here an *active* point of view: a vector is transformed into another, different vector under the rotation. It is not a passive change of basis for a fixed vector. Fix an orthonormal basis $e_1, e_2, e_3 \in T_p\mathbb{E}^3$. Referring to that basis, the action of $SO(3)$ is, if $\mathbf{v} = v^i e_i$ and R is the matrix of coefficients $R^i{}_j$:

$$\mathbf{v} \mapsto \rho_R \mathbf{v}$$

where

$$(\rho_R \mathbf{v})^i := R^i{}_j v^j .$$

A direct verification shows that

$$\rho_R \rho_{R'} = \rho_{RR'} , \quad \text{for all } R, R' \in SO(3) ,$$

therefore $SO(3) \ni R \mapsto \rho_R$ defines, in fact, a simple representation of $SO(3)$ on $T_p\mathbb{E}^3$ called the **fundamental representation** of $SO(3)$. This representation acts both on $\mathbf{n}$ and $\mathbf{s}$ and enjoys a straightforward physical meaning.

Let us pass to the tensor $\sigma \in T_p^*\mathbb{E}^3 \otimes T_p\mathbb{E}^3$. In the bases $\{e_i\}_{i=1,2,3}$ and the associated dual one $\{e^{*j}\}_{j=1,2,3}$, the relation (4.2) reads

$$s^i = \sigma^i{}_j n^j ,$$

and thus, by action of a rotation R, one gets

$$(R\mathbf{s})^i = R^i{}_k \sigma^k{}_j n^j ,$$

or also,

$$(R\mathbf{s})^i = R^i{}_k \sigma^k{}_h (R^{-1})^h{}_j (R\mathbf{n})^j \ .$$

We conclude that, if we require that the relation (4.2) is preserved under rotations, we must define the action of $R \in SO(3)$ on the tensors $\sigma \in T_p^*\mathbb{E}^3 \otimes T_p\mathbb{E}^3$ by

$$\sigma \mapsto \gamma_R \sigma$$

where, *in the said bases*,

$$(\gamma_R \sigma)^i{}_j = R^i{}_k \, (R^{-1})^h{}_j \, \sigma^k{}_h \, . \tag{4.3}$$

It follows by direct verification that

$$\gamma_R \gamma_{R'} = \gamma_{RR'} , \quad \text{for all } R, R' \in SO(3) ,$$

therefore $SO(3) \ni R \mapsto \gamma_R$ defines a representation of $SO(3)$ on $T_p^*\mathbb{E}^3 \otimes T_p\mathbb{E}^3$. This representation is the tensor product of ρ acting on $T_p\mathbb{E}^3$ and another representation of $SO(3)$, ρ', acting on $T_p^*\mathbb{E}^3$ and defined as follows. If $\mathbf{w} = w_j e^{*j} \in T_p^*\mathbb{E}^3$ and $R \in SO(3)$ is the matrix of coefficients $R^i{}_j$:

$$\mathbf{w} \mapsto \rho'_R \mathbf{w}$$

where, referring to the dual basis $e^{*1}, e^{*2}, e^{*3} \in T_p^*\mathbb{E}^3$,

$$(\rho'_R \mathbf{w})_j := w_i (R^{-1})^i{}_j = w_i (R^t)^i{}_j \, .$$

Indeed one has

$$\gamma_R = \rho_R \otimes \rho'_R\,, \quad \text{for all } R \in SO(3)\,.$$

It is simply proved that ρ' is nothing but the dual representation associated with ρ, as defined in Definition 4.7. The proof is immediate.

4.2.3 The Schur Lemma and an Application to Continuum Mechanics

Schur's lemma actually consists of a number of different but similar statements which characterize irreducible linear representations of groups on vector spaces. Here we deal with a basic version of the lemma. There are many proofs; the proof we present here can be easily extended to the theory of unitary-projective representations in Hilbert spaces [1, 2].

Theorem 4.13 (Schur's Lemma in $\mathbb{C}^n$ and a Real Counterpart) *Let $\rho : G \ni g \mapsto GL(n, \mathbb{K})$ be a group representation in terms of isometric matrices (i.e., orthogonal if $\mathbb{K} = \mathbb{R}$ or unitary if $\mathbb{K} = \mathbb{C}$). Define the* **commutant** *of ρ by*

$$\rho' := \{A \in M(n, \mathbb{K}) \mid A\rho_g = \rho_g A\,, \quad \forall g \in G\}\,.$$

Then:

(a) *if $\mathbb{K} = \mathbb{C}$, the representation ρ is irreducible if and only if*

$$\rho' = \mathbb{C}I := \{cI \mid c \in \mathbb{C}\}\,;$$

(b) *if $\mathbb{K} = \mathbb{R}$ and*

$$\rho' = \mathbb{R}I := \{cI \mid c \in \mathbb{R}\}\,,$$

then the representation ρ is irreducible.

Proof We prove statement **(a)** in the complex case. Equip $\mathbb{C}^n$ with the standard Hermitian scalar product

$$\langle u|v\rangle := \sum_{k=1}^{n} \overline{u^k} v^k\,.$$

The Hermitian adjoint operator of $A \in M(n, \mathbb{C})$ will be denoted by

$$A^\dagger := \overline{A}^t\,,$$

where the bar indicates complex conjugation. By definition

$$\langle u|Av\rangle = \langle A^\dagger u|v\rangle \quad \text{for all } u, v \in \mathbb{C}^n .$$

By hypothesis $\rho_g^\dagger = (\rho_g)^{-1} = \rho_{g^{-1}}$, since the representation is made of unitary operators. Notice that, if $A \in \rho'$, taking the Hermitian adjoint of $A\rho_g = \rho_g A$ we get

$$\rho_{g^{-1}} A^\dagger = A^\dagger \rho_{g^{-1}} ,$$

so that $\rho_h A^\dagger = A^\dagger \rho_h$ for all $h \in G$ (since g^{-1} ranges throughout G when g does). Thus $A \in \rho'$ implies $A^\dagger \in \rho'$. The commutant is closed under Hermitian conjugation. As a consequence, since the commutant is a complex vector space, if $A \in \rho'$, then also

$$B := A + A^\dagger \in \rho' \quad \text{and} \quad C := i(A - A^\dagger) \in \rho' ,$$

and both B and C are Hermitian.

We now prove that, if ρ' is not trivial, then ρ is not irreducible. Suppose that $A \in \rho'$ is not of the form aI. It is not possible that both $B = bI$ and $C = cI$ for some $b, c \in \mathbb{C}$, since this would imply $A = aI$ for some $a \in \mathbb{C}$, contrary to assumption. Suppose that B is not of the said form, the other case being similar. As $B = B^\dagger$, it admits an orthogonal decomposition in terms of eigenspaces. Let λ be one of its eigenvalues. Notice that the corresponding eigenspace H_λ is non-trivial, but also proper; otherwise $B = \lambda I$, which we have excluded. Hence H_λ is a proper non-trivial subspace. On the other hand, if $u \in H_\lambda$, one has

$$B\rho_g u = \rho_g B u = \lambda \rho_g u ,$$

so that $\rho_g u \in H_\lambda$ and $\rho_g(H_\lambda) \subset H_\lambda$ for every $g \in G$. In summary, ρ cannot be irreducible since it admits a proper non-trivial invariant subspace.

We now prove that, if ρ is not irreducible, then the commutant ρ' is not trivial. Let $H \subset \mathbb{C}^n$ be a proper non-trivial subspace which is invariant under ρ: $\rho_g(H) \subset H$. If $u_1, \ldots, u_m$ is an orthonormal basis of H, define

$$P(\cdot) := \sum_{j=1}^{m} \langle u_j|\cdot\rangle u_j .$$

This operator is different from I (because $\dim(H) = m < n$) and $P \neq 0$ because $\dim(H) > 0$. P is the orthogonal projector onto H. By construction $P^\dagger = P$, $PP = P$, and $P(\mathbb{C}^n) = H$. Saying that $x \in H$ is equivalent to writing $Px = x$. Since H is invariant, $\rho_g P u \in H$ for every $u \in \mathbb{C}^n$, so that

$$P\rho_g P u = \rho_g P u .$$

Hence $P\rho_g P = \rho_g P$ because u is arbitrary. Taking the Hermitian adjoint of both sides, and using $P^\dagger = P$ and $\rho_g^\dagger = \rho_{g^{-1}}$, we get

$$P\rho_{g^{-1}}P = P\rho_{g^{-1}} .$$

Since g^{-1} is arbitrary, this boils down to

$$P\rho_g P = P\rho_g .$$

Comparing with $P\rho_g P = \rho_g P$, we conclude that

$$\rho_g P = P\rho_g ,$$

where P is not of the form cI by construction. This proves the complex case.

Finally, statement **(b)** in the real case is immediate: if ρ were reducible, the orthogonal projector onto a proper non-trivial invariant subspace would belong to the commutant and would not be of the form cI, contradicting the assumption $\rho' = \mathbb{R}I$. □

Let us come back to the previous example of the representation (4.3) γ_R, for $R \in SO(3)$, on the stress tensor of a continuous body. We assume that not only space but also the body is *isotropic*, in the sense that

$$R\sigma_p(\mathbf{n}) = \sigma_p(R\mathbf{n})$$

for every point p in the body, every $R \in SO(3)$ and every unit vector $\mathbf{n}$. *Notice that R acts here around p, which is viewed as the origin of the rotation of the vectors* $\mathbf{n}$ *applied at p*. All *ideal fluids at rest in a reference frame* satisfy (by definition) this hypothesis. If we represent everything in an orthonormal basis of $\mathbb{R}^3$, taking into account the linearity of σ_p and R, we can extend the above identity to

$$R\sigma_p(\mathbf{v}) = \sigma_p(R\mathbf{v})$$

for every $\mathbf{v} \in \mathbb{R}^3$ and $R \in SO(3)$. Namely

$$R\sigma_p = \sigma_p R , \quad \forall R \in SO(3) .$$

Now observe that the fundamental representation $\rho : SO(3) \ni R \mapsto R \in GL(3, \mathbb{R})$ is irreducible. This is because from every fixed unit vector we produce the entire 2-sphere by applying all rotations of $SO(3)$, so no proper non-trivial $SO(3)$-invariant subspaces can exist. Moreover, in this specific case the commutant is trivial:

$$\rho' = \mathbb{R}I .$$

Indeed, if $A \in M(3, \mathbb{R})$ commutes with every $R \in SO(3)$, fix a unit vector u and let $R_u(\theta)$ be any rotation around the axis spanned by u. Then

$$R_u(\theta)(Au) = A(R_u(\theta)u) = Au ,$$

hence Au is fixed by every rotation around the axis u, and therefore Au must be parallel to u. Since this holds for every unit vector u, every vector is an eigenvector of A, which implies that $A = \lambda I$ for some $\lambda \in \mathbb{R}$. We conclude that

$$\sigma^i_{p\,j} = -\mathcal{P}(p)\delta^i{}_j ,$$

for some scalar function of the point $\mathcal{P}$ (the sign above is conventional and $\mathcal{P}$ is positive if a volume of the continuum is compressed). This function is the usual *pressure* at p.

4.2.4 An Example from Quantum Mechanics

Hilbert spaces provide the standard mathematical setting whenever one handle quantum mechanics. A Hilbert space is nothing but a complex vector space equipped with a Hermitian scalar product (see the next chapter) such that it is complete with respect to the norm topology induced by that scalar product. However, here we analyze only the structure of vector spaces. Physically speaking, the vectors of the Hilbert space represent the states of the considered physical system (actually things are more complicated, but we do not discuss that here). To consider the simplest case we assume that the vector space which describes the states of the system is finite-dimensional (that is the case for the spin part of a quantum particle). Moreover, physics implies that the space $\mathcal{H}$ of the states of a composite system S made of two systems S_1 and S_2 associated with Hilbert spaces $\mathcal{H}_1$ and $\mathcal{H}_2$ respectively, is the Hilbert tensor product (see Sect. 2.2.4 for more details)

$$\mathcal{H} = \mathcal{H}_1 \otimes_H \mathcal{H}_2 .$$

As far as this example is concerned, the reader may omit the adjective "Hilbert" and think of $\mathcal{H}$ as the standard algebraic tensor product

$$\mathcal{H} = \mathcal{H}_1 \otimes \mathcal{H}_2 ,$$

which is correct when $\mathcal{H}_1, \mathcal{H}_2$ are finite-dimensional. Let the system S_1 be described by a state $\psi \in \mathcal{H}_1$ and suppose we transform the system by the action of an element R of some physical group of transformations $\mathcal{G}$ (e.g. $SO(3)$). The transformed state ψ' is given by $U^{(1)}_R\psi$ where $\mathcal{G} \ni R \mapsto U^{(1)}_R$ is a representation of $\mathcal{G}$ in terms of linear transformations $U^{(1)}_R : \mathcal{H}_1 \to \mathcal{H}_1$. Actually, physics, and Wigner's theorem in particular, requires that every $U_R^{(1)}$ be a *unitary* (or *anti-unitary*) transformation, but this is not relevant for our present purposes. The natural question concerning the representation of the action of $\mathcal{G}$ on the composite system S is:

> *If we know the representations* $\mathcal{G} \ni R \mapsto U_R^{(1)}$ *and* $\mathcal{G} \ni R \mapsto U_R^{(2)}$*, what is the representation of the action of* $\mathcal{G}$ *on* S *in terms of linear transformations in the Hilbert space* $\mathcal{H} = \mathcal{H}_1 \otimes \mathcal{H}_2$*?*

The answer given by physics, at least when the systems S_1 and S_2 do not interact, is that

$$U_R := U_R^{(1)} \otimes U_R^{(2)} .$$

4.3 Permutation Group and Symmetry of Tensors

We recall the definition of the *group of permutations of n objects* and give some well-known results of basic group theory whose proofs may be found in any group-theory textbook.

Definition 4.14 (Group of Permutations) Consider the set $I_n := \{1, \dots, n\}$. The **group of permutations of** n **objects**, $\mathscr{P}_n$, is the set of the bijective maps $\sigma : I_n \to I_n$ equipped with the composition rule given by the usual composition rule of functions. Moreover,

(a) the elements of $\mathscr{P}_n$ are called **permutations** (of n objects),
(b) a permutation of $\mathscr{P}_n$ with $n \geq 2$ is said to be a **transposition** if it differs from the identity map and reduces to the identity map when restricted to some subset of I_n containing $n - 2$ elements. ■

Remarks 4.15

(1) $\mathscr{P}_n$ contains $n!$ elements.
(2) Each permutation $\sigma \in \mathscr{P}_n$ can be represented by the corresponding string $(\sigma(1), \dots, \sigma(n))$.
(3) If, for instance, $n = 5$, then with the notation above $(1, 2, 3, 5, 4)$, $(5, 2, 3, 4, 1)$, $(1, 2, 4, 3, 5)$ are transpositions, while $(2, 3, 4, 5, 1)$ and $(5, 4, 3, 2, 1)$ are not.
(4) It is possible to show that each permutation $\sigma \in \mathscr{P}_n$ can be decomposed as a product of transpositions $\sigma = \tau_1 \circ \dots \circ \tau_k$. In general there are several different transposition-product decompositions for each permutation, however it is possible to show that if $\sigma = \tau_1 \circ \dots \circ \tau_k = \tau'_1 \circ \dots \circ \tau'_r$, where τ_i and τ'_j are transpositions, then $r + k$ is even. Equivalently, r is even or odd if and only if k is. This defines the **parity**, $\epsilon_\sigma \in \{-1, +1\}$, of a permutation σ, where $\epsilon_\sigma = +1$ if σ can be decomposed as a product of an *even* number of transpositions and $\epsilon_\sigma = -1$ if σ can be decomposed as a product of an *odd* number of transpositions.

(5) If $A = [A_{ij}]$ is a real or complex $n \times n$ matrix, it is possible to show (by induction) that

$$\det A = \sum_{\sigma \in \mathscr{P}_n} \epsilon_\sigma A_{1\sigma(1)} \cdots A_{n\sigma(n)} .$$

Alternatively, the identity above may be used to define the determinant of a matrix. ■

We pass to consider the action of $\mathscr{P}_n$ on tensors. Fix $n > 1$, consider the tensor algebra $\mathscr{A}_{\mathbb{K}}(V)$ and single out the tensor space

$$V^{n\otimes} := \underbrace{V \otimes \ldots \otimes V}_{n \text{ times}} .$$

Then consider the following action of $\mathscr{P}_n$ on

$$V^{n\times} := \underbrace{V \times \ldots \times V}_{n \text{ times}} .$$

For each $\sigma \in \mathscr{P}_n$ consider the map

$$\hat{\sigma} : V^{n\times} \to V^{n\otimes} \quad \text{such that} \quad \hat{\sigma} : (v_1, \ldots, v_n) \mapsto v_{\sigma^{-1}(1)} \otimes \ldots \otimes v_{\sigma^{-1}(n)} .$$

It is quite straightforward to show that $\hat{\sigma}$ is multilinear. Therefore, let

$$\sigma^{\otimes} : V^{n\otimes} \to V^{n\otimes} ,$$

be the linear map uniquely determined by $\hat{\sigma}$ by means of the universality theorem. By definition, it is completely determined by linearity and by the requirement

$$\sigma^{\otimes} : v_1 \otimes \ldots \otimes v_n \mapsto v_{\sigma^{-1}(1)} \otimes \ldots \otimes v_{\sigma^{-1}(n)} .$$

Theorem 4.16 *The above-defined linear map*

$$\sigma \mapsto \sigma^{\otimes}$$

with $\sigma \in \mathscr{P}_n$ is a group representation of $\mathscr{P}_n$ on $V^{n\otimes}$.

Proof First we show that if $\sigma, \sigma' \in \mathscr{P}_n$ then

$$\sigma^{\otimes}(\sigma'^{\otimes}(v_1 \otimes \ldots \otimes v_n)) = (\sigma \circ \sigma')^{\otimes}(v_1 \otimes \ldots \otimes v_n) .$$

This follows from the definition:

$$\sigma'^{\otimes}(v_1 \otimes \ldots \otimes v_n) = v_{\sigma'^{-1}(1)} \otimes \ldots \otimes v_{\sigma'^{-1}(n)} .$$

Re-defining $u_i := v_{\sigma'^{-1}(i)}$ so that

$$u_{\sigma^{-1}(j)} = v_{\sigma'^{-1}(\sigma^{-1}(j))} ,$$

one finds

$$\sigma^{\otimes}(\sigma'^{\otimes}(v_1 \otimes \ldots \otimes v_n)) = u_{\sigma^{-1}(1)} \otimes \ldots \otimes u_{\sigma^{-1}(n)}$$

$$= v_{\sigma'^{-1}\circ\sigma^{-1}(1)} \otimes \ldots \otimes v_{\sigma'^{-1}\circ\sigma^{-1}(n)}$$

$$= v_{(\sigma\circ\sigma')^{-1}(1)} \otimes \ldots \otimes v_{(\sigma\circ\sigma')^{-1}(n)} = (\sigma \circ \sigma')^{\otimes}(v_1 \otimes \ldots \otimes v_n) .$$

In other words

$$\sigma^{\otimes}(\sigma'^{\otimes}(v_1 \otimes \ldots \otimes v_n)) = (\sigma \circ \sigma')^{\otimes}(v_1 \otimes \ldots \otimes v_n) .$$

In particular, that identity also holds for a canonical basis of elements $e_{i_1} \otimes \ldots \otimes e_{i_n}$:

$$\sigma^{\otimes}(\sigma'^{\otimes}(e_{i_1} \otimes \ldots \otimes e_{i_n})) = (\sigma \circ \sigma')^{\otimes}(e_{i_1} \otimes \ldots \otimes e_{i_n}) .$$

By linearity such an identity holds for all arguments in $V^{n\otimes}$, and thus

$$\sigma^{\otimes}\sigma'^{\otimes} = (\sigma \circ \sigma')^{\otimes} .$$

The maps $\sigma^{\otimes}$ are linear by construction and are bijective because

$$\sigma^{\otimes}\sigma^{-1\otimes} = \sigma^{-1\otimes}\sigma^{\otimes} = e^{\otimes} = I .$$

The identity $e^{\otimes} = I$ can be proved by noticing that $e^{\otimes} - I$ is linear and vanishes when evaluated on any canonical basis of $V^{n\otimes}$. We have shown that $\sigma^{\otimes} \in GL(V^{n\otimes})$ for all $\sigma \in \mathscr{P}_n$ and that the map $\sigma \mapsto \sigma^{\otimes}$ is a homomorphism. This concludes the proof. □

Let us pass to the abstract-index notation and give a representation of the action of $\sigma^{\otimes}$ within that picture.

Theorem 4.17 *If t is a tensor in $V^{n\otimes} \in \mathscr{A}_{\mathbb{K}}(V)$ with $n \geq 2$ and $\sigma \in \mathscr{P}_n$, then the components of t with respect to any canonical basis of $V^{n\otimes}$ satisfy*

$$(\sigma^{\otimes}t)^{i_1\ldots i_n} = t^{i_{\sigma(1)}\ldots i_{\sigma(n)}} .$$

An analogous statement holds for tensors $s \in V^{*n\otimes}$:

$$(\sigma^{\otimes}s)_{i_1\ldots i_n} = s_{i_{\sigma(1)}\ldots i_{\sigma(n)}} .$$

Proof

$$\sigma^{\otimes} t = \sigma^{\otimes}(t^{j_1 \cdots j_n} e_{j_1} \otimes \ldots \otimes e_{j_n}) = t^{j_1 \cdots j_n} e_{j_{\sigma^{-1}(1)}} \otimes \ldots \otimes e_{j_{\sigma^{-1}(n)}} .$$

Since $\sigma : I_n \to I_n$ is bijective, if we define $i_k := j_{\sigma^{-1}(k)}$, it follows that $j_k = i_{\sigma(k)}$. Using this identity above we find

$$\sigma^{\otimes} t = t^{i_{\sigma(1)} \cdots i_{\sigma(n)}} e_{i_1} \otimes \ldots \otimes e_{i_n} .$$

That is nothing but the thesis. The final statement is a direct consequence of the initial one just replacing the vector space V with the vector space V^* (the fact that V^* is a dual space is immaterial; the statement relies only on the vector-space structure.) □

To conclude we introduce the notion of symmetric or anti-symmetric tensor.

Definition 4.18 (Symmetric and Antisymmetric Tensors) Let V be a finite-dimensional vector space over the field $\mathbb{K} = \mathbb{C}$ or $\mathbb{R}$. Consider the space $V^{n\otimes} \in \mathscr{A}_{\mathbb{K}}(V)$ of tensors of order $(n, 0)$ with $n \geq 2$.

(a) $t \in V^{n\otimes}$ is said to be **symmetric** if

$$\sigma^{\otimes} t = t ,$$

for all $\sigma \in \mathscr{P}_n$, or equivalently, using the abstract index notation,

$$t^{j_1 \cdots j_n} = t^{j_{\sigma(1)} \cdots j_{\sigma(n)}} ,$$

for all $\sigma \in \mathscr{P}_n$.

(b) $t \in V^{n\otimes}$ is said to be **antisymmetric** if

$$\sigma^{\otimes} t = \epsilon_\sigma \, t ,$$

for all $\sigma \in \mathscr{P}_n$, or equivalently, using the abstract index notation,

$$t^{j_1 \cdots j_n} = \epsilon_\sigma \, t^{j_{\sigma(1)} \cdots j_{\sigma(n)}} ,$$

for all $\sigma \in \mathscr{P}_n$. ■

Remarks 4.19

(1) Concerning the definition in (b), notice that

$$\epsilon_\sigma = \epsilon_{\sigma^{-1}} .$$

(1) Notice that the sets of symmetric and antisymmetric tensors in $V^{n\otimes}$ are separately closed under linear combinations. In other words each of them is a subspace of $V^{n\otimes}$. ■

Examples 4.20

1. Suppose $n = 2$. Then a symmetric tensor $s \in V \otimes V$ satisfies $s^{ij} = s^{ji}$ and an antisymmetric tensor $a \in V \otimes V$ satisfies $a^{ij} = -a^{ji}$.
2. Suppose $n = 3$. Then one checks that $\sigma \in \mathscr{P}_3$ has parity $+1$ if and only if σ is a **cyclic permutation**, i.e.,

$$(\sigma(1), \sigma(2), \sigma(3)) = (1, 2, 3)\,, \quad (2, 3, 1)\,, \quad \text{or} \quad (3, 1, 2)\,.$$

Now consider the vector space V with $dim V = 3$. It turns out that a tensor $e \in V \otimes V \otimes V$ is antisymmetric if and only if

$$e^{ijk} = 0\,,$$

if (i, j, k) is *not* a permutation of $(1, 2, 3)$ and, otherwise,

$$e^{ijk} = \pm e^{123}\,,$$

where the sign $+$ occurs if the permutation $(\sigma(1), \sigma(2), \sigma(3)) = (i, j, k)$ is cyclic and the sign $-$ occurs otherwise. That relation between the parity of a permutation and cyclicity does *not* hold for $n > 3$. ■

Remarks 4.21

(1) Consider a generic tensor space $S \in \mathscr{A}_{\mathbb{K}}(V)$ which contains $n \geq 2$ spaces V as factors. We may suppose, for the sake of simplicity, that

$$S = S_1 \otimes V^{n\otimes} \otimes S_2\,,$$

where

$$S_1 = U_1 \otimes \ldots \otimes U_k\,, \quad S_2 = U_{k+1} \otimes \ldots \otimes U_m$$

and U_i is either V or V^*. Anyway, all that we are going to say holds also if the considered n spaces V do not define a unique block $V^{n\otimes}$. We may define the action of $\sigma \in \mathscr{P}_n$ on the whole space S starting from a multilinear map

$$\hat{\sigma} : U_1 \times \ldots \times U_k \times V^{n\times} \times U_{k+1} \times \ldots \times U_m$$
$$\to U_1 \otimes \ldots \otimes U_k \otimes V^{n\otimes} \otimes U_{k+1} \otimes \ldots \otimes U_m\,,$$

such that it reduces to the tensor-product map on $U_1 \times \ldots \times U_k$ and $U_{k+1} \times \ldots \times U_m$:

$$\hat{\sigma} : (u_1, \ldots, u_k, v_1, \ldots, v_n, u_{k+1}, \ldots, u_m)$$
$$\mapsto u_1 \otimes \ldots \otimes u_k \otimes v_{\sigma^{-1}(1)} \otimes \ldots \otimes v_{\sigma^{-1}(n)} \otimes u_{k+1} \otimes \ldots \otimes u_m \, .$$

Using the universality theorem as above, we build up a representation of $\mathscr{P}_n$ on S,

$$\sigma \mapsto \sigma^{\otimes} \, ,$$

which "acts on $V^{n\otimes}$ only". Using the abstract index notation, the action of $\sigma^{\otimes}$ is well represented by

$$\sigma^{\otimes} : t^{A \; i_1 \ldots i_n \; B} \mapsto t^{A \; i_{\sigma(1)} \ldots i_{\sigma(n)} \; B} \, .$$

This allows one to define and study the symmetry of a tensor with respect to a few indices singled out among the complete set of indices of the tensor. For instance, a tensor $t^{ij}{}_k{}^r$ may be symmetric or antisymmetric with respect to the indices i and j, or j and r, or i, j, r.

(2) Notice that no discussion on the symmetry of indices of different kinds (one covariant and the other contravariant) is possible. ■

Exercises 4.22

1. Let t be a tensor in $V^{n\otimes}$ (or $V^{*n\otimes}$). Show that t is symmetric or antisymmetric if there is a canonical basis where the components have symmetric or antisymmetric indices, i.e.,

$$t^{i_1 \ldots i_n} = t^{i_{\sigma(1)} \ldots i_{\sigma(n)}}$$

or respectively

$$t^{i_1 \ldots i_n} = \epsilon_\sigma \, t^{i_{\sigma(1)} \ldots i_{\sigma(n)}}$$

for all $\sigma \in \mathscr{P}_n$.

Remarks 4.23 The result implies that, to show that a tensor is symmetric or antisymmetric, it is sufficient to verify the symmetry or antisymmetry of its components within a *single* canonical basis. ■

2. Show that the sets of symmetric tensors of order $(n, 0)$ and $(0, n)$ are vector subspaces of $V^{n\otimes}$ and $V^{*n\otimes}$ respectively.
3. Show that the subspace of antisymmetric tensors of order $(0, n)$ in $V^{*n\otimes}$ has dimension $\binom{dimV}{n}$ if $n \leq dimV$. What about $n > dimV$?

4. Consider a tensor $t^{i_1 \ldots i_n}$. Show that the tensor is symmetric if and only if it is symmetric with respect to each arbitrarily chosen pair of indices, i.e.

$$t^{\ldots i_k \ldots i_p \ldots} = t^{\ldots i_p \ldots i_k \ldots} ,$$

for all $p, k \in \{1, \ldots, n\}$, $p \neq k$.
5. Consider a tensor $t^{i_1 \ldots i_n}$. Show that the tensor is antisymmetric if and only if it is antisymmetric with respect to each arbitrarily chosen pair of indices, i.e.

$$t^{\ldots i_k \ldots i_p \ldots} = -t^{\ldots i_p \ldots i_k \ldots} ,$$

for all $p, k \in \{1, \ldots, n\}$, $p \neq k$.
6. Show that $V \otimes V = A \oplus S$, where $\oplus$ denotes the direct sum and A and S are respectively the spaces of antisymmetric and symmetric tensors in $V \otimes V$. Does such a direct decomposition hold if considering $V^{n\otimes}$ with $n > 2$?

4.4 Grassmann Algebra, Also Known as Exterior Algebra (and Something About the Space of Symmetric Tensors)

p-forms play a crucial role in several physical applications of tensor calculus. Let us review, very quickly, the most important features of p-forms.

Definition 4.24 (Spaces $\Lambda^p(V)$ and $\Lambda^p(V^*)$) If V denotes a linear space over the field $\mathbb{K}$ and has finite dimension n, $\Lambda^p(V)$ and $\Lambda^p(V^*)$ indicate respectively the linear spaces of *antisymmetric* tensors of order $(p, 0)$ and of *antisymmetric* tensors of order $(0, p)$. We assume the convention that $\Lambda^0(V) := \mathbb{K}$ and $\Lambda^0(V^*) := \mathbb{K}$, and furthermore, if $p > n$ is an integer, $\Lambda^p(V) := \{0\}$, $\Lambda^p(V^*) := \{0\}$, $\{0\}$ being the trivial vector space made of the zero vector only. The elements of $\Lambda^p(V)$ are called p-**vectors**. The elements of $\Lambda^p(V^*)$ are called p-**forms**. ■

In the following we will be mainly concerned with the spaces $\Lambda^p(V)$ and $\Lambda^p(V^*)$. However, for some applications to physics (especially quantum mechanics), it is useful to define the spaces $S^p(V)$ and $S^p(V^*)$, respectively the linear spaces of *symmetric* tensors of order $(p, 0)$ and of *symmetric* tensors of order $(0, p)$. As before, we assume $S^0(V) := \mathbb{K}$, $S^0(V^*) := \mathbb{K}$.

Remarks 4.25 $\Lambda^p(V) = \{0\}$ if $p > n$ is a natural definition. Indeed, in the considered case, if $s \in \Lambda^p(V)$, in every component $s^{i_1 \ldots i_p}$ there must be at least two indices, say i_1 and i_2, with the same value so that

$$s^{i_1 i_2 \ldots i_p} = s^{i_2 i_1 \ldots i_p} .$$

Theorem 4.17 implies

$$s^{i_1 i_2 \ldots i_p} = -s^{i_2 i_1 \ldots i_p} ,$$

and thus

$$s^{i_1 i_2 \ldots i_p} = s^{i_2 i_1 \ldots i_p} = 0 .$$

■

Before proceeding, we recall some basic notions about *projectors*.

If U is a vector space over the field $\mathbb{K} = \mathbb{R}$ or $\mathbb{C}$, a **projector onto** the subspace $S \subset U$ is a linear operator $P : U \to U$ such that it is **idempotent**, i.e. $PP = P$, and $P(U) = S$.

We have the following well-known results concerning projectors.

(1) If P is a projector onto S, then $Q := I - P$ is a projector onto another subspace $S' := Q(U)$. Moreover it holds

$$U = S \oplus S' ,$$

where $\oplus$ denotes the direct sum of S and S' (i.e. $S \oplus S'$ is the space generated by linear combinations of elements in S and elements in S', and it holds $S \cap S' = \{0\}$).

(2) Conversely, if S and S' are subspaces of U such that $S \oplus S' = U$—and thus, for every $v \in U$ there is a pair of uniquely determined elements $v_S \in S$ and $v_{S'} \in S'$ such that $v = v_S + v_{S'}$—the maps

$$P : U \ni v \mapsto v_S \quad \text{and} \quad Q : U \ni v \mapsto v_{S'}$$

are projectors onto S and S' respectively and $Q = I - P$.

(3) If (a) P is a projector onto S, (b) P' is a projector onto S', and (c) $PP' = P'P$, then PP' is a projector as well, and it is a projector onto $S \cap S'$.

These results are valid also in the infinite-dimensional case. However, in that case, if U is a Banach space it is more convenient to specialize the definition of projectors by requiring their continuity too. With this restricted definition the statements above are valid anyway provided that the involved subspaces are closed (i.e. they are Banach subspaces).

Proposition 4.26 *Consider the linear operators* $\mathcal{A} : V^{\otimes p} \to V^{\otimes p}$ *and* $\mathcal{S} : V^{\otimes p} \to V^{\otimes p}$*, where, respectively,*

$$\mathcal{A} := \frac{1}{p!} \sum_{\sigma \in \mathscr{P}_p} \epsilon_\sigma \sigma^{\otimes} , \tag{4.4}$$

and

$$\mathcal{S} := \frac{1}{p!} \sum_{\sigma \in \mathscr{P}_p} \sigma^{\otimes} . \tag{4.5}$$

Then $\mathcal{A}$ is a projector onto $\Lambda^p(V)$, i.e. $\mathcal{A}\mathcal{A} = \mathcal{A}$ and $\mathcal{A}\left(V^{\otimes p}\right) = \Lambda^p(V)$ and, similarly, $\mathcal{S}$ is a projector onto $S^p(V)$, i.e. $\mathcal{S}\mathcal{S} = \mathcal{S}$ and $\mathcal{S}\left(V^{\otimes p}\right) = S^p(V)$. If $p \geq 2$, one also has

$$\mathcal{A}\mathcal{S} = \mathcal{S}\mathcal{A} = 0 \tag{4.6}$$

where 0 is the zero operator, which implies $\Lambda^p(V) \cap S^p(V) = \{0\}$.

Proof Let us prove that $\mathcal{A}\mathcal{A} = \mathcal{A}$. (Actually it would be a trivial consequence of the two facts proved immediately afterwards, but this proof is interesting in its own right because similar procedures will be used later in more complicated proofs.) Using Theorem 4.16 and the fact that $\epsilon_\sigma \epsilon_\tau = \epsilon_{\sigma \circ \tau}$ for $\sigma, \tau \in \mathscr{P}_p$, one has

$$\mathcal{A}\mathcal{A} = \frac{1}{p!} \sum_{\sigma \in \mathscr{P}_p} \epsilon_\sigma \sigma^{\otimes} \frac{1}{p!} \sum_{\tau \in \mathscr{P}_p} \epsilon_\tau \tau^{\otimes} = \frac{1}{p!} \sum_{\sigma \in \mathscr{P}_p} \frac{1}{p!} \sum_{\tau \in \mathscr{P}_p} \epsilon_{\sigma \circ \tau} \left(\sigma \circ \tau\right)^{\otimes}$$

$$= \frac{1}{p!} \sum_{\sigma \in \mathscr{P}_p} \frac{1}{p!} \sum_{\sigma \circ \tau \in \mathscr{P}_p} \epsilon_{\sigma \circ \tau} \left(\sigma \circ \tau\right)^{\otimes} .$$

In the last step, we have taken advantage of the fact that, for fixed σ, $\sigma \circ \tau$ ranges over the whole $\mathscr{P}_p$ if τ ranges over the whole $\mathscr{P}_p$. Therefore summing over τ is equivalent to summing over $\sigma \circ \tau$. In other words, if $\sigma' := \sigma \circ \tau$,

$$\mathcal{A}\mathcal{A} = \frac{1}{p!} \sum_{\sigma \in \mathscr{P}_p} \frac{1}{p!} \sum_{\sigma' \in \mathscr{P}_p} \epsilon_{\sigma'} \sigma'^{\otimes} = \frac{1}{p!} p! \, \mathcal{A} = \mathcal{A} .$$

To prove $\mathcal{A}\left(V^{\otimes p}\right) = \Lambda^p(V)$ it is sufficient to verify that, if $s \in V^{\otimes p}$, then $\sigma^{\otimes}\mathcal{A}s = \epsilon_\sigma \mathcal{A}s$, so that $\mathcal{A}(V^{\otimes p}) \subset \Lambda^p(V)$, and furthermore that $t \in \Lambda^p(V)$ entails $\mathcal{A}t = t$, so that $\mathcal{A}(V^{\otimes p}) \supset \Lambda^p(V)$. Using, in particular, the fact that

$$\epsilon_\sigma \epsilon_{\tau \circ \sigma} = \epsilon_\sigma \epsilon_\tau \epsilon_\sigma = \left(\epsilon_\sigma\right)^2 \epsilon_\tau ,$$

we have

$$\sigma^{\otimes}\mathcal{A}s = \frac{1}{p!} \sum_{\tau \in \mathscr{P}_p} \epsilon_\tau \, \sigma^{\otimes}(\tau^{\otimes}s) = \frac{1}{p!} \sum_{\tau \in \mathscr{P}_p} \epsilon_\sigma \epsilon_{\tau \circ \sigma} (\sigma \circ \tau)^{\otimes} s$$

$$= \epsilon_\sigma \frac{1}{p!} \sum_{\sigma \circ \tau \in \mathscr{P}_p} \epsilon_{\tau \circ \sigma} (\sigma \circ \tau)^{\otimes} s = \epsilon_\sigma \mathcal{A}s .$$

To conclude we prove that $t \in \Lambda^p(V)$ entails $\mathcal{A}t = t$.

$$\mathcal{A}t = \frac{1}{p!} \sum_{\sigma \in \mathscr{P}_p} \epsilon_\sigma \sigma^\otimes t = \frac{1}{p!} \sum_{\sigma \in \mathscr{P}_p} (\epsilon_\sigma)^2 t = \frac{1}{p!} \sum_{\sigma \in \mathscr{P}_p} t = \frac{p!}{p!} t = t\,.$$

We have used the fact that $\sigma^\otimes t = \epsilon_\sigma t$ since $t \in \Lambda^p(V)$.

The proof for $\mathcal{S}$ and the spaces $S^p(V)$ is essentially identical, with the only difference that the coefficients ϵ_τ, ϵ_σ, etc. do not appear anywhere in the proof. If $p \geq 2$, the last statement can be proved as follows. First notice that, if τ is a fixed transposition,

$$-\sum_{\sigma \in \mathscr{P}_p} \epsilon_\sigma = \sum_{\sigma \in \mathscr{P}_p} \epsilon_\tau \epsilon_\sigma = \sum_{\sigma \in \mathscr{P}_p} \epsilon_{\tau\circ\sigma} = \sum_{\tau\circ\sigma \in \mathscr{P}_p} \epsilon_{\tau\circ\sigma} = \sum_{\sigma' \in \mathscr{P}_p} \epsilon_{\sigma'} = \sum_{\sigma \in \mathscr{P}_p} \epsilon_\sigma\,,$$

so that $\sum_{\sigma \in \mathscr{P}_p} \epsilon_\sigma = 0$. Next, notice that if $t \in V^{\otimes p}$, then $\mathcal{S}t = s \in S^p(V)$ is symmetric and thus

$$p!\mathcal{A}\mathcal{S}t = p!\mathcal{A}s = \sum_{\sigma \in \mathscr{P}_p} \epsilon_\sigma \sigma^\otimes s = \left(\sum_{\sigma \in \mathscr{P}_p} \epsilon_\sigma \right) s = 0$$

since $\sum_{\sigma \in \mathscr{P}_p} \epsilon_\sigma = 0$. If $t \in V^{\otimes p}$, then $\mathcal{A}t = a \in \Lambda^p(V)$ is antisymmetric and thus

$$p!\mathcal{S}\mathcal{A}t = p!\mathcal{S}a = \sum_{\sigma \in \mathscr{P}_p} \sigma^\otimes a = \left(\sum_{\sigma \in \mathscr{P}_p} \epsilon_\sigma \right) a = 0\,,$$

again, since $\sum_{\sigma \in \mathscr{P}_p} \epsilon_\sigma = 0$. Using property (3) of projectors mentioned above, it follows that $\Lambda^p(V) \cap S^p(V) = \{0\}$. □

Remarks 4.27 In view of the last statement, one may wonder whether the space $V^{\otimes p}$ is the direct sum of $\Lambda^p(V)$ and $S^p(V)$. This is not the case if $p > 2$, and $\Lambda^p(V) \oplus S^p(V)$ is a proper subspace of $V^{\otimes p}$ for $p > 2$. The proof is omitted. ■

Examples 4.28 This example is very important in Quantum Mechanics. Consider a set of p identical quantum systems S, each represented on the Hilbert space $\mathcal{H}$. As we know, in principle, the states of the overall system are represented by unit-norm vectors in the Hilbert tensor product of p copies of the space $\mathcal{H}$, $\mathcal{H} \otimes_H \cdots \otimes_H \mathcal{H}$ (see Definition 2.32). This is not the whole story, because the axioms of Quantum Mechanics (but the following statement turns out to be a theorem in relativistic quantum field theory, called the *spin-statistics theorem*) assume that only some vectors are permitted to describe states when the system S is a quantum particle. There are only two categories of quantum particles: *bosons* and *fermions*. The first ones are those with integer spin (or helicity) and the others are those with semi-integer spin (or helicity). The states of a system of p identical bosons are allowed to be represented by unit-norm vectors in $S^p(\mathcal{H})$ only, whereas the states of a system of p identical fermions are allowed to be represented by unit-norm vectors in $\Lambda^p(\mathcal{H})$. All the possible transformations of states by means of operators, representing some physical action on the system, must respect that requirement.

4.4.1 The Exterior Product and Grassmann Algebra

We are in a position to give the central definition, that of the *exterior product* of k-vectors.

Definition 4.29 (Exterior Product) If $s \in \Lambda^p(V)$ and $t \in \Lambda^q(V)$, the **exterior product** of s and t is the $(p+q)$-vector (the 0-vector when $p+q > dim V$):

$$s \wedge t := \mathcal{A}(s \otimes t)\,.$$

Theorem 4.30 *Consider the exterior product $\wedge$ in the spaces $\Lambda^r(V)$ where V is a linear space over the field $\mathbb{K}$ and has finite dimension n. It fulfills the following properties for $s \in \Lambda^p(V)$, $t \in \Lambda^q(V)$, $w \in \Lambda^r(V)$ and $a, b \in \mathbb{K}$:*

(a) $s \wedge (t \wedge w) = (s \wedge t) \wedge w = \mathcal{A}(s \otimes t \otimes w)$,
(b) $s \wedge t = (-1)^{pq} t \wedge s$, *in particular* $u \wedge v = -v \wedge u$ *if* $u, v \in V$,
(c) $s \wedge (at + bw) = a(s \wedge t) + b(s \wedge w)$.
Furthermore, if $u_1, \ldots, u_m \in V$, these vectors are linearly independent if and only if

$$u_1 \wedge \cdots \wedge u_m \neq 0\,,$$

where the exterior product of several vectors is well defined (without parentheses) in accordance with the associativity property **(a)**.

Proof

(a) Let us prove that $s \wedge (t \wedge w) = \mathcal{A}(s \otimes t \otimes w)$. Exploiting a similar procedure one proves that $(s \wedge t) \wedge w = \mathcal{A}(s \otimes t \otimes w)$, and this concludes the proof. To prove that $s \wedge (t \wedge w) = \mathcal{A}(s \otimes t \otimes w)$, we notice that, if $s \in \Lambda^p(V)$, $t \in \Lambda^q(V)$, $w \in \Lambda^r(V)$, then

$$s \wedge (t \wedge w)$$

$$= \frac{1}{(p+q+r)!} \sum_{\sigma \in \mathscr{P}_{p+q+r}} \epsilon_\sigma \sigma^\otimes \left(\frac{1}{(q+r)!} \sum_{\tau \in \mathscr{P}_{q+r}} \epsilon_\tau\, s \otimes \tau^\otimes (t \otimes w) \right)\,.$$

That is,

$$s \wedge (t \wedge w) = \frac{1}{(q+r)!} \sum_{\tau \in \mathscr{P}_{q+r}} \left(\frac{1}{(p+q+r)!} \sum_{\sigma \in \mathscr{P}_{p+q+r}} \epsilon_\tau \epsilon_\sigma \, \sigma^\otimes (s \otimes \tau^\otimes (t \otimes w)) \right) .$$

Now we can view $\mathscr{P}_{q+r}$ as the subgroup of $\mathscr{P}_{p+q+r}$ which leaves unchanged the first p objects. With this interpretation one has

$$s \wedge (t \wedge w) = \frac{1}{(q+r)!} \sum_{\tau \in \mathscr{P}_{q+r}} \left(\frac{1}{(p+q+r)!} \sum_{\sigma \in \mathscr{P}_{p+q+r}} \epsilon_\tau \epsilon_\sigma \sigma^\otimes (\tau^\otimes (s \otimes t \otimes w)) \right) .$$

Equivalently,

$$\begin{aligned} s \wedge (t \wedge w) &= \frac{1}{(q+r)!} \sum_{\tau \in \mathscr{P}_{q+r}} \left(\frac{1}{(p+q+r)!} \sum_{\sigma \in \mathscr{P}_{p+q+r}} \epsilon_{\sigma \circ \tau} \, (\sigma \circ \tau)^\otimes (s \otimes t \otimes w) \right) \\ &= \frac{1}{(q+r)!} \sum_{\tau \in \mathscr{P}_{q+r}} \left(\frac{1}{(p+q+r)!} \sum_{\sigma' \in \mathscr{P}_{p+q+r}} \epsilon_{\sigma'} \, \sigma'^\otimes (s \otimes t \otimes w) \right) \\ &= \frac{(q+r)!}{(q+r)!} \mathcal{A}(s \otimes t \otimes w) = \mathcal{A}(s \otimes t \otimes w) . \end{aligned}$$

(b) One may pass from $u_1 \otimes \cdots \otimes u_q \otimes v_1 \otimes \cdots \otimes v_p$ to $v_1 \otimes \cdots \otimes v_p \otimes u_1 \otimes \cdots \otimes u_q$ by means of the following procedure. First, using $p+q-1$ transpositions one achieves

$$u_2 \otimes \cdots \otimes u_q \otimes v_1 \otimes \cdots \otimes v_p \otimes u_1 .$$

In this case, due to the definition of $\mathcal{A}$, one has

$$\mathcal{A}(u_1 \otimes u_2 \otimes \cdots \otimes u_q \otimes v_1 \otimes \cdots \otimes v_p) = (-1) \mathcal{A}(u_2 \otimes u_1 \otimes \cdots \otimes u_q \otimes v_1 \otimes \cdots \otimes v_p),$$

so that, using $p+q-1$ iterated transpositions:

$$\mathcal{A}(u_1 \otimes \cdots \otimes u_q \otimes v_1 \otimes \cdots \otimes v_p) = (-1)^{p+q-1} \mathcal{A}(u_2 \otimes \cdots \otimes u_q \otimes v_1 \otimes \cdots \otimes v_p \otimes u_1).$$

Using again the same procedure one gets also

$$\begin{aligned}&\mathcal{A}(u_1 \otimes \cdots \otimes u_q \otimes v_1 \otimes \cdots \otimes v_p)\\&\quad = (-1)^{p+q-1}(-1)^{p+q-1}\mathcal{A}(u_3 \otimes \cdots \otimes u_q \otimes v_1 \otimes \cdots \otimes v_p \otimes u_1 \otimes u_2)\,.\end{aligned}$$

By iterating this prescription one finally gets

$$\mathcal{A}(u_1\otimes\cdots\otimes u_q\otimes v_1\otimes\cdots\otimes v_p) = (-1)^{q(p+q-1)}\mathcal{A}(v_1\otimes\cdots\otimes v_p\otimes u_1\otimes\cdots\otimes u_q).$$

That is,

$$u_1 \wedge\cdots\wedge u_q \wedge v_1 \wedge\cdots\wedge v_p = (-1)^{q(q-1)}(-1)^{pq} v_1 \wedge\cdots\wedge v_p \wedge u_1 \wedge\cdots\wedge u_q\,.$$

Since $(-1)^{q(q-1)} = 1$, by linearity the obtained identity yields the thesis.

[**(c)**] is a straightforward consequence of the definition of $\mathcal{A}$.

Let us come to the last statement. If $u_1, \ldots, u_m$ are linearly dependent, there must be one of them, say u_1, which can be represented as a linear combination of the others. In this case $u_1 \wedge \cdots \wedge u_m$ can be rewritten as

$$\sum_{i=2}^{m} c^i\, u_i \wedge u_2 \wedge \cdots \wedge u_m\,.$$

Since, whenever $c^i \neq 0$, u_i appears twice in $u_i \wedge u_2 \wedge \cdots \wedge u_m = \mathcal{A}(u_i \otimes u_2 \otimes \cdots \otimes u_m)$, this term vanishes (interchanging the two copies of u_i amounts on the one hand to doing nothing, and on the other hand to performing a transposition, which produces a minus sign), and thus $u_1 \wedge \cdots \wedge u_m = 0$. If $u_1, \ldots, u_m$ are linearly independent, one can add further vectors $u_{m+1}, \ldots, u_n$ in order to obtain a basis for V, which we denote by $\{e_i\}_{i=1,\ldots,n}$. In this case

$$u_1 \wedge \cdots \wedge u_n = \mathcal{A}(e_1 \otimes \cdots \otimes e_n) \neq 0\,.$$

Indeed, by direct inspection one finds that if $\{e^{*i}\}_{i=1,\ldots,n}$ is the dual basis, then

$$\begin{aligned}&\langle e^{*1} \otimes \cdots \otimes e^{*n}, \mathcal{A}(e_1 \otimes \cdots \otimes e_n)\rangle\\&\quad = \frac{1}{n!}\langle e^{*1} \otimes \cdots \otimes e^{*n}, e_1 \otimes \cdots \otimes e_n\rangle + \text{vanishing terms} = \frac{1}{n!} \neq 0\,,\end{aligned}$$

and so $u_1 \wedge \cdots \wedge u_n \neq 0$. As a consequence, the factor $u_1 \wedge \cdots \wedge u_m$ cannot vanish either: if it did, then using **(a)** and **(c)** one would have

$$u_1\wedge\cdots\wedge u_n = (u_1\wedge\cdots\wedge u_m)\wedge(u_{m+1}\wedge\cdots\wedge u_n) = \mathbf{0}\wedge(u_{m+1}\wedge\cdots\wedge u_n) = \mathbf{0},$$

which is impossible.

□

Remarks 4.31 It is very important to stress that there is another *inequivalent* definition of wedge product which gives rise to a naturally isomorphic structure. If $s \in \Lambda^p(V)$ and $t \in \Lambda^q(V)$, the **exterior product** of s and t can also be defined as the $(p+q)$-vector (the 0-vector when $p+q > dimV$):

$$s\tilde{\wedge}t := \frac{(p+q)!}{p!q!}\mathcal{A}(s \otimes t) = \frac{(p+q)!}{p!q!} s \wedge t\,. \tag{4.7}$$

With this different definition, *all* the statements of Theorem 4.30 and of the next corollary are nevertheless *true*. In particular $\tilde{\wedge}$ is associative. The new structure is as natural as the one we have introduced and is constructed by taking suitable quotients in the tensor product space instead of exploiting the operator $\mathcal{A}$. In symplectic geometry and Hamiltonian mechanics the second definition of wedge product is preferred. This ambiguity is often the source of mathematical disasters. ■

Theorem 4.30 has an immediate and important corollary.

Corollary *Each space $\Lambda^p(V)$, with $0 \le p \le n = dim(V)$, has dimension $\binom{n}{p}$. Moreover, if $\{e_i\}_{i=1,\dots,n}$ is a basis of V, the elements*

$$e_{i_1} \wedge \cdots \wedge e_{i_p}\,, \quad \textit{with } 1 \le i_1 < i_2 < \cdots < i_p \le n$$

form a basis of $\Lambda^p(V)$.

Proof Define $I := \{1, 2, \dots, n\}$. Notice that, with the given definitions,

$$e_{i_1} \wedge \cdots \wedge e_{i_p} = \mathcal{A}(e_{i_1} \otimes \cdots \otimes e_{i_p})\,.$$

Since $e_{i_1} \otimes \cdots \otimes e_{i_p}$ with $i_1, \dots, i_p \in I$ define a basis of $V^{p\otimes}$ and $\mathcal{A} : V^{p\otimes} \to \Lambda^p(V)$ is a projector, the elements $e_{i_1} \wedge \cdots \wedge e_{i_p} = \mathcal{A}(e_{i_1} \otimes \cdots \otimes e_{i_p})$ are generators of $\Lambda^p(V)$. Moreover

$$e_{i_1} \wedge \cdots \wedge e_{i_p} = \pm e_{j_1} \wedge \cdots \wedge e_{j_p}$$

if $(i_1, \dots, i_p)$ and $(j_1, \dots, j_p)$ differ by a permutation, and

$$e_{i_1} \wedge \cdots \wedge e_{i_p} = 0$$

if there are repeated indices. All that implies that

$$S := \{e_{i_1} \wedge \cdots \wedge e_{i_p}\,, \quad \text{with } 1 \le i_1 < i_2 < \cdots < i_p \le n\}$$

is still a set of generators of $\Lambda^p(V)$. Finally, the elements of S are linearly independent. Indeed, suppose that

$$c^{i_1\cdots i_p} e_{i_1} \wedge \cdots \wedge e_{i_p} = 0\,, \quad \text{with } 1 \le i_1 < i_2 < \cdots < i_p \le n\,.$$

If we apply the functional $e^{*j_1} \wedge \cdots \wedge e^{*j_p}$ to this identity, with $1 \le j_1 < j_2 < \cdots < j_p \le n$ fixed, only the term $e_{j_1} \wedge \cdots \wedge e_{j_p}$ gives a non-vanishing contribution:

$$\begin{aligned} 0 &= c^{i_1\cdots i_p} \langle e_{i_1} \wedge \cdots \wedge e_{i_p},\, e^{*j_1} \wedge \cdots \wedge e^{*j_p} \rangle \\ &= c^{i_1\cdots i_p} \left\langle \mathcal{A}(e_{i_1} \otimes \cdots \otimes e_{i_p}), e^{*j_1} \otimes \cdots \otimes e^{*j_p} \right\rangle \\ &= c^{i_1\cdots i_p} \frac{1}{p!} \delta^{i_1}_{j_1} \cdots \delta^{i_p}_{j_p} = c^{j_1\cdots j_p} \frac{1}{p!}\,. \end{aligned}$$

Thus we conclude that $c^{j_1\cdots j_p} = 0$ for all $j_1, \ldots, j_p$ with $1 \le j_1 < j_2 < \cdots < j_p \le n$. Therefore the elements of S are linearly independent. By elementary combinatorics, those elements are exactly $n!/(p!(n-p)!) = \binom{n}{p}$. □

Remarks 4.32

(1) Let $dim V = n$ and fix a basis $\{e_i\}_{i=1,\ldots,n}$ in V. If $w \in \Lambda^p(V)$, it can be represented in two different ways:

$$w = w^{i_1\cdots i_p} e_{i_1} \otimes \cdots \otimes e_{i_p}\,,$$

or

$$w = \sum_{1 \le i_1 < \cdots < i_p \le n} \tilde{w}^{i_1\cdots i_p} e_{i_1} \wedge \cdots \wedge e_{i_p}\,.$$

The relation between the components with $1 \le i_1 < \cdots < i_p \le n$ is

$$\tilde{w}^{i_1\cdots i_p} = p! w^{i_1\cdots i_p}\,. \tag{4.8}$$

(2) Every $u^{*1} \wedge \cdots \wedge u^{*p} \in \Lambda^p(V^*) \subset V^{*p\otimes}$ acts on $V^{p\otimes}$ and thus, in particular, it acts on every tensor $v_1 \wedge \cdots \wedge v_p \in \Lambda^p(V) \subset V^{p\otimes}$. Let us focus on that action:

$$\begin{aligned} \langle v_1 \wedge \cdots \wedge v_p, u^{*1} \wedge \cdots \wedge u^{*p} \rangle &= \langle \mathcal{A}(v_1 \otimes \cdots \otimes v_p), \mathcal{A}(u^{*1} \otimes \cdots \otimes u^{*p}) \rangle \\ &= \sum_{\tau \in \mathscr{P}_p} \frac{\epsilon_\tau}{p!} \langle \mathcal{A}(v_1 \otimes \cdots \otimes v_p), u^{*\tau^{-1}(1)} \otimes \cdots \otimes u^{*\tau^{-1}(p)} \rangle \end{aligned}$$

$$= \sum_{\tau \in \mathscr{P}_p} \frac{(\epsilon_\tau)^2}{p!} \langle \mathcal{A}(v_{\tau^{-1}(1)} \otimes \cdots \otimes v_{\tau^{-1}(p)}), u^{*\tau^{-1}(1)} \otimes \cdots \otimes u^{*\tau^{-1}(p)} \rangle$$

$$= \frac{p!}{p!} \langle \mathcal{A}(v_1 \otimes \cdots \otimes v_p), u^{*1} \otimes \cdots \otimes u^{*p} \rangle ,$$

since, as one can prove by direct inspection,

$$\langle \mathcal{A}(v_{\tau^{-1}(1)} \otimes \cdots \otimes v_{\tau^{-1}(p)}), u^{*\tau^{-1}(1)} \otimes \cdots \otimes u^{*\tau^{-1}(p)} \rangle$$
$$= \langle \mathcal{A}(v_1 \otimes \cdots \otimes v_p), u^{*1} \otimes \cdots \otimes u^{*p} \rangle .$$

To conclude, we have found that

$$\langle v_1 \wedge \cdots \wedge v_p, u^{*1} \wedge \cdots \wedge u^{*p} \rangle = \frac{1}{p!} \sum_{\sigma \in \mathscr{P}_p} \epsilon_\sigma \langle v_{\sigma^{-1}(1)}, u^{*1} \rangle \cdots \langle v_{\sigma^{-1}(p)}, u^{*p} \rangle .$$

Since summing over $\sigma \in \mathscr{P}_p$ is completely equivalent to summing over $\sigma^{-1} \in \mathscr{P}_p$ and the right-hand side is nothing but

$$(1/p!) \det \Big([\langle v_i, u^{*j} \rangle]_{i,j=1,\ldots,p} \Big) ,$$

we have proved the following formula:

$$\langle v_1 \wedge \cdots \wedge v_p, u^{*1} \wedge \cdots \wedge u^{*p} \rangle = \frac{1}{p!} \det \Big([\langle v_i, u^{*j} \rangle]_{i,j=1,\ldots,p} \Big) . \tag{4.9}$$

We also recall that, if $V_0, \cdots , V_k$ are vector spaces over the field $\mathbb{K}$, $\bigoplus_{p=0}^{k} V_p$ denotes the **external direct sum**, i.e. the vector space defined on $V_0 \times \cdots \times V_k$ with composition rule

$$\alpha(w_0, \ldots , w_k) + \beta(u_0, \ldots , u_k) := (\alpha w_0 + \beta u_0, \ldots , \alpha w_k + \beta u_k)$$

for all $\alpha, \beta \in \mathbb{K}$ and $(w_0, \ldots , w_k), (u_0, \ldots , u_k) \in V_0 \times \cdots \times V_k$.

Definition 4.33 (Grassmann Algebra) Let V be a linear space over the field $\mathbb{K}$ with finite dimension n and let $\wedge$ be the exterior product defined above. The **Grassmann algebra** on V (also called the **exterior algebra** on V) is the pair $(\Lambda(V), \wedge)$ where the former is the external direct sum

$$\Lambda(V) = \bigoplus_{p=0}^{n} \Lambda^p(V) ,$$

and the product $\wedge$ denotes the obvious bi-linear extension of the standard exterior product to the above direct sum. ■

Let us finally focus on the space of forms $\Lambda^p(V^*)$, where, as usual, V has finite dimension $n \geq p$. We show that it is naturally isomorphic to $(\Lambda^p(V))^*$. This is stated in the following proposition.

Proposition 4.34 *If V is a linear space over the field $\mathbb{K}$ and has finite dimension n, and if $0 \leq p \leq n$ is an integer, then there is a natural isomorphism F_p between $\Lambda^p(V^*)$ and $(\Lambda^p(V))^*$ obtained by restricting any functional in $\Lambda^p(V^*)$ to the space $\Lambda^p(V)$. In other words,*

$$F_p : \Lambda^p(V^*) \ni f \mapsto f\restriction_{\Lambda^p(V)} \in (\Lambda^p(V))^* .$$

Therefore $\Lambda(V^*)$ is naturally isomorphic to $(\Lambda(V))^*$.

Proof By construction the map F_p is linear. Let us prove that it is injective and surjective. Concerning injectivity, it is sufficient to show that $f\restriction_{\Lambda^p(V)}= 0$ entails $f = 0$. Let us prove it. Fix a basis in V and generate the associated canonical bases in every relevant tensor space. Therefore, using the corollary of Theorem 4.30,

$$f = c_{i_1\ldots i_p} e^{*i_1} \wedge \cdots \wedge e^{*i_p} ,$$

where the sum is extended over the set $1 \leq i_1 < \cdots < i_p \leq n$. If $f\restriction_{\Lambda^p(V)}= 0$, one has in particular:

$$f(e_{j_1} \wedge \cdots \wedge e_{j_p}) = 0 ,$$

so that

$$c_{i_1\ldots i_p} \langle e_{j_1} \wedge \cdots \wedge e_{j_p}, e^{*i_1} \wedge \cdots \wedge e^{*i_p} \rangle = 0 .$$

Remembering that $1 \leq i_1 < \cdots < i_p \leq n$ and $1 \leq j_1 < \cdots < j_p \leq n$, and using the definition of $\wedge$, one concludes that

$$\langle e_{j_1} \wedge \cdots \wedge e_{j_p}, e^{*i_1} \wedge \cdots \wedge e^{*i_p} \rangle = 0$$

unless $i_1 = j_1, i_2 = j_2, \ldots, i_p = j_p$. In this case formula (4.9) produces

$$\langle e_{i_1} \wedge \cdots \wedge e_{i_p}, e^{*i_1} \wedge \cdots \wedge e^{*i_p} \rangle = (p!)^{-1} \det I = 1/p! .$$

We conclude that

$$\langle e_{j_1} \wedge \cdots \wedge e_{j_p}, e^{*i_1} \wedge \cdots \wedge e^{*i_p} \rangle = \frac{1}{p!} \delta^{i_1}_{j_1} \cdots \delta^{i_p}_{j_p} , \tag{4.10}$$

and thus

$$c_{j_1 \ldots j_p} = 0 .$$

Since $j_1, \ldots, j_p$ are arbitrary, this implies that

$$f = c_{i_1 \ldots i_p} e^{*i_1} \wedge \cdots \wedge e^{*i_p} = 0 .$$

Let us pass to the proof of surjectivity. Actually surjectivity follows from injectivity since $\Lambda^p(V^*)$ and $(\Lambda^p(V))^*$ have the same dimension, however we also give a direct proof. By direct inspection one verifies straightforwardly that, if $h \in (\Lambda^p(V))^*$, the functional of $\Lambda^p(V^*)$ (where, as usual, the sum is extended over the set $1 \leq j_1 < \cdots < j_p \leq n$)

$$f_h := p!\, h(e_{j_1} \wedge \cdots \wedge e_{j_p}) e^{*j_1} \wedge \cdots \wedge e^{*j_p}$$

satisfies

$$f_h \restriction_{\Lambda^p(V)} = h .$$

The natural isomorphism from $\Lambda(V^*)$ to $(\Lambda(V))^*$ is defined by

$$F : \Lambda(V^*) \ni (u_0^*, u_1^*, \ldots, u_n^*) \mapsto \big(F_0(u_0^*), F_1(u_1^*), \ldots, F_n(u_n^*)\big) \in (\Lambda(V))^* .$$

This concludes the proof. □

4.4.2 *Interior Product*

Consider the exterior algebras $\Lambda(V)$ and $\Lambda(V^*)$ with $dim V = n$ and field $\mathbb{K}$, and the pairing $\langle \cdot | \cdot \rangle : \Lambda(V) \times \Lambda(V^*) \to \mathbb{K}$, defined by

$$\langle w | z^* \rangle := \sum_{p=0}^{n} p! \langle w_p, z^{*p} \rangle , \quad \text{for all } w = (w_0, \ldots, w_n) \in \Lambda(V)$$
$$\text{and } z^* = (z^{*0}, \ldots, z^{*n}) \in \Lambda(V^*). \tag{4.11}$$

By direct inspection one proves that the pairing $\langle \cdot | \cdot \rangle$ is bilinear. Moreover one has the following proposition.

Proposition 4.35 *Consider the exterior algebras $\Lambda(V)$ and $\Lambda(V^*)$ with $dim V = n$ and field $\mathbb{K}$, and the pairing $\langle \cdot | \cdot \rangle : \Lambda(V) \times \Lambda(V^*) \to \mathbb{K}$ defined in (4.11).*

If $T : \Lambda(V) \to \Lambda(V)$ is a linear operator, there is a unique linear operator $T^* : \Lambda(V^*) \to \Lambda(V^*)$, called the ***dual operator*** *of T with respect to $\langle\cdot|\cdot\rangle$, such that*

$$\langle Tw|z^*\rangle = \langle w|T^*z^*\rangle\,, \quad \text{for all } w \in \Lambda(V) \text{ and } z^* \in \Lambda(V^*)\,.$$

Proof Fix a basis $\{e_i\}_{i=1,\ldots,n}$ in V and consider its dual one $\{e^{*i}\}_{i=1,\ldots,n}$ in V^*. These bases induce bases $\{E_{p,i_p}\}_{i_p\in I_p}$ in each space $\Lambda^p(V)$ and, taking advantage of Proposition 4.34, induce associated dual bases $\{E_p^{*i_p}\}_{i_p\in I_p}$ in $\Lambda^p(V^*)$. Each element E_{p,i_p} has the form $e_{j_1} \wedge \cdots \wedge e_{j_p}$ whereas each element $E_p^{*i_p}$ has the form $e^{*j_1} \wedge \cdots \wedge e^{*j_p}$. By (4.10),

$$\langle E_{p,i_p}|E_q^{*j_q}\rangle = \delta_{pq}\delta_{i_p}^{j_q}\,.$$

Obviously $\{E_{p,i_p}\}_{i_p\in I_p,\ p=0,\ldots,n}$ is a basis of $\Lambda(V)$ and $\{E_p^{*i_p}\}_{i_p\in I_p,\ p=0,\ldots,n}$ is a basis of $\Lambda(V^*)$. Consider the operator $T^* : \Lambda(V^*) \to \Lambda(V^*)$ defined by

$$T^*z^* := \sum_{p=0}^{n}\sum_{i_p\in I_p} \langle TE_{p,i_p}|z^*\rangle E_p^{*i_p}\,.$$

With the given definitions one finds by direct inspection that

$$\langle Tw|z^*\rangle = \langle w|T^*z^*\rangle\,.$$

Now suppose that there is another operator T'^* satisfying the requirement above. As a consequence, one finds

$$\langle E_{p,i_p}|(T^* - T'^*)z^*\rangle = 0$$

for all p and i_p. Since

$$(T^* - T'^*)z^* = \sum_{p=0}^{n}\sum_{i_p\in I_p} c_{p,i_p}E_p^{*i_p},$$

where

$$c_{p,i_p} = \langle E_{p,i_p}|(T^* - T'^*)z^*\rangle,$$

we conclude that every c_{p,i_p} vanishes and thus $(T^*-T'^*)z^* = 0$ for all $z^* \in \Lambda(V^*)$. In other words $T^* = T'^*$. □

In view of the proved proposition we can give the following definition when, for $u \in \Lambda^p(V)$,

$$u\wedge : \Lambda(V) \to \Lambda(V)$$

is defined by the linear extension to the whole $\Lambda(V)$ of the analogous operator $u\wedge : \Lambda^q(V) \to \Lambda^{p+q}(V)$ and the spaces $\Lambda^r(V)$ with $r > n$ are identified with $\{0\}$, 0 being the zero vector of $\bigoplus_{p=0}^n \Lambda^p(V)$.

Definition 4.36 (Interior Product) Consider the exterior algebras $\Lambda(V)$ and $\Lambda(V^*)$ with $dimV = n$ and field $\mathbb{K}$. If $u \in \Lambda^p(V)$,

$$u\lrcorner : \Lambda(V^*) \to \Lambda(V^*)$$

denotes the adjoint $(u\wedge)^*$ of the operator $u\wedge : \Lambda(V) \to \Lambda(V)$ with respect to $\langle\cdot|\cdot\rangle$. If $z^* \in \Lambda(V^*)$, $u\lrcorner\, z^*$ is called the **interior product** of u and z^*. ■

The most important properties of the interior product are encompassed by the following theorem.

Theorem 4.37 *Consider the exterior algebras $\Lambda(V)$ and $\Lambda(V^*)$ with $dimV = n$ and field $\mathbb{K}$. The interior product enjoys the following properties.*

(a) *If $w \in \Lambda^p(V)$ and $z^* \in \Lambda^q(V^*)$, then $w\lrcorner\, z^* \in \Lambda^{q-p}(V^*)$ (so that $w\lrcorner\, z^* = 0$ if $p > q$).*

(b) *If $w_1 \in \Lambda^p(V)$, $w_2 \in \Lambda^q(V)$ and $z^* \in \Lambda^r(V^*)$, then*

$$(w_1 \wedge w_2)\lrcorner\, z^* = w_2\lrcorner\, (w_1\lrcorner\, z^*)\,.$$

(c) *If $x \in V$ and $y^{*1}, \ldots, y^{*p} \in V^*$, then*

$$x\lrcorner\, (y^{*1}\wedge\cdots\wedge y^{*p}) = \sum_{k=1}^{p}(-1)^{k+1}\langle x, y^{*k}\rangle y^{*1}\wedge\ldots\wedge y^{*k-1}\wedge y^{*k+1}\wedge\ldots\wedge y^{*p}.$$

(d) *If $x \in V$, $x\lrcorner$ is an **anti-derivation** on $\Lambda(V^*)$ with respect to $\wedge$, i.e.*

$$x\lrcorner\, (z^{*1}\wedge z^{*2}) = (x\lrcorner\, z^{*1})\wedge z^{*2} + (-1)^p z^{*1}\wedge (x\lrcorner\, z^{*2}),$$

$$\textit{for } z^{*1} \in \Lambda^p(V^*) \textit{ and } z^{*2} \in \Lambda^q(V^*)\,.$$

Proof Statement **(a)** is valid by definition. Statement **(b)** can be proved as follows:

$$\langle y|(w_1 \wedge w_2)\lrcorner\, z^*\rangle = \langle w_1 \wedge w_2 \wedge y|z^*\rangle = \langle w_2 \wedge y|w_1\lrcorner\, z^*\rangle = \langle y|w_2\lrcorner\, (w_1\lrcorner\, z^*)\rangle\,.$$

The arbitrariness of y implies the thesis.
Let us pass to prove **(c)**:

$$\langle x_2\wedge\cdots\wedge x_p|x_1\lrcorner\, (y^{*1}\wedge\cdots\wedge y^{*p})\rangle = \langle x_1\wedge\cdots\wedge x_p|(y^{*1}\wedge\cdots\wedge y^{*p})\rangle = \det[\langle x_i, y^{*j}\rangle],$$

where we have used (4.9). But

$$\det[\langle x_i, y^{*j}\rangle] = \sum_{k=1}^{p}(-1)^{k+1}\langle x_1, y^{*k}\rangle \det[\langle x_r, y^{*s}\rangle]\,, \quad \text{for } r \neq 1 \text{ and } s \neq k\,.$$

The right-hand side can be rewritten as

$$\sum_{k=1}^{p}(-1)^{k+1}\langle x_1, y^{*k}\rangle\langle x_2 \wedge \cdots \wedge x_p | y^{*1} \wedge \ldots \wedge y^{*k-1} \wedge y^{*k+1} \wedge \ldots \wedge y^{*p}\rangle\,.$$

The arbitrariness of $x_2 \wedge \cdots \wedge x_p$ implies the validity of the thesis with $x = x_1$. The proof of **(d)** is a straightforward consequence of **(c)**. □

Remarks 4.38

(1) The use of linearity and of **(b)** and **(c)** of Theorem 4.37 allows one to compute the action of $x \lrcorner$ explicitly. For instance,

$$\begin{aligned}
&(u^{ij} e_i \wedge e_j) \lrcorner \left(\omega_{pqr} e^{*p} \wedge e^{*q} \wedge e^{*r}\right) \\
&\quad = u^{ij}\omega_{pqr} e_i \lrcorner \left(\delta^p_j e^{*q} \wedge e^{*r} - \delta^q_j e^{*p} \wedge e^{*r} + \delta^r_j e^{*p} \wedge e^{*q}\right)
\end{aligned}$$

$$\begin{aligned}
&= u^{ij}\left(\omega_{ijr} e^{*r} - \omega_{iqr} e^{*q} - \omega_{jir} e^{*r} + \omega_{pij} e^{*p} + \omega_{jqi} e^{*q} - \omega_{pji} e^{*p}\right) \\
&= 6u^{ij}\omega_{ijr} e^{*r}\,.
\end{aligned}$$

(2) With the alternative definition of $\wedge$ (4.7), it is possible to define a corresponding notion of interior product $\tilde{\lrcorner}$ which satisfies all properties of the theorem above with respect to $\tilde{\wedge}$. That is the unique linear extension of

$$x \tilde{\lrcorner} z^* := \frac{1}{p} x \lrcorner z^* \quad \text{for } x \in V \text{ and } z^* \in \Lambda^p(V^*).$$

■

References

1. Moretti, V.: Spectral Theory and Quantum Mechanics: Mathematical Foundations of Quantum Theories, Symmetries and Introduction to the Algebraic Formulation, 2nd edn. Springer, Berlin (2017)
2. Moretti, V.: Fundamental Mathematical Structures of Quantum Theory Spectral Theory: Foundational Issues, Symmetries, Algebraic Formulation, Quantum Measurement Theory, 2nd revised and enlarged edn. Springer, Berlin (2026)

Chapter 5
Scalar Products and Metric Tools

This chapter introduces scalar products and their tensorial consequences. The treatment includes ordinary positive scalar products, Hermitian scalar products, pseudo-scalar products, and semi-scalar products, because all of these structures occur in mathematical physics. In particular, pseudo-scalar products are indispensable in relativistic theories, where the metric is non-degenerate but not positive definite.

The chapter then explains how a non-degenerate scalar product identifies a vector space with its dual and how this identification is expressed by the metric tensor. This leads to the operations of raising and lowering indices, to the notion of signature, to pseudo-orthonormal bases, and to pseudo-orthogonal groups. The final sections discuss orientation, pseudo-tensors, Ricci's pseudo-tensor, and tensor densities, thereby preparing the algebraic tools needed to write invariant volume elements, integration formulae, and relativistic tensor equations.

5.1 Scalar Products

First of all we give the definitions of a *pseudo scalar product* and of *semi scalar products*, which differ from the notion of *scalar product* by the positivity and the non-degeneracy requirement respectively. In fact, a pseudo scalar product is a generalization of the usual definition of scalar product which has many applications in mathematical physics, relativistic theories in particular. Semi scalar products are used in several applications of quantum field theory (for instance in the celebrated *GNS theorem* [1]).

Definition 5.1 (Pseudo Scalar Product) Let V be a vector space over the field either $\mathbb{K} = \mathbb{R}$ or $\mathbb{C}$.

V. Moretti, *Geometric Methods in Mathematical Physics I*, La Matematica per il 3+2 184, https://doi.org/10.1007/978-3-032-31512-0_5

(a) A **pseudo scalar product** is a map $(\,|\,) : V \times V \to \mathbb{K}$ which is:

(i) **bilinear**, i.e., for all $u \in V$ both $(u|\cdot) : v \mapsto (u|v)$ and $(\cdot|u) : v \mapsto (v|u)$ are linear functionals on V,
(ii) **symmetric**, i.e., $(u|v) = (v|u)$ for all $u, v \in V$,
(iii) **non-degenerate**, i.e., $(u|v) = 0$ for all $v \in V$ implies $u = 0$.

(b) If $\mathbb{K} = \mathbb{C}$, a **Hermitian pseudo scalar product** is a map $(\,|\,) : V \times V \to \mathbb{K}$ which is:

(i) **sesquilinear**, i.e., for all $u \in V$, $(u|\cdot)$ and $(\cdot|u)$ are a linear functional and an anti-linear functional on V respectively,
(ii) **Hermitian**, i.e., $(u|v) = \overline{(v|u)}$ for all $u, v \in V$,
(iii) **non-degenerate**.

■

Definition 5.2 (Semi Scalar Product) Let V be a vector space over the field either $\mathbb{K} = \mathbb{R}$ or $\mathbb{C}$.

(a) If $\mathbb{K} = \mathbb{R}$, a **semi scalar product** is a map $(\,|\,) : V \times V \to \mathbb{R}$ which satisfies **(a)(i)**, **(a)(ii)** above and is

(iv) **positive semi-definite**, i.e.,

$$(u|u) \geq 0 \quad \text{for all} \quad u \in V\,.$$

(b) If $\mathbb{K} = \mathbb{C}$, a **Hermitian semi scalar product** is a map $(\,|\,) : V \times V \to \mathbb{K}$ which satisfies **(b)(i)**, **(b)(ii)** above and is

(iv) **positive semi-definite**.

■

Finally we give the definition of scalar product.

Definition 5.3 (Scalar Product) Let V be a vector space over the field $\mathbb{K} = \mathbb{R}$ (resp. $\mathbb{C}$) endowed with a pseudo scalar product (resp. Hermitian pseudo scalar product) $(\,|\,)$. Then $(\,|\,)$ is called a **scalar product** (resp. **Hermitian scalar product**) if $(\,|\,)$ is also a semi scalar product, i.e., if it is **positive semi-definite**. ■

Remarks 5.4

(1) Notice that all given definitions do not require that V is finite dimensional.
(2) If $\mathbb{K} = \mathbb{C}$ and $(\,|\,)$ is not Hermitian, in general any requirement on the positivity of $(u|u)$ does not make sense because $(u|u)$ may *not* be real. If instead Hermiticity holds, we have $(u|u) = \overline{(u|u)}$, which assures that $(u|u) \in \mathbb{R}$ and thus positivity may be investigated.

(3) Actually, a (Hermitian) scalar product is *positive definite*, i.e.

$$(u|u) > 0 \quad \text{if} \quad u \in V \setminus \{0\} ,$$

because of Cauchy-Schwarz' inequality

$$|(u|v)|^2 \leq (u|u)(v|v) ,$$

which we shall prove below for semi scalar products.

(4) A **semi norm** on a vector space V over the field $\mathbb{K} = \mathbb{C}$ or $\mathbb{R}$ is a map

$$|| \ || : V \to \mathbb{R}$$

such that the following properties hold:

(i) **(semi positivity)** $||v|| \geq 0$ for all $v \in V$,
(ii) **(homogeneity)** $||\alpha v|| = |\alpha| \, ||v||$ for all $\alpha \in \mathbb{K}$ and $v \in V$,
(iii) **(triangular inequality)** $||u + v|| \leq ||u|| + ||v||$ for all $u, v \in V$.
A semi norm $|| \ || : V \to \mathbb{R}$ is a **norm** if
(iv) $||v|| = 0$ implies $v = 0$.
Notice that for semi norms it holds $||0|| = 0$ because of **(ii)** above. With the given definitions, one checks that if V with field $\mathbb{R}$ (resp. $\mathbb{C}$) is equipped with a (Hermitian) semi scalar product, then

$$||v|| := \sqrt{(v|v)} \quad \text{for all } v \in V$$

defines a semi norm. Furthermore, if $(\ | \ )$ is a scalar product, then the associated semi norm is a norm.

(5) If a vector space V is equipped with a norm $|| \ ||$, it becomes a *metric space* by defining the *distance*

$$d(u, v) := ||u - v|| \quad \text{for all } u, v \in V.$$

A **Banach space** $(V, || \ ||)$ [2] is a vector space equipped with a norm such that the associated metric space is *complete*, i.e., all Cauchy sequences converge. A **Hilbert space** [2] is a Banach space with norm given by a (Hermitian, if the field is $\mathbb{C}$) scalar product as above. Hilbert spaces are the central mathematical objects used in Quantum Mechanics. ■

Exercises 5.5

Show that if $(\, | \,)$ is a (Hermitian) semi scalar product on V with field $\mathbb{R}$ (resp. $\mathbb{C}$), then the map on V

$$v \mapsto ||v|| := \sqrt{(v|v)}$$

satisfies $||u + v|| \leq ||u|| + ||v||$ as a consequence of Cauchy-Schwarz' inequality

$$|(u|v)|^2 \leq (u|u)(v|v) ,$$

which holds for every (Hermitian) semi scalar product.
(*Hint*. Compute $||u + v||^2 = (u + v|u + v)$ using bilinearity or sesquilinearity of $(\,|\,)$, then use Cauchy-Schwarz' inequality.)

Theorem 5.6 (Cauchy-Schwarz' Inequality) *Let V be a vector space with field $\mathbb{R}$ (resp. $\mathbb{C}$) equipped with a (Hermitian) semi scalar product $(\,|\,)$. Then, for all $u, v \in V$, Cauchy-Schwarz' inequality holds:*

$$|(u|v)|^2 \leq (u|u)(v|v) .$$

Proof Consider the complex case with a Hermitian semi scalar product. Take $u, v \in V$. For all $z \in \mathbb{C}$ it must hold $(zu + v|zu + v) \geq 0$ by definition of Hermitian semi scalar product. Using sesquilinearity and Hermiticity:

$$0 \leq \overline{z}z(u|u) + (v|v) + \overline{z}(u|v) + z(v|u) = |z|^2(u|u) + (v|v) + \overline{z}(u|v) + z\overline{(u|v)} ,$$

which can be rewritten as

$$|z|^2(u|u) + (v|v) + 2\Re\{\overline{z}(u|v)\} \geq 0 . \tag{5.1}$$

Then we pass to the polar representation of z, namely

$$z = re^{i\alpha}$$

with $r \in \mathbb{R}$ and $\alpha \in \mathbb{R}$ arbitrarily and independently fixed. Decompose also $(u|v)$ as

$$(u|v) = |(u|v)|e^{i\arg(u|v)} .$$

Inserting above we get

$$F(r, \alpha) := r^2(u|u) + 2r|(u|v)|\Re\left[e^{i(\arg(u|v)-\alpha)}\right] + (v|v) \geq 0 ,$$

for all $r \in \mathbb{R}$ when $\alpha \in \mathbb{R}$ is fixed arbitrarily. Since the right-hand side above is a second-order polynomial in r, the inequality implies that, for all $\alpha \in \mathbb{R}$,

$$\left\{2|(u|v)|\Re\left[e^{i(\arg(u|v)-\alpha)}\right]\right\}^2 - 4(v|v)(u|u) \leq 0 ,$$

which is equivalent to

$$|(u|v)|^2\cos^2(\arg(u|v) - \alpha) - (u|u)(v|v) \leq 0 ,$$

for all $\alpha \in \mathbb{R}$. Choosing $\alpha = \arg(u|v)$, we get Cauchy-Schwarz' inequality:

$$|(u|v)|^2 \leq (u|u)(v|v) .$$

The real case can be treated similarly, replacing $z \in \mathbb{C}$ with $x \in \mathbb{R}$. By hypothesis it must hold $(xu + v|xu + v) \geq 0$, which implies the analogue of inequality (5.1),

$$x^2(u|u) + 2x(u|v) + (v|v) \geq 0 .$$

That inequality must be valid for every $x \in \mathbb{R}$. Since $2x(u|v)$ may have arbitrary sign and value if x is chosen opportunely, if $(u|u) = 0$ then the inequality above can hold for every x only if $(u|v) = 0$. In that case $(u|v)^2 \leq (u|u)(v|v)$ is trivially true. If $(u|u) \neq 0$, and thus $(u|u) > 0$, the inequality

$$x^2(u|u) + 2x(u|v) + (v|v) \geq 0$$

is valid for every $x \in \mathbb{R}$ if and only if

$$[2(u|v)]^2 - 4(u|u)(v|v) \leq 0 ,$$

that is,

$$(u|v)^2 \leq (u|u)(v|v) .$$

□

Corollary *A bilinear symmetric (resp. sesquilinear Hermitian) map*

$$(\,|\,) : V \times V \to \mathbb{R} \quad (\textit{resp. } \mathbb{C})$$

is a scalar product (resp. Hermitian scalar product) if and only if it is ***positive definite****, that is* $(u|u) > 0$ *for all* $u \in V \setminus \{\mathbf{0}\}$.

Proof Assume that $(\,|\,)$ is a (Hermitian) scalar product. Hence $(u|u) \geq 0$ by definition and $(\,|\,)$ is non-degenerate. Moreover it holds

$$|(u|v)|^2 \leq (u|u)(v|v) .$$

As a consequence, if $(u|u) = 0$ then $(u|v) = 0$ for all $v \in V$, and thus $u = 0$ because $(\,|\,)$ is non-degenerate. We have proved that $(u|u) > 0$ if $u \neq 0$. That is, a scalar product is a positive-definite bilinear symmetric (resp. sesquilinear Hermitian) map

$$(\,|\,) : V \times V \to \mathbb{R} \quad (\text{resp. } \mathbb{C}) .$$

Now assume that $(\,|\,)$ is positive definite and bilinear symmetric (resp. sesquilinear Hermitian). By definition it is a semi scalar product since positive definiteness implies positive semi-definiteness. Let us prove that $(\cdot|\cdot)$ is non-degenerate, and this concludes the proof. If $(u|v) = 0$ for all $v \in V$, then choosing $v = u$, positive definiteness implies $u = 0$. □

5.2 Natural Isomorphism Between V and V^* and Metric Tensor

Let us show that if V is a finite-dimensional vector space endowed with a pseudo scalar product, then V is isomorphic to V^*. That isomorphism is *natural* because it is built up using only the structure of vector space with scalar product, specifying nothing further.

Theorem 5.7 (Natural (Anti)isomorphism Between V **and** V^***)** *Let V be a finite-dimensional vector space over the field $\mathbb{K} = \mathbb{R}$ or $\mathbb{C}$.*

(a) *If $\mathbb{K} = \mathbb{R}$ and V is endowed with a pseudo scalar product $(\,|\,)$, then*

(i) *the map defined on V,*

$$h : u \mapsto (u|\cdot) \in V^*$$

(where $(u|\cdot)$ is the linear functional $(u|\cdot) : v \mapsto (u|v)$) is an isomorphism,

(ii) *the formula*

$$(u|v)^* := (h^{-1}(u)|h^{-1}(v)) \quad \textit{for } u, v \in V^*$$

defines a pseudo scalar product on V^*.

(b) *If $\mathbb{K} = \mathbb{C}$ and V is endowed with a Hermitian pseudo scalar product $(|)$, then*

(i) *the map defined on V,*

$$h : u \mapsto (u|\cdot) \in V^*$$

(where $(u|\cdot)$ is the linear functional $(u|\cdot) : v \mapsto (u|v)$) is an anti-isomorphism,

(ii) *the formula*

$$(u|v)^* := \overline{(h^{-1}(u)|h^{-1}(v))} = (h^{-1}(v)|h^{-1}(u)) \quad \textit{for } u, v \in V^*$$

defines a Hermitian pseudo scalar product on V^*.

Proof First consider (i) in cases (a) and (b). It follows immediately that $(u|\cdot) \in V^*$ in both cases. Moreover the linearity or anti-linearity of the map $u \mapsto (u|\cdot)$ is a trivial consequence of the definition of pseudo scalar product and Hermitian pseudo scalar product respectively.

Then recall the well-known theorem

$$dim(Kerf) + dimf(V) = dimV ,$$

which holds for linear and anti-linear maps from some finite-dimensional vector space V to some vector space V'. Since $dimV = dimV^*$, it is sufficient to show that $h : V \to V^*$ defined by $u \mapsto (u|\cdot)$ has trivial kernel, i.e., is injective: this also assures surjectivity. Therefore, we have to show that

$$(u|\cdot) = (u'|\cdot)$$

implies $u = u'$. This is equivalent to showing that

$$(u - u'|v) = 0 \quad \text{for all } v \in V$$

implies $u - u' = 0$. This is nothing but the non-degeneracy property, which holds by definition of (Hermitian) pseudo scalar product.

Statements (ii) in the two cases are immediate from the definition of (Hermitian) pseudo scalar products and from the fact that h is a (anti-)isomorphism. □

Remarks 5.8

(1) Notice that $(u|v)^* = (h^{-1}(u)|h^{-1}(v))$ and, in the Hermitian case,

$$(u|v)^* = \overline{(h^{-1}(u)|h^{-1}(v))}$$

mean that h and h^{-1} (anti)preserve the scalar products.

(2) The theorem above also holds when considering a Hilbert space and its topological dual space (i.e., the subspace of the algebraic dual space consisting of continuous linear functionals on the Hilbert space). That is the mathematical content of the celebrated *Riesz representation theorem*.

(3) From Theorem 5.7 it follows immediately that, if $u, v \in V$, then in the real case

$$(u|v) = \langle u, h(v)\rangle = (h(u)|h(v))^* ,$$

whereas in the Hermitian case

$$(u|v) = \langle v, h(u)\rangle = (h(v)|h(u))^* .$$

■

From now on we specialize to the pseudo-scalar-product case, dropping the Hermitian one. Suppose $(\ |\ )$ is a pseudo scalar product on a finite-dimensional vector space V over the field $\mathbb{K}$ either $\mathbb{R}$ or $\mathbb{C}$. The map $(u, v) \mapsto (u|v)$ belongs to $\mathscr{L}(V, V) = V^* \otimes V^*$ and thus it is a tensor $\mathbf{g} \in V^* \otimes V^*$. Fixing a canonical basis in $V^* \otimes V^*$ induced by a basis $\{e_i\}_{i\in I} \subset V$, we can write

$$\mathbf{g} = g_{ij} e^{*i} \otimes e^{*j} ,$$

where, by Theorem 2.22,

$$g_{ij} = (e_i|e_j) .$$

Definition 5.9 ((Pseudo) Metric Tensor) A pseudo scalar product $(\,|\,) = \mathbf{g} \in V^* \otimes V^*$ on a finite-dimensional vector space V over the field $\mathbb{K}$ either $\mathbb{R}$ or $\mathbb{C}$ is called a **pseudo-metric tensor**. If $\mathbb{K} = \mathbb{R}$, a pseudo-metric tensor is called a **metric tensor** if it defines a scalar product. ■

Remarks 5.10

(1) By Theorem 3.9, the isomorphism $h : V \to V^*$ is represented by a tensor of $V^* \otimes V^*$ which acts on elements of V by means of a product of tensors and a contraction. The introduction of the pseudo-metric tensor allows us to represent the isomorphism $h : V \to V^*$ by means of the abstract index notation, determining the tensor representing h as we now illustrate. Since $h : u \mapsto (u|\cdot) = \mathbf{g}(u, \cdot)$ and

$$(u|v) = (e_i|e_j)u^i v^j = g_{ij} u^i v^j ,$$

we trivially have

$$(hu)_j = g_{ij} u^i .$$

$h(u)$ is obtained by the product $\mathbf{g} \otimes u$ followed by a contraction. Hence, in the sense of Theorem 3.9, the linear map $h : V \to V^*$ is represented by the tensor $\mathbf{g}$ itself.

(2) Pseudo-metric tensors are *symmetric* because of the symmetry of pseudo scalar products:

$$\mathbf{g}(u, v) = (u|v) = (v|u) = \mathbf{g}(v, u) .$$

(3) The symmetry requirement on the metric tensor is not necessary to define an isomorphism between V and V^*. In Weyl spinor theory [3, 4], the space V is a two-dimensional complex vector space whose elements are called Weyl *spinors*. V is equipped with a fixed *antisymmetric* tensor $\epsilon \in V^* \otimes V^*$ (the so-called *metric spinor*) defining a non-degenerate linear map from V to V^* by contraction. Using the abstract index notation

$$V \ni \xi^A \mapsto \epsilon_{AB}\xi^B \in V^* .$$

Notice that, differently from the metric tensor case, now $\epsilon_{AB}\xi^B = -\epsilon_{BA}\xi^B$. Non-degeneracy of ϵ entails that the map $\xi^A \mapsto \epsilon_{AB}\xi^B$ is an isomorphism from V to V^*.

■

Components of pseudo-metric tensors with respect to canonical bases enjoy some simple but important properties which are listed below.

Theorem 5.11 (Properties of the Metric Tensor) *Referring to Definition 5.9, the components of any pseudo-metric tensor* **g**,

$$g_{ij} := \mathbf{g}(e_i, e_j)$$

with respect to the canonical basis induced in $V^* \otimes V^*$ by any basis $\{e_i\}_{i=1,\dots,n} \subset V$, enjoy the following properties:

(1) *they define a symmetric matrix $[g_{ij}]$, i.e.,*

$$g_{ij} = g_{ji} \,,$$

(2) *$[g_{ij}]$ is non-singular, i.e., it satisfies*

$$\det[g_{ij}] \neq 0 \,,$$

(3) *if $\mathbb{K} = \mathbb{R}$ and* **g** *is a scalar product, then the matrix $[g_{ij}]$ is positive definite.*

Proof

(1) This follows from the symmetry of the scalar product:

$$g_{ij} = (e_i|e_j) = (e_j|e_i) = g_{ji} \,.$$

(2) Suppose $\det[g_{ij}] = 0$ and define $n = dim V$. The linear map $\mathbb{K}^n \to \mathbb{K}^n$ determined by the matrix $g := [g_{ij}]$ has a non-trivial kernel. In other words, there are n scalars u^j, $j = 1, \dots, n$, defining a $\mathbb{K}^n$ vector $[u] := (u^1, \dots, u^n)^t$ with $g[u] = 0$ and $[u] \neq 0$. In particular,

$$[v]^t g[u] = 0$$

for every choice of $[v] \in \mathbb{K}^n$. Defining $u := u^j e_j$, the obtained result implies that there is $u \in V \setminus \{0\}$ with

$$(u|v) = (v|u) = 0 \quad \text{for all } v \in V.$$

This is impossible because $(\,|\,)$ is non-degenerate by hypothesis.

(3) The statement $(u|u) > 0$ if $u \in V \setminus \{\mathbf{0}\}$ reads, in the considered canonical basis,

$$[u]^t g[u] > 0 \quad \text{for } [u] \in \mathbb{R}^n \setminus \{\mathbf{0}\}.$$

That is one of the equivalent definitions of a positive definite matrix g.

□

The following theorem shows that a (pseudo) scalar product can be assigned by means of a suitable tensor which satisfies some properties when represented in some canonical basis. The important point is that one does not need to check these properties for *all* canonical bases: verification for a single canonical basis is sufficient.

Theorem 5.12 (Assignment of a (Pseudo) Scalar Product) *Let V be a finite-dimensional vector space over the field $\mathbb{K}$ either $\mathbb{R}$ or $\mathbb{C}$. Suppose $\mathbf{g} \in V^* \otimes V^*$ is a tensor such that there is a canonical basis of $V^* \otimes V^*$ where the components g_{ij} of $\mathbf{g}$ define a symmetric matrix $g := [g_{ij}]$ with non-vanishing determinant. Then $\mathbf{g}$ is a pseudo-metric tensor, i.e. a pseudo scalar product.*

Furthermore, if $\mathbb{K} = \mathbb{R}$ and $[g_{ij}]$ is positive definite, the pseudo scalar product is a scalar product.

Proof If $\mathbf{g}$ is represented by a symmetric matrix of components in one canonical basis, then the same holds in all remaining bases and the tensor is symmetric (see Exercise 4.221). This implies that

$$(u|v) := \mathbf{g}(u, v)$$

is a bilinear symmetric functional. Suppose $(\,|\,)$ is degenerate. Then there is $u \in V$ such that $u \neq \mathbf{0}$ and

$$(u|v) = 0 \quad \text{for all } v \in V.$$

Using the notation of the proof of item **(2)** of Theorem 5.11, this reads in components

$$[u]^t g[v] = 0 \quad \text{for all } [v] = (v^1, \ldots, v^n)^t \in \mathbb{K}^n,$$

where $n = dim V$. Hence

$$[u]^t g = 0\,,$$

and therefore, since $g = g^t$,

$$g[u] = g^t[u] = 0\,.$$

Since $[u] \neq \mathbf{0}$, g cannot be injective and thus $\det g = 0$, contrary to the hypothesis. Therefore $(\,|\,)$ is non-degenerate. We conclude that

$$(u|v) := \mathbf{g}(u, v)$$

defines a pseudo scalar product.

Finally, if $\mathbb{K} = \mathbb{R}$ and g is also positive definite, then $(|)$ itself is positive definite, i.e., it is a scalar product, since

$$(u|u) = [u]^t g[u] > 0$$

if $[u] \neq \mathbf{0}$, which is equivalent to $u \neq \mathbf{0}$. □

5.2.1 *Signature of a Pseudo-Metric Tensor, Pseudo-Orthonormal Bases and Pseudo-Orthogonal Groups*

Let us introduce the notion of *signature* of a pseudo-metric tensor in a vector space over the field $\mathbb{R}$ by recalling Sylvester's theorem, whose proof can be found in any linear algebra textbook. The definition is intertwined with those of *pseudo-orthonormal basis* and *pseudo-orthogonal group*.

Theorem 5.13 (Sylvester's Theorem) *Let G be a real symmetric $n \times n$ matrix.*

(a) *There is a non-singular (i.e., with non-vanishing determinant) real $n \times n$ matrix D such that*

$$DGD^t = diag(0, \ldots, 0, -1, \ldots, -1, +1, \ldots, +1)\,,$$

where the reals $0, -1, +1$ appear $v \geq 0$ times, $m \geq 0$ times and $p \geq 0$ times respectively, with $v + m + p = n$.

(b) *The triple (v, m, p) does not depend on D. In other words, if, for some non-singular real $n \times n$ matrix $E \neq D$, EGE^t is diagonal and the diagonal contains only the reals $0, -1, +1$ (in whatever order), then $0, -1, +1$ respectively appear v times, m times and p times.*

If $\mathbf{g} \in V^* \otimes V^*$ is a pseudo-metric tensor on the finite-dimensional vector space V over the field $\mathbb{R}$, the transformation rule of the components of $\mathbf{g}$ with respect to canonical bases (see Theorem 3.6) induced by bases $\{e_i\}_{i\in I}$, $\{e'_j\}_{j\in I}$ of V is

$$g'_{pq} = B_p{}^i B_q{}^j g_{ij}\,.$$

Defining $g' := [g'_{pq}]$, $g := [g_{ij}]$, $B := [B_h{}^k]$, this can be rewritten as

$$g' = BgB^t .$$

Recall from Theorem 3.6 that the non-singular matrices B are defined by

$$B = A^{-1t} ,$$

where $A = [A^i{}_j]$ and

$$e_m = A^l{}_m e'_l .$$

Notice that the specification of B is completely equivalent to the specification of A, because $A = B^{-1t}$.

Hence, since g is real and symmetric by Theorem 5.11, Sylvester's theorem implies that, starting from any basis $\{e_i\}_{i\in I} \subset V$, one can find another basis $\{e'_j\}_{j\in I}$, which induces a canonical basis in $V^* \otimes V^*$ where the pseudo-metric tensor is represented by a diagonal matrix. It is sufficient to pick a transformation matrix B as specified in **(a)** of Theorem 5.13. In particular, one can find B such that each diagonal entry of g' turns out to be either -1 or $+1$ only. The value 0 is not allowed, because it would imply that the matrix has vanishing determinant, and this is impossible by Theorem 5.11. Moreover the pair (m, p), where (m, p) are defined in Theorem 5.13, does not depend on the basis $\{e'_j\}_{j\in I}$. In other words, it is an *intrinsic* property of the pseudo-metric tensor: that is the *signature* of the pseudo-metric tensor.

Definition 5.14 (Pseudo-Orthonormal Bases and Signature) Let $\mathbf{g} \in V^* \otimes V^*$ be a pseudo-metric tensor on the finite-dimensional vector space V over the field $\mathbb{R}$.

(a) A basis $\{e_i\}_{i\in I} \subset V$ is called **pseudo-orthonormal** with respect to $\mathbf{g}$ if the components of $\mathbf{g}$ with respect to the canonical basis induced in $V^* \otimes V^*$ form a diagonal matrix with diagonal entries in $\{-1, +1\}$. In other words, $\{e_i\}_{i\in I}$ is pseudo-orthonormal if

$$(e_i, e_j) = \pm\delta_{ij} .$$

If the pseudo-metric tensor is a metric tensor, the pseudo-orthonormal bases are called **orthonormal bases**.

(b) The pair (m, p), where m is the number of diagonal entries equal to -1 and p is the number of diagonal entries equal to $+1$ of a matrix representing the components of $\mathbf{g}$ in a pseudo-orthonormal basis, is called the **signature** of $\mathbf{g}$.

(c) $\mathbf{g}$ and its signature are said **elliptic** or **Euclidean** or **Riemannian** if $m = 0$, **hyperbolic** if $m > 0$ and $p \neq 0$, and **Lorentzian** or **normally hyperbolic** if $m = 1$ and $p \neq 0$.

(d) If $\mathbf{g}$ is hyperbolic, a pseudo-orthonormal basis $\{e_i\}_{i\in I}$ is said to be **canonical** if the matrix of the components of $\mathbf{g}$ takes the form

$$diag(-1, ..., -1, +1, ..., +1) .$$

■

Exercises 5.15 Show that a *pseudo-metric* tensor $\mathbf{g}$ is a *metric* tensor if and only if its signature is elliptic.

Remarks 5.16 If $\{e_i\}_{i\in I}$ is a pseudo-orthonormal basis with respect to a hyperbolic pseudo-metric tensor $\mathbf{g}$, one can trivially reorder the vectors of the basis obtaining a canonical pseudo-orthonormal basis. ■

Let us consider a pseudo-metric tensor $\mathbf{g}$ in V over the field $\mathbb{R}$. Let (m, p) be the signature of $\mathbf{g}$ and let $\mathcal{N}_{\mathbf{g}}$ be the class of all canonical pseudo-orthonormal bases in V with respect to $\mathbf{g}$. In the following we shall indicate by η the matrix $diag(-1, ..., -1, +1, ..., +1)$ which represents the components of $\mathbf{g}$ with respect to each basis of $\mathcal{N}_{\mathbf{g}}$. If A is a matrix corresponding to a change of basis in $\mathcal{N}_{\mathbf{g}}$, and $B := A^{-1t}$ is the associated matrix describing the induced change of basis in V^*, it has to hold

$$\eta = B\eta B^t .$$

Conversely, each real $n \times n$ matrix B satisfying the identity above determines $A := B^{-1t}$ which represents a change of basis in $\mathcal{N}_{\mathbf{g}}$. That A is well defined because every B satisfying $\eta = B\eta B^t$ is *non-singular* and, more precisely, satisfies $\det B = \pm 1$. Indeed, taking the determinant of both sides in $\eta = B\eta B^t$, using the fact that $\det \eta = (-1)^m$ and $\det B = \det B^t$, we conclude that

$$(\det B)^2 = 1$$

and thus $\det B = \pm 1$. As a consequence, since $\det A = (\det B)^{-1}$, one also has $\det A = \pm 1$.

The identity $\eta = B\eta B^t$ can equivalently be rewritten in terms of the matrix A. Since $A = B^{-1t}$, one has $B^{-1} = A^t$. Thus, applying A^t on the left and A on the right of $\eta = B\eta B^t$, we get the equivalent relation in terms of the matrix A:

$$\eta = A^t \eta A .$$

That equation completely determines the set $O(m, p) \subset GL(n, \mathbb{R})$ $(n = m + p)$ of all real non-singular $n \times n$ matrices which correspond to changes of bases in $\mathcal{N}_{\mathbf{g}}$.

It is simply proved that $O(m, p)$ is a subgroup of $GL(n, \mathbb{R})$. We can state the following definition.

Definition 5.17 If η is the matrix $diag(-1, ..., -1, +1, ..., +1)$ where -1 occurs m times and $+1$ occurs p times, the subgroup of $GL(m+p, \mathbb{R})$

$$O(m, p) := \{A \in GL(m+p, \mathbb{R}) \mid \eta = A^t \eta A\}$$

is called the **pseudo-orthogonal** group of order (m, p). ■

Notice that, if $m = 0$, $O(0, p) = O(n)$ reduces to the usual orthogonal group of order n. $O(1, 3)$ is the celebrated **Lorentz group**, which is the central mathematical object in relativistic theories. We shall come back to those issues in the last two chapters, focusing on the Lorentz group in particular.

Exercises 5.18

1. Show that if $A \in O(m, p)$ then A^{-1} exists and

$$A^{-1} = \eta A^t \eta \, .$$

2. Show that $O(m, p)$ is a group with respect to the usual multiplication of matrices.

Remarks 5.19 This implies that $O(m, p)$ is a subgroup of $GL(n, \mathbb{R})$ with $n = p + m$.

(*Hint*. You have to prove that (1) the identity matrix I belongs to $O(m, p)$, (2) if A and A' belong to $O(m, p)$, then AA' belongs to $O(m, p)$, (3) if A belongs to $O(m, p)$, then A^{-1} exists and belongs to $O(m, p)$.)

3. Show that $SO(m, p) := \{A \in O(m, p) \mid \det A = 1\}$ is non-empty and is a subgroup of $O(m, p)$. $SO(m, p)$ is called the *special pseudo-orthogonal group* of order (m, p).

4. Consider the special Lorentz group $SO(1, 3)$ and show that the set (called the *special orthochronous Lorentz group*)

$$SO(1, 3)\!\uparrow := \{A \in SO(1, 3) \mid A^1{}_1 > 0\}$$

is a non-empty subgroup. ■

5.2.2 *Raising and Lowering Indices of Tensors*

Consider a finite-dimensional vector space V over the field $\mathbb{K} = \mathbb{R}$ or $\mathbb{C}$ endowed with a pseudo-metric tensor $\mathbf{g}$. As we said above, there is a natural isomorphism $h : V \to V^*$ defined by

$$h : u \mapsto (u|\cdot) = \mathbf{g}(u, \cdot) \, .$$

This isomorphism may be extended to the whole tensor algebra $\mathscr{A}_{\mathbb{K}}(V)$ using the universality theorem and Definition 4.11.

Indeed, consider a space $S \in \mathscr{A}_{\mathbb{K}}(V)$ of the form $A \otimes V \otimes B$, where A and B are tensor spaces of the form $U_1 \otimes \ldots \otimes U_k$ and $U_{k+1} \otimes \ldots \otimes U_m$ respectively, with each U_i being either V or V^*. We may define the operators

$$h^{\otimes} := I_1 \otimes \ldots \otimes I_k \otimes h \otimes I_{k+1} \otimes \ldots \otimes I_m : A \otimes V \otimes B \to A \otimes V^* \otimes B\,,$$

and

$$(h^{-1})^{\otimes} := I_1 \otimes \ldots \otimes I_k \otimes h^{-1} \otimes I_{k+1} \otimes \ldots \otimes I_m : A \otimes V^* \otimes B \to A \otimes V \otimes B\,,$$

where $I_j : U_j \to U_j$ is the identity operator. Using the remark after Definition 4.11, one finds

$$(h^{-1})^{\otimes} h^{\otimes} = I_1 \otimes \ldots \otimes I_k \otimes (h^{-1}h) \otimes I_{k+1} \otimes \ldots \otimes I_m = Id_{A \otimes V \otimes B}\,,$$

and

$$h^{\otimes}(h^{-1})^{\otimes} = I_1 \otimes \ldots \otimes I_k \otimes (hh^{-1}) \otimes I_{k+1} \otimes \ldots \otimes I_m = Id_{A \otimes V^* \otimes B}\,.$$

Therefore $h^{\otimes}$ is an isomorphism with inverse $(h^{-1})^{\otimes}$.
The action of $h^{\otimes}$ and $(h^{-1})^{\otimes}$ is that of **lowering** and **raising indices** respectively. In fact, in abstract-index notation one has

$$h^{\otimes} : t^{AiB} \mapsto t^{A}{}_{j}{}^{B} := t^{AiB} g_{ij}\,,$$

and

$$(h^{-1})^{\otimes} : u^{A}{}_{i}{}^{B} \mapsto u^{AjB} := u^{A}{}_{i}{}^{B} \tilde{g}^{ij}\,.$$

Above g_{ij} represents the pseudo-metric tensor as specified in Remark **(1)** after Definition 5.9. What about the tensor $\tilde{g} \in V \otimes V$ representing h^{-1} via Theorem 3.9?

Theorem 5.20 *Let $h : V \to V^*$ be the isomorphism determined by a pseudo scalar product, i.e. by a pseudo-metric tensor $\mathbf{g}$ on the finite-dimensional vector space V over the field $\mathbb{K} = \mathbb{R}$ or $\mathbb{C}$.*

(a) *The inverse map $h^{-1} : V^* \to V$ is represented via Theorem 3.9 by a* ***symmetric*** *tensor $\tilde{\mathbf{g}} \in V \otimes V$ such that, if $\{e_i\}_{i \in I}$ is a basis of V, $\tilde{g}^{rs} := \tilde{\mathbf{g}}(e^{*r}, e^{*s})$ and $g_{ij} := \mathbf{g}(e_i, e_j)$, then the matrix $[\tilde{g}^{ij}]$ is the* inverse matrix *of $[g_{ij}]$.*

(b) *The tensor $\tilde{\mathbf{g}}$ coincides with the pseudo-metric tensor* ***with both indices raised****.*

Proof

(a) By Theorem 3.9, h^{-1} determines a tensor $\tilde{\mathbf{g}} \in V \otimes V$ with

$$h^{-1}(u^*) = \tilde{\mathbf{g}}(u^*, \cdot)\,.$$

In components,

$$(h^{-1}u^*)^i = u^*_k \tilde{g}^{ki} .$$

On the other hand it must be

$$h(h^{-1}u^*) = u^* ,$$

or

$$u^*_k \tilde{g}^{ki} g_{ir} = u^*_r ,$$

for all $u^* \in V^*$. This can be rewritten as

$$[u^*]^t(\tilde{g}g - I) = 0 ,$$

for all $\mathbb{K}^n$ vectors

$$[u^*] = (u^*_1, \ldots, u^*_n) .$$

Then the matrix $(\tilde{g}g - I)^t$ is the null matrix. This implies that

$$\tilde{g}g = I ,$$

which is the thesis. $\tilde{g}$ is symmetric because it is the inverse of a symmetric matrix, and thus also the tensor $\tilde{\mathbf{g}}$ is symmetric.

(b) Let g^{ij} be the pseudo-metric tensor with both indices raised, i.e.

$$g^{ij} := g_{rk}\ \tilde{g}^{kj}\tilde{g}^{ri} .$$

By **(a)**, the right-hand side is equal to

$$\delta^j_r \tilde{g}^{ri} = \tilde{g}^{ji} = \tilde{g}^{ij} .$$

That is the thesis.

□

Remarks 5.21

(1) The isomorphisms $h : V \to V^*$ and $h^{-1} : V^* \to V$ are often respectively denoted by $\flat : V \to V^*$ and $\sharp : V^* \to V$ and are called **musical isomorphisms**. In particular, the following notations are used:

$$\omega^\sharp := \sharp(\omega) \quad (\omega \in V^*) , \qquad u^\flat := \flat(u) \quad (u \in V)$$

respectively corresponding to the process of raising and lowering an index.

(2) Another result which arises from the proof of the second part of the theorem is that

$$g_i^j = \delta_i^j \, .$$

(3) When a vector space is endowed with a pseudo scalar product, tensors can be viewed as abstract objects which may be represented either as covariant or contravariant concrete tensors using the procedure of raising and lowering indices. For instance, a tensor t^{ij} of $V \otimes V$ may be viewed as a covariant tensor when represented in its covariant form

$$t_{pq} := g_{pi} \, g_{qj} \, t^{ij} \, .$$

Also, it can be viewed as a mixed tensor

$$t_p{}^j := g_{pi} \, t^{ij} \quad \text{or} \quad t^i{}_q := g_{qj} \, t^{ij} \, .$$

(4) Now consider a finite-dimensional vector space over $\mathbb{R}$, V, endowed with a metric tensor $\mathbf{g}$, i.e. with *elliptic signature*. In orthonormal bases the contravariant and covariant components numerically coincide because $g_{ij} = \delta_{ij} = g^{ij}$. This is the reason why, using the usual scalar product of vector spaces isomorphic to $\mathbb{R}^n$ and working in orthonormal bases, the difference between covariant and contravariant vectors does not arise.
Conversely, in relativistic theories, where a Lorentzian scalar product is necessary, the difference between covariant and contravariant vectors turns out to be evident also in pseudo-orthonormal bases, since the diagonal matrix $[g_{ij}]$ has an eigenvalue -1.

■

This section is devoted to introducing very important tools in both theoretical/mathematical physics and pure mathematics: pseudo tensors and tensor densities.

5.3 Orientation and Pseudo Tensors

The first example of a "pseudo" object we are going to discuss is an *orientation* of a *real* vector space.

Consider a finite-dimensional vector space V over the field $\mathbb{R}$. In the following, $\mathcal{B}$ indicates the set of all vector bases of V. Consider two bases $\{e_i\}_{i\in I}$ and $\{e'_j\}_{j\in I}$ in $\mathcal{B}$. Concerning the determinant of the transformation matrix $A := [A^r{}_s]$, with

$e_i = A^j{}_i e'_j$, there are only two possibilities: $\det A > 0$ or $\det A < 0$. It follows by direct verification that the relation in $\mathcal{B}$

$$\{e_i\}_{i\in I} \sim \{e'_j\}_{j\in I} \quad \text{iff} \quad \det A > 0$$

where A denotes the transformation matrix as above, is an *equivalence relation*. Since the only two possibilities are those above, the partition of $\mathcal{B}$ induced by $\sim$ is made of two *equivalence classes* $\mathcal{B}_1$ and $\mathcal{B}_2$. Hence, if a basis belongs to $\mathcal{B}_1$ or $\mathcal{B}_2$, any other basis belongs to the same set if and only if the corresponding transformation matrix has positive determinant.

Definition 5.22 (Orientation of a Vector Space) Consider a finite-dimensional vector space V over the field $\mathbb{R}$. An **orientation** of V is a bijective map $\mathcal{O} : \{\mathcal{B}_1, \mathcal{B}_2\} \to \{-1, +1\}$. If V is equipped with an orientation $\mathcal{O}$, it is said to be **oriented**, and a basis $\{e_i\}_{i\in I} \in \mathcal{B}_k$ is said to be **positively oriented** if $\mathcal{O}(\mathcal{B}_k) = +1$, or **negatively oriented** if $\mathcal{O}(\mathcal{B}_k) = -1$. ■

Remarks 5.23 The usual physical vector space can be oriented "by hand" using the natural basis given by our own right hand. When we use the right hand to assign an orientation, we determine $\mathcal{O}^{-1}(+1)$ by exhibiting a basis contained in it. ■

The given definition can be, in some sense, generalized by introducing the notion of a *pseudo tensor*.

Definition 5.24 (Pseudotensors) Let V be a finite-dimensional vector space over the field $\mathbb{R}$. Let S be a tensor space of $\mathscr{A}_{\mathbb{R}}(V)$. A **pseudo tensor of** S is a map $t_s : \{\mathcal{B}_1, \mathcal{B}_2\} \to S$ of the form

$$t_s(\mathcal{B}_1) = s\,, \qquad t_s(\mathcal{B}_2) = -s\,,$$

where $s \in S$. If $s \neq 0$, this is equivalently a bijection from $\{\mathcal{B}_1, \mathcal{B}_2\}$ onto $\{s, -s\}$; if $s = 0$, it is the zero pseudo tensor. Moreover:

(a) the various tensorial properties enjoyed by both s and $-s$ are attributed to t_s. (So, for instance, if s, and thus $-s$, is symmetric, t_s is said to be symmetric);
(b) if $\{e_i\}_{i\in I} \in \mathcal{B}_i$, the **components of** t_s with respect to the canonical bases induced by that basis are the components of $t_s(\mathcal{B}_i)$.

■

Remarks 5.25

(1) The given definition encompasses the definition of a *pseudo scalar*.
(2) It follows immediately that the assignment of a pseudo tensor t_s of, for instance, $V^{n\otimes} \otimes V^{*m\otimes}$, is equivalent to the assignment of components

$$t^{i_1\ldots i_n}{}_{j_1\ldots j_m}$$

for each canonical basis

$$\{e_{i_1} \otimes \ldots \otimes e_{i_n} \otimes e^{*j_1} \otimes \cdots \otimes e^{*j_m}\}_{i_1,\ldots,i_n,j_1,\ldots,j_m \in I}$$

such that the transformation rules, when passing to the basis

$$\{e'_{r_1} \otimes \ldots \otimes e'_{r_n} \otimes e'^{*l_1} \otimes \cdots \otimes e'^{*l_m}\}_{r_1,\ldots,r_n,l_1,\ldots,l_m \in I} ,$$

are given by

$$t'^{k_1\ldots k_n}{}_{h_1\ldots h_m} = \frac{\det A}{|\det A|} A^{k_1}{}_{i_1} \cdots A^{k_n}{}_{i_n} B_{h_1}{}^{j_1} \cdots B_{h_m}{}^{j_m} t^{i_1\ldots i_n}{}_{j_1\ldots j_m} ,$$

where $e_l = A^m{}_l e'_m$ and $B = A^{-1t}$ with $B := [B_k{}^j]$ and $A := [A^p{}_q]$.

In fact, $t^{i_1\ldots i_n}{}_{j_1\ldots j_m} = s^{i_1\ldots i_n}{}_{j_1\ldots j_m}$ if the considered basis is in $t_s^{-1}(+1)$, or $t^{i_1\ldots i_n}{}_{j_1\ldots j_m} = (-s)^{i_1\ldots i_n}{}_{j_1\ldots j_m}$ if the considered basis is in $t_s^{-1}(-1)$.

(3) Evidently the pseudotensors of a fixed type (e.g. (p, q)) form a vector space.

■

Examples 5.26

1. Consider the **magnetic field** $B = B^i e_i$, where e_1, e_2, e_3 is a right-handed orthonormal basis of the space V_3 of the vectors with origin at a point of the physical space E_3. Actually, as every physicist knows, when changing basis the components of B transform as usual only if the new basis is right-handed; otherwise a sign $-$ appears in front of each component. This is a physical requirement due to the Lorentz law. This means that the magnetic field has to be represented in terms of **pseudo vectors**.

5.4 Ricci's Pseudo Tensor

A particular pseudo tensor is Ricci's one, which is very important in physical applications. Its definition requires a preliminary discussion.

Consider an n-dimensional vector space V endowed with a pseudo-metric tensor **g**. We know that, changing basis $\{e_i\}_{i\in I} \to \{e'_j\}_{j\in I}$, the components of the pseudo-metric tensor referred to the corresponding canonical bases transform as

$$g' = BgB^t ,$$

where $g = [g_{ij}]$, $g' = [g'_{pq}]$ and $B = A^{-1t}$, with $A := [A^p{}_q]$, and $e_l = A^m{}_l e'_m$. This implies that

$$\det g' = (\det B)^2 \ \det g , \tag{5.2}$$

which is equivalent to

$$\sqrt{|\det g'|} = |\det A|^{-1} \sqrt{|\det g|}\,. \tag{5.3}$$

Now fix a basis $E := \{e_i\}_{i\in I}$ of V in the set of bases $\mathcal{B}$, and consider the canonical basis induced in $V^{*n\otimes}$, where $n = dimV$,

$$\{e^{*i_1} \otimes \ldots \otimes e^{*i_n}\}_{i_1,\ldots,i_n\in I}\,.$$

Finally define the n-form associated with that basis:

$$\mathcal{E}^{(E)} := n!\sqrt{|\det g|}e^{*1} \wedge \cdots \wedge e^{*n}\,. \tag{5.4}$$

We want to prove that the map

$$\mathcal{B} \ni E \mapsto \mathcal{E}^{(E)} \in \Lambda^n(V^*)$$

defines an antisymmetric pseudotensor of maximal order. In other words, we intend to prove that, if $E, E' \in \mathcal{B}$ with $E = \{e_i\}_{i=1,\ldots,n}$, $E' = \{e'_j\}_{j=1,\ldots,n}$, with $e_i = A^j{}_i e'_j$, then

$$\mathcal{E}^{(E')} = \frac{\det A}{|\det A|}\mathcal{E}^{(E)}\,. \tag{5.5}$$

To prove (5.5), we expand $\mathcal{E}^{(E)}$ in terms of the corresponding objects associated with the basis E':

$$\mathcal{E}^{(E)} = n!\sqrt{|\det g|}e^{*1}\wedge\cdots\wedge e^{*n} = n!\sqrt{|\det g'|}\frac{1}{|\det B|}B_{j_1}{}^1 \cdots B_{j_n}{}^n e'^{*j_1}\wedge\cdots\wedge e'^{*j_n}.$$

Hence

$$\mathcal{E}^{(E)} = n!\frac{\det A}{|\det A|}\frac{\sqrt{|\det g'|}}{\det B}B_{j_1}{}^1 \cdots B_{j_n}{}^n e'^{*j_1} \wedge \cdots \wedge e'^{*j_n}\,. \tag{5.6}$$

Now focus attention on the last term on the right-hand side:

$$B_{j_1}{}^1 \cdots B_{j_n}{}^n e'^{*j_1} \wedge \cdots \wedge e'^{*j_n}\,.$$

In view of the properties of the wedge product, if some indices among $j_1, \ldots, j_n$ are repeated, we have $e'^{*j_1} \wedge \cdots \wedge e'^{*j_n} = 0$. So the only possibility is that $(j_1, \ldots, j_n)$ is a permutation of $(1, 2, \ldots, n)$, that is, $j_k = \sigma(k)$ for some $\sigma \in \mathcal{P}_n$. Since

$$e'^{*\sigma(1)} \wedge \cdots \wedge e'^{*\sigma(n)} = \epsilon_\sigma e'^{*1} \wedge \cdots \wedge e'^{*n}\,,$$

we finally have

$$B_{j_1}{}^1 \cdots B_{j_n}{}^n e'^{*j_1} \wedge \cdots \wedge e'^{*j_n} = \sum_{\sigma \in \mathcal{P}_n} \epsilon_\sigma B_{\sigma(1)}{}^1 \cdots B_{\sigma(n)}{}^n e'^{*1} \wedge \cdots \wedge e'^{*n}$$
$$= (\det B) e'^{*1} \wedge \cdots \wedge e'^{*n} .$$

Inserting this result into (5.6), we conclude that

$$\mathcal{E}^{(E)} = n! \frac{\det A}{|\det A|} \frac{\sqrt{|\det g'|}}{\det B} (\det B) e'^{*1} \wedge \cdots \wedge e'^{*n} = \frac{\det A}{|\det A|} \mathcal{E}^{(E')} ,$$

namely

$$\mathcal{E}^{(E')} = \frac{|\det A|}{\det A} \mathcal{E}^{(E)}$$

so that (5.5) is proved because $\frac{|\det A|}{\det A} = \frac{\det A}{|\det A|}$.

Definition 5.27 (Ricci's Pseudo Tensor) Let V be a vector space over the field $\mathbb{R}$ with dimension $n < +\infty$, endowed with a pseudo-metric tensor $\mathbf{g}$. **Ricci's pseudo tensor** is the antisymmetric $(0, n)$ pseudotensor $\mathcal{E}$ represented in each canonical basis by

$$\mathcal{E} = n! \sqrt{|\det g|}\, e^{*1} \wedge \cdots \wedge e^{*n} , \tag{5.7}$$

where $g = [g_{ij}]$, with g_{ij} the components of $\mathbf{g}$ in the considered basis. ∎

Remarks 5.28

(1) $\mathcal{E}$ can also be written in standard components:

$$\mathcal{E} = \mathcal{E}_{i_1 \cdots i_n} e^{*i_1} \otimes \cdots \otimes e^{*i_n} ,$$

where $i_k \in \{1, 2, \ldots, n\}$ for $k = 1, 2, \ldots, n$ and, as usual, we use the summation convention over repeated indices. It is easy to prove that

$$\mathcal{E}_{i_1 \cdots i_n} = \sqrt{|\det g|} \eta_{i_1 \ldots i_n} , \tag{5.8}$$

where

$$\eta_{i_1 \ldots i_n} = \epsilon_{\sigma_{i_1 \ldots i_n}} ,$$

whenever $(i_1, \ldots, i_n)$ is a permutation of $(1, 2, \ldots, n)$; otherwise

$$\eta_{i_1 \ldots i_n} = 0 .$$

Above, $\epsilon_{\sigma_{i_1\ldots i_n}}$ is the sign of the permutation $\sigma_{i_1\ldots i_n} \in \mathscr{P}_n$ defined by

$$(\sigma(1), \ldots, \sigma(n)) = (i_1, \ldots, i_n) .$$

(2) If V is oriented, it is possible to define $\mathcal{E}$ as a *proper tensor* instead of a pseudotensor. In this case one defines the components in a canonical basis associated with a positively oriented basis of V as

$$\mathcal{E}_{i_1\ldots i_n} := \sqrt{|\det g|}\ \eta_{i_1\ldots i_n}\ , \text{that is}\ \ \mathcal{E} = n!\,\sqrt{|\det g|}\, e^{*1} \wedge \cdots \wedge e^{*n}$$

and

$$\mathcal{E}_{i_1\ldots i_n} := -\sqrt{|\det g|}\ \eta_{i_1\ldots i_n}\ , \text{that is}\ \ \mathcal{E} = -n!\,\sqrt{|\det g|}\, e^{*1} \wedge \cdots \wedge e^{*n}$$

if the chosen basis is associated with a negatively oriented basis of V. One can prove straightforwardly that the defined components give rise to a tensor of $V^{*n\otimes}$ called the **Ricci tensor**. This alternative point of view is equivalent in practice to the point of view corresponding to Definition 5.27.

■

Ricci's pseudotensor has various applications in mathematical physics, in particular when it is used as a linear operator producing pseudotensors when it acts on tensors. In fact, consider $t \in V^{m\otimes}$ with $m \leq n$. Fix a basis in V and, in the canonical bases induced in the tensor spaces by that basis, consider the action of $\mathcal{E}$ on t:

$$t^{i_1\ldots i_m} \mapsto \tilde{t}_{j_1\ldots j_{n-m}} := \mathcal{E}_{j_1\ldots j_{n-m} i_1\ldots i_m} t^{i_1\ldots i_m} .$$

A direct verification shows that the components $\tilde{t}_{j_1,\ldots,j_{n-m}}$ define an antisymmetric pseudotensor of order $(0, n-m)$, called the **conjugate** pseudotensor of t.

Let us restrict ourselves to antisymmetric tensors and pseudotensors. The correspondence above from antisymmetric *tensors* of order $(m, 0)$ to antisymmetric *pseudo*tensors of order $(0, n-m)$ is surjective, due to the formula (with $p = n-m$)

$$t_{i_1\ldots i_p} = \frac{|\det g|}{\det g}\,\frac{(-1)^{p(n-p)}}{p!(n-p)!}\,\mathcal{E}_{i_1\ldots i_p j_1\ldots j_m}\mathcal{E}^{j_1\ldots j_m r_1\ldots r_p}\ t_{r_1\ldots r_p} .$$

However, since the space of antisymmetric tensors of order $(m, 0)$ and that of antisymmetric pseudotensors of order $(0, n-m)$ have the same dimension,

$$\binom{n}{m} = \binom{n}{n-m} ,$$

we conclude that the linear map associating antisymmetric tensors of order $(m, 0)$ with their antisymmetric dual pseudotensors of order $(0, n-m)$ is an isomorphism of vector spaces. This implies that, if the tensor t is antisymmetric, then its conjugate pseudotensor $\tilde{t}$ carries the same information as t itself.

Examples 5.29 **1**. As a trivial but very important example, consider the **vector product** in a three-dimensional vector space V over the field $\mathbb{R}$ endowed with a metric tensor g. If $u, v \in V$ we may define the pseudovector of order $(1, 0)$:

$$(u \times v)^r := g^{ri}\mathcal{E}_{ijk}u^j v^k .$$

If $\{e_i\}_{i=1,2,3}$ is an orthonormal basis in V, everything strongly simplifies. In fact, the Ricci tensor is represented by components

$$\mathcal{E}_{ijk} := 0$$

if (i, j, k) is not a permutation of $(1, 2, 3)$ and, otherwise,

$$\mathcal{E}_{ijk} := \pm 1 ,$$

where $+1$ corresponds to *cyclic* permutations of $1, 2, 3$ and -1 to *non-cyclic* permutations (see **Examples 3.1.2**). In such a basis,

$$(u \times v)_i = \mathcal{E}_{ijk}\, u^j v^k ,$$

because $g^{ij} = \delta^{ij}$ in every orthonormal basis. We have obtained that

$$\begin{aligned}(u \times v)_i &= \mathcal{E}_{ijk}\frac{1}{2}(u^j v^k - u^k v^j)\\ &= \mathcal{E}_{ijk}(u \wedge v)^{jk} ,\end{aligned}$$

where we have used

$$\begin{aligned}u \wedge v &= u^j e_j \wedge v^k e_k = u^j v^k e_j \wedge e_k\\ &= u^j v^k \frac{1}{2!}(e_j \otimes e_k - e_k \otimes e_j) = \frac{1}{2}(u^j v^k - u^k v^j) e_j \otimes e_k .\end{aligned}$$

The identity

$$(u \times v)_i = \mathcal{E}_{ijk}(u \wedge v)^{jk} ,$$

proves that *the vector product of two vectors is nothing but the conjugate pseudotensor of the wedge product of those vectors. It also proves that the vector product is a pseudo (covariant) vector.*

Exercises 5.30

1. Often, the definition of the vector product in $\mathbb{R}^3$ is given, in an orthonormal basis, as

$$(u \times v)_i = \mathcal{E}_{ijk}\, u^j v^k \,,$$

where it is assumed that the basis is *right-oriented*. Show that this defines a proper vector (and not a pseudovector) if a convenient definition of $\times$ is given in a *left-oriented* basis.

2. Is it possible to define a sort of vector product (mapping pairs of vectors to vectors) in $\mathbb{R}^4$ generalizing the vector product in $\mathbb{R}^3$?
3. In the physics literature one may find the statement: "Differently from the momentum $\vec{p}$, which is a **polar vector**, the angular momentum $\vec{l}$ is an **axial vector**." What does it mean?
(Solution. Polar vector = vector, Axial vector = pseudovector.)
4. Consider the **parity inversion**, $P \in O(3)$, as the active transformation of vectors of physical space defined by $P := -I$ when acting on the components of vectors in any orthonormal basis. What do physicists mean when saying: "Axial vectors transform differently from polar vectors under parity inversion"?
(*Hint*. Interpret P as a passive transformation, i.e. a change of basis, and extend the result to the active interpretation.)
5. Can the formula defining the conjugate pseudotensor of t,

$$t^{i_1\ldots i_m} \mapsto \tilde{t}_{j_1\ldots j_{n-m}} := \mathcal{E}_{j_1\ldots j_{n-m} i_1\ldots i_m} t^{i_1\ldots i_m} \,,$$

be generalized to the case where t is a pseudotensor? If yes, what sort of geometric object is $\tilde{t}$?

6. Consider a vector product $u \times v$ in $\mathbb{R}^3$ using an orthonormal basis. In that basis there is an antisymmetric matrix carrying the same information as $u \times v$ and which can be written down using the components of the vectors u and v. Determine that matrix and explain its tensorial meaning.
7. Prove the formula introduced above:

$$t_{i_1\ldots i_p} = \frac{|\det g|}{\det g} \frac{(-1)^{p(n-p)}}{p!(n-p)!} \mathcal{E}_{i_1\ldots i_p j_1\ldots j_m} \mathcal{E}^{j_1\ldots j_m r_1\ldots r_p} \, t_{r_1\ldots r_p} \,,$$

for antisymmetric tensors $t \in V^{*p\otimes}$, $0 \leq p \leq n = dim V$, $m := n - p$.

5.5 Tensor Densities

In the previous section we have seen that the determinant of the matrix representing a pseudo-metric tensor g transforms, under a change of basis, according to the rule

$$\det g' = |\det A|^{-2} \det g$$

where the pseudo-metric tensor is

$$g'_{ij} e'^{*i} \otimes e'^{*j} = g_{pq} e^{*p} \otimes e^{*q}$$

and $A = [A^i{}_j]$ is the matrix used in the change of basis for contravariant vectors

$$t^i e_i = t'^p e'_p \,,$$

that is

$$t'^i = A^i{}_p t^p \,.$$

Thus the assignment of the numbers $\det g$ to each basis in $\mathcal{B}$ does not define a scalar because of the presence of the factor $|\det A|^{-2}$. Similar mathematical objects play a relevant role in mathematical/theoretical physics and therefore deserve a precise definition.

Definition 5.31 (Tensor Densities) Let V be a finite-dimensional vector space over the field $\mathbb{R}$ and let $\mathcal{B}$ be the class of all bases of V.

If $S \in \mathscr{A}_{\mathbb{R}}(V)$, a **tensor density of** S **with weight** $w \in \mathbb{Z}$ is a map $d : \mathcal{B} \to S$ such that, if $B = \{e_i\}_{i\in I}$ and $B' = \{e'_j\}_{j\in I}$ are two bases in $\mathcal{B}$ with $e_k = A^i{}_k e'_i$, then

$$d(B') = |\det A|^w d(B) \,,$$

where $A = [A^i{}_k]$. Furthermore:

(a) the various tensorial properties enjoyed by all $d(B)$ are attributed to d. (So, for instance, if a $d(B)$ is symmetric, and thus all $d(B)$ with $B \in \mathcal{B}$ are symmetric, then d is said to be symmetric);

(b) if $B \in \mathcal{B}$, the **components of** d with respect to the canonical bases induced by B are the components of $d(B)$ in those bases.

■

If $\mathbf{g}$ is a pseudo-metric tensor on V, a trivial example of a density with weight w in, for instance, $S = V \otimes V^* \otimes V$, can be built up as follows. Take $t \in V \otimes V^* \otimes V$ and define

$$d_t(\{e_i\}_{i\in I}) := (\sqrt{|\det g|})^{-w} t \,,$$

where g is the matrix of the coefficients of $\mathbf{g}$ in the canonical basis associated with $\{e_i\}_{i\in I}$. In components, in the sense of **(b)** of the definition above:

$$(d_t)^i{}_j{}^k = (\sqrt{|\det g|})^{-w} t^i{}_j{}^k\,.$$

To conclude, we give the definition of pseudo-tensor density, which is the straightforward extension of the definition given above.

Definition 5.32 (Pseudo-Tensor Densities) Let V be a finite-dimensional vector space over the field $\mathbb{R}$ and let $\mathcal{B}$ be the class of all bases of V.

If $S \in \mathscr{A}_{\mathbb{R}}(V)$, a **pseudo-tensor density of** S **with weight** $w \in \mathbb{Z}$ is a map $d : \mathcal{B} \to S$ such that, if $B = \{e_i\}_{i\in I}$ and $B' = \{e'_j\}_{j\in I}$ are two bases in $\mathcal{B}$ with $e_k = A^i{}_k e'_i$, then

$$d(B') = \frac{\det A}{|\det A|} |\det A|^w d(B)\,.$$

where $A = [A^i{}_k]$. Furthermore:

(a) the various tensorial properties enjoyed by all $d(B)$ are attributed to d. (So, for instance, if a $d(B)$ is symmetric, and thus all $d(B)$ with $B \in \mathcal{B}$ are symmetric, then d is said to be symmetric);

(b) if $B \in \mathcal{B}$, the **components of** d with respect to the canonical bases induced by B are the components of $d(B)$ in those bases. ■

Remarks 5.33 It follows immediately that the sets of tensor densities and pseudo-tensor densities of a fixed space S and fixed weight form vector spaces, with the composition rule reducing to the usual linear composition rule of components. ■

There is an important property of densities with weight -1 which is very useful in integration theory on manifolds. The property is stated in the following theorem.

Theorem 5.34 *Let V be a vector space over $\mathbb{R}$ with dimension $n < +\infty$. There is a natural isomorphism G from the space of scalar densities of weight -1 to the space of antisymmetric covariant tensors of order n. In components, using the notation introduced above,*

$$G : \alpha \mapsto \alpha\eta_{i_1\ldots i_n}\,.$$

Proof Fix a canonical basis of $V^{*n\otimes}$ associated with a basis of V. Any non-vanishing tensor t in the space $\Lambda^n(V^*)$ of antisymmetric covariant tensors of order $n = dim V$ must have the form

$$t_{i_1\ldots i_n} = \alpha\eta_{i_1\ldots i_n}$$

in components, because different non-vanishing components can differ only by a sign due to the antisymmetry of t. Therefore the dimension of $\Lambda^n(V^*)$ is 1, which is also the dimension of the space of scalar densities of weight -1. The application G defined above in components is linear and surjective, and thus injective as well. Finally, by straightforwardly adapting the relevant part of the discussion used to define $\mathcal{E}$, one finds that the coefficient α in $t_{i_1\ldots i_n} = \alpha\eta_{i_1\ldots i_n}$ transforms as a scalar density of weight -1 under a change of basis. □

References

1. Moretti, V.: Spectral Theory and Quantum Mechanics: Mathematical Foundations of Quantum Theories, Symmetries and Introduction to the Algebraic Formulation, 2nd edn. Springer, Berlin (2017)
2. Rudin, W.: Real and Complex Analysis. McGraw-Hill, New York (1987)
3. Streater, R.F., Wightman, A.S.: PCT, Spin and Statistics, and All That with Revised Preface and Corrections. Princeton University Press, Princeton (2000)
4. Wald, R.M.: General Relativity. University of Chicago Press, Chicago (1984)

Chapter 6
Polar Decomposition Theorem in the Finite-Dimensional Case

This chapter is devoted to the polar decomposition theorem in finite-dimensional spaces with scalar product. The theorem states, in essence, that a linear operator can be decomposed into a metric part and an isometric part, in a way analogous to decomposing a complex number into its modulus and phase. Although the result is purely algebraic, it has many applications in geometry and physics.

The chapter reviews adjoint operators, positive operators, complexification, and square roots of positive operators before proving the polar decomposition. This preparation is important because the theorem will later be used to analyze the structure of the Lorentz group and to separate boosts from rotations. Thus the chapter is both an independent piece of linear algebra and a technical tool for the relativistic applications that follow.

6.1 Operators in Spaces with Scalar Product

We recall here some basic definitions and results which should be known to the reader from elementary courses in linear algebra [1, 2].

If $A \in \mathscr{L}(V|V)$, that is, $A : V \to V$ is a (linear) operator on a finite-dimensional vector space V over the field $\mathbb{K} = \mathbb{R}$ or $\mathbb{C}$, an **eigenvalue** of A is an element $\lambda \in \mathbb{K}$ such that

$$(A - \lambda I)u = 0$$

for some $u \in V \setminus \{0\}$. In that case u is called an **eigenvector** associated with λ. The set $\sigma(A)$ containing all eigenvalues of A is called the **spectrum** of A. The **eigenspace** E_λ associated with $\lambda \in \sigma(A)$ is the subspace of V spanned by the eigenvectors associated with λ.

V. Moretti, *Geometric Methods in Mathematical Physics I*, La Matematica per il 3+2 184, https://doi.org/10.1007/978-3-032-31512-0_6

Proposition 6.1 *Let V be a real (respectively complex) finite-dimensional vector space equipped with a scalar product (respectively a Hermitian scalar product) $(\cdot|\cdot)$. For every operator $A : V \to V$ there exists exactly one operator $A^\dagger : V \to V$, called the* ***adjoint operator*** *of A, such that*

$$(A^\dagger u|v) = (u|Av)\,,$$

for all $u, v \in V$.

Proof Fix $u \in V$. The map $v \mapsto (u|Av)$ is a linear functional and thus an element of V^*. By Theorem 5.7 there is a unique element $w_{u,A} \in V$ such that

$$(u|Av) = (w_{u,A}|v)$$

for all $v \in V$. Consider the map $u \mapsto w_{u,A}$. If a, b are scalars in the field of V and $u, u' \in V$, then

$$\begin{aligned}(w_{au+bu',A}|v) &= (au + bu'|Av) = \overline{a}(u|Av) + \overline{b}(u'|Av)\\ &= \overline{a}(w_{u,A}|v) + \overline{b}(w_{u',A}|v) = (aw_{u,A} + bw_{u',A}|v)\,.\end{aligned}$$

Hence, for all $v \in V$:

$$(w_{au+bu',A} - aw_{u,A} - bw_{u',A}|v) = 0\,.$$

The scalar product is non-degenerate by definition, and this implies

$$w_{au+bu',A} = aw_{u,A} + bw_{u',A}\,.$$

We have obtained that the map $A^\dagger : u \mapsto w_{u,A}$ is linear; in other words, it is an operator. The uniqueness is trivial: if another operator B satisfies $(Bu|v) = (u|Av)$ for all $u, v \in V$, then

$$((B - A^\dagger)u|v) = 0$$

for all $u, v \in V$, which entails

$$(B - A^\dagger)u = 0$$

for all $u \in V$. In other words, $B = A^\dagger$. □

Remarks 6.2 There is an interesting interplay between this notion of adjoint operator and that given in Definition 2.15. Let $h : V \to V^*$ be the natural (anti)isomorphism induced by the scalar product as in Theorem 5.7. If $T \in$

$\mathscr{L}(V|V)$ and $T^*, T^\dagger$ denote, respectively, the two adjoint operators of T defined in Definition 2.15 and in Proposition 6.1, one has

$$h \circ T^\dagger = T^* \circ h\,. \tag{6.1}$$

Indeed, for every $u, v \in V$,

$$\langle v, h(T^\dagger u)\rangle = (T^\dagger u|v) = (u|Tv) = \langle Tv, h(u)\rangle = \langle v, T^* h(u)\rangle\,,$$

where in the Hermitian case the same computation is read with the anti-isomorphism h and the corresponding convention for the dual adjoint. Hence $h(T^\dagger u) = T^* h(u)$ for every $u \in V$. ■

There are a few simple properties of the adjoint operator whose proofs are straightforward. Below A, B are operators in a complex (respectively real) finite-dimensional vector space V equipped with a Hermitian (respectively real) scalar product $(\cdot|\cdot)$, and a, b belong to the field of V.

(1) $(A^\dagger)^\dagger = A$,
(2) $(aA + bB)^\dagger = \overline{a}A^\dagger + \overline{b}B^\dagger$ (respectively $(aA + bB)^\dagger = aA^\dagger + bB^\dagger$),
(3) $(AB)^\dagger = B^\dagger A^\dagger$,
(4) $(A^\dagger)^{-1} = (A^{-1})^\dagger$ (if A^{-1} exists).

Take a finite-dimensional vector space V equipped with a scalar product $(\cdot|\cdot)$ and consider $A \in \mathscr{L}(V|V)$. We distinguish two cases: the *real case*, where the field of V and the scalar product are real, and the *complex case*, where the field of V is $\mathbb{C}$ and the scalar product is Hermitian. We have the following definitions.

(1) In the real case, A is said to be **symmetric** if $(Au|v) = (u|Av)$ for all $u, v \in V$. In the complex case, A is said to be **Hermitian** if $(Au|v) = (u|Av)$ for all $u, v \in V$. It is simply proved that, in the respective cases, A is symmetric or Hermitian if and only if $A = A^\dagger$. In both cases $\sigma(A) \subset \mathbb{R}$.
(2) In the real case, A is said to be **orthogonal** if $(Au|Av) = (u|v)$ for all $u, v \in V$. In the complex case, A is said to be **unitary** if $(Au|Av) = (u|v)$ for all $u, v \in V$. It is simply proved that, in the respective cases, A is orthogonal or unitary if and only if A is bijective and $A^{-1} = A^\dagger$. In both cases, if $\lambda \in \sigma(A)$, then $|\lambda| = 1$.
(3) In both cases, A is said to be **normal** if $AA^\dagger = A^\dagger A$. It follows immediately that symmetric, Hermitian, unitary, and orthogonal operators are normal.
An important and straightforwardly proved result is that, if $A \in \mathscr{L}(V|V)$ is normal, then:

(a) $u \in V$ is an eigenvector of A with eigenvalue λ if and only if the same u is an eigenvector of $A^\dagger$ with eigenvalue $\overline{\lambda}$ (which coincides with λ in the real case);
(b) if $\lambda, \mu \in \sigma(A)$ and $\lambda \neq \mu$, then $(u_\lambda|u_\mu) = 0$ if u_λ and u_μ are eigenvectors of A with eigenvalues λ and μ, respectively.

In either the real or complex case, if $U \subset V$ is a subspace, the **orthogonal** of U, $U^\perp$, is the subspace of V made of all vectors orthogonal to U, i.e., $v \in U^\perp$ if and only if $(u|v) = 0$ for all $u \in U$. By the standard projection theorem in finite-dimensional spaces (see [1, 2]), if $w \in V$, the decomposition

$$w = u + v$$

with $u \in U$ and $v \in U^\perp$ is uniquely determined, and the map $P_U : w \mapsto u$ is linear; it is called the **orthogonal projector** onto U.

If V is a finite-dimensional vector space (either real or complex), and $U \subset V$ is a subspace, the subspace $U^*_\perp \subset V^*$ defined as

$$U^*_\perp := \{v \in V^* \mid \langle u, v\rangle = 0\,, \quad \text{for all } u \in U\}$$

provides another and more general notion of "orthogonal space", in the absence of a scalar product on V. It is simply proved that, in the presence of a (pseudo)scalar product, if $h : V \to V^*$ is the natural (anti)isomorphism induced by the scalar product as in Theorem 5.7, then

$$h^{-1}(U^*_\perp) = U^\perp\,.$$

A direct verification shows that an operator $P : V \to V$ is an orthogonal projector onto some subspace $U \subset V$ if and only if both conditions below hold:

(1) $PP = P$,
(2) $P = P^\dagger$.

In that case P is the orthogonal projector onto $U = \{Pv \mid v \in V\}$.

Another pair of useful results concerning orthogonal projectors is the following. Let V be as above, let U, U' be subspaces of V, and let P, P' be the corresponding orthogonal projectors.

(a) U and U' are **orthogonal** to each other, i.e., $U' \subset U^\perp$ (which is equivalent to $U \subset U'^\perp$), if and only if $PP' = P'P = 0$.
(b) $U \subset U'$ if and only if $PP' = P'P = P$.

If V is as above, has finite dimension n, and $A : V \to V$ is normal, there is a well-known spectral decomposition theorem [1, 2] (the finite-dimensional version of the "spectral theorem").

Proposition 6.3 (Spectral Decomposition for Normal Operators in Complex Spaces) *Let V be a complex finite-dimensional vector space equipped with a Hermitian scalar product $(\cdot|\cdot)$. If $A \in \mathscr{L}(V|V)$ is normal (i.e., $A^\dagger A = AA^\dagger$), the following expansion holds:*

$$A = \sum_{\lambda \in \sigma(A)} \lambda P_\lambda\,,$$

where P_λ *is the orthogonal projector onto the eigenspace associated with* λ*. Moreover the map* $\sigma(A) \ni \lambda \mapsto P_\lambda$ *satisfies the following two properties:*

(1) $I = \sum_{\lambda\in\sigma(A)} P_\lambda$,
(2) $P_\lambda P_\mu = P_\mu P_\lambda = 0$ *for* $\mu \neq \lambda$.

A ***spectral measure****, i.e. a map* $B \ni \mu \mapsto P'_\mu$ *with* $B \subset \mathbb{C}$ *finite,* P'_μ *non-vanishing orthogonal projectors, and:*

(1)' $I = \sum_{\mu\in B} P'_\mu$,
(2)' $P'_\lambda P'_\mu = P'_\mu P'_\lambda = 0$ *for* $\mu \neq \lambda$*, coincides with* $\sigma(A) \ni \lambda \mapsto P_\lambda$ *if*
(3)' $A = \sum_{\mu\in B} \mu P'_\mu$.

Proof The equation $\det(A-\lambda I) = 0$ must have a solution $\lambda \in \mathbb{C}$ by the fundamental theorem of algebra. As a consequence, A admits an eigenspace $E_{\lambda_1} \subset V$. From properties **(a)** and **(b)** of normal operators, one obtains that

$$A(E^\perp_{\lambda_1}) \subset E^\perp_{\lambda_1} .$$

Moreover $A\restriction_{E^\perp_{\lambda_1}}$ is normal. Therefore the procedure can be iterated, obtaining a new eigenspace $E_{\lambda_2} \subset E^\perp_{\lambda_1}$. Notice that $\lambda_2 \neq \lambda_1$, because every eigenvector of $A\restriction_{E^\perp_{\lambda_1}}$ is also an eigenvector of A with the same eigenvalue. Since eigenspaces with different eigenvalues are orthogonal, one finally gets a sequence of pairwise orthogonal eigenspaces

$$E_{\lambda_1}, E_{\lambda_2}, \ldots, E_{\lambda_k},$$

with $E_{\lambda_l} \subset E^\perp_{\lambda_{l'}}$ if $l' < l$. This sequence must be finite because the dimension of V is finite, and orthogonal eigenvectors are linearly independent. Therefore

$$E^\perp_{\lambda_k} = \{0\}$$

and so

$$\oplus_{l=1}^k E_{\lambda_l} = V .$$

On the other hand, the set $\{\lambda_1, \lambda_2, \ldots, \lambda_k\}$ must coincide with the whole $\sigma(A)$, because eigenspaces with different eigenvalues are orthogonal, and thus any other eigenspace E_{λ_0} with $\lambda_0 \notin \{\lambda_1, \lambda_2, \ldots, \lambda_k\}$ would be included in

$$(\oplus_{l=1}^k E_{\lambda_l})^\perp = \{0\} .$$

If $\{e_j^{(l)}\}_{j\in I_l}$ is an orthonormal basis of E_{λ_l}, the orthogonal projector onto E_{λ_l} is

$$P_{\lambda_l} := \sum_{j\in I_l} (e_j^{(l)}|\ \ )e_j^{(l)} .$$

Property **(1)** is nothing but the fact that

$$\{e_j^{(l)}\}_{j \in I_l,\ l=1,\ldots,k}$$

is an orthonormal basis of V. Property **(2)** is another way to say that eigenspaces with different eigenvalues are orthogonal. Finally, since $e_j^{(l)}$ is an eigenvector of A with eigenvalue λ_l, it holds

$$A = \sum_{l=1}^{k} \sum_{j \in I_l} \lambda_l (e_j^{(l)} | \quad) e_j^{(l)} ,$$

which can be rewritten as

$$A = \sum_{\lambda \in \sigma(A)} \lambda P_\lambda .$$

The uniqueness property of spectral measures is proved as follows. It holds

$$\sum_{\mu \in B} \mu P'_\mu = \sum_{\lambda \in \sigma(A)} \lambda P_\lambda .$$

Applying P'_ν on the left and P_τ on the right, and using properties **(1)**' and **(1)**, one finds

$$\nu P'_\nu P_\tau = \tau P'_\nu P_\tau ,$$

so that

$$(\nu - \tau) P'_\nu P_\tau = 0 .$$

This is possible only if $\nu = \tau$ or $P'_\nu P_\tau = 0$. For a fixed P'_ν, it is not possible that $P'_\nu P_\tau = 0$ for all P_τ, because, using **(1)**, it would imply

$$0 = P'_\nu \sum_{\tau \in \sigma(A)} P_\tau = P'_\nu I = P'_\nu$$

but $P'_\nu \neq 0$ by hypothesis. Therefore $\nu = \tau$ for some $\tau \in \sigma(A)$ and hence $B \subset \sigma(A)$. By the analogous argument one sees that $\sigma(A) \subset B$. We conclude that $B = \sigma(A)$. Finally, the decomposition

$$A = \sum_{\lambda \in \sigma(A)} \lambda P'_\lambda$$

implies that each P'_λ is an orthogonal projector onto a subspace of E_λ. As a consequence, if u_λ is an eigenvector of A with eigenvalue λ, it must be

$$P'_{\lambda'} u_\lambda = 0$$

for $\lambda' \neq \lambda$. Since **(1)**' is valid, one concludes that $P'_\lambda u_\lambda = u_\lambda$, and thus not only $P'_\lambda(V) \subset E_\lambda$, but, more strongly, $P'_\lambda(V) = E_\lambda$. Since the orthogonal projector onto a subspace is uniquely determined by the subspace itself, we have proved that

$$P'_\lambda = P_\lambda$$

for every $\lambda \in \sigma(A)$. □

6.1.1 Positive Operators

Definition 6.4 If V is a real (respectively complex) vector space equipped with a scalar product (respectively a Hermitian scalar product) $(\cdot|\cdot)$, an operator $A : V \to V$ is said to be **positive** (or *positive semidefinite*) if

$$(u|Au) \geq 0 \quad \text{for all } u \in V\,.$$

A positive operator A is said to be **strictly positive** (or *positive definite*) if

$$(u|Au) = 0 \quad \text{entails } u = 0\,.$$

■

A straightforward consequence of the definition is the following lemma.

Lemma 6.5 *Let V be a complex vector space equipped with a Hermitian scalar product $(\cdot|\cdot)$.*

Any positive operator $A : V \to V$ is Hermitian.

Moreover, if $\dim V < \infty$, a normal operator $A : V \to V$ satisfies:

(a) *A is positive if and only if $\sigma(A) \subset [0, +\infty)$;*
(b) *A is strictly positive if and only if $\sigma(A) \subset (0, +\infty)$.*

Proof Since $(v|Av) \geq 0$, by complex conjugation we have

$$(Av|v) = \overline{(v|Av)} = (v|Av)\,,$$

and thus, setting $B := A^\dagger - A$, we get

$$(Bv|v) = 0$$

for all $v \in V$. Now use the polarization identity:

$$4(Bu|w) = (B(u+w)|u+w) - (B(u-w)|u-w) \\ - i(B(u+iw)|u+iw) + i(B(u-iw)|u-iw) \,.$$

Since each quadratic term on the right-hand side vanishes, it follows that

$$(Bu|w) = 0$$

for all $u, w \in V$. Hence $B = 0$, that is, $A^\dagger = A$.

Let us prove **(a)**. Suppose A is positive. We already know that $\sigma(A) \subset \mathbb{R}$. Suppose there is $\lambda < 0$ in $\sigma(A)$. Let $u \neq 0$ be an eigenvector associated with λ. Then

$$(u|Au) = \lambda(u|u) < 0$$

because $(u|u) > 0$. This is impossible.

Now assume that A is normal with $\sigma(A) \subset [0, +\infty)$. By Proposition 6.3:

$$A = \sum_{\lambda\in\sigma(A)} \lambda P_\lambda \,.$$

Therefore

$$Au = \sum_{\lambda\in\sigma(A)} \lambda P_\lambda u \,,$$

and hence

$$(u|Au) = \sum_{\lambda\in\sigma(A)} \lambda(u|P_\lambda u) \,.$$

Since $P_\lambda = P_\lambda^\dagger = P_\lambda^2$, we have

$$(u|P_\lambda u) = (P_\lambda u|P_\lambda u) \geq 0 \,,$$

so that

$$(u|Au) = \sum_{\lambda\in\sigma(A)} \lambda(P_\lambda u|P_\lambda u) \geq 0$$

because $\lambda \geq 0$ for every $\lambda \in \sigma(A)$. Thus A is positive.

Concerning **(b)**, assume that A is strictly positive. Then A is positive, hence Hermitian, and by **(a)**

$$\sigma(A) \subset [0, +\infty) .$$

If $0 \in \sigma(A)$, there would exist $u \neq 0$ such that $Au = 0$, and thus

$$(u|Au) = 0,$$

contradicting strict positivity. Therefore

$$\sigma(A) \subset (0, +\infty) .$$

Conversely, if A is normal with $\sigma(A) \subset (0, +\infty)$, then A is positive by **(a)**. If A were not strictly positive, there would exist $u \neq 0$ such that $(u|Au) = 0$, and thus

$$0 = (u|Au) = \sum_{\lambda \in \sigma(A)} \lambda (P_\lambda u | P_\lambda u) .$$

Since $\lambda > 0$ and $(P_\lambda u | P_\lambda u) \geq 0$, it follows that

$$(P_\lambda u | P_\lambda u) = 0$$

for all $\lambda \in \sigma(A)$, hence $P_\lambda u = 0$ for all λ. This is impossible because, using Property **(1)** in Proposition 6.3,

$$0 \neq u = Iu = \sum_\lambda P_\lambda u = 0 .$$

□

6.1.2 Complexification Procedure

There is a standard and useful way to associate a complex vector space endowed with a Hermitian scalar product to a given real vector space equipped with a scalar product. This correspondence allows one to exploit some results valid for complex vector spaces also in the case of real vector spaces. We shall use that possibility shortly.

Definition 6.6 If V is a real vector space equipped with the scalar product $(\cdot|\cdot)$, the **complexification** of V, denoted by $V \oplus iV$, is the complex vector space with

Hermitian scalar product $(\cdot|\cdot)_{\mathbb{C}}$ defined as follows. The elements of $V \oplus iV$ are the pairs $(u, v) \in V \times V$, denoted by

$$u + iv := (u, v) .$$

Moreover,

(1) the product of a scalar in $\mathbb{C}$ and a vector in $V \oplus iV$ is defined as

$$(a+ib)(u+iv) := au - bv + i(bu + av) \quad \text{for } a+ib \in \mathbb{C} \text{ and } u+iv \in V \oplus iV ,$$

(2) the sum of vectors is defined as

$$(u + iv) + (x + iy) := (u + x) + i(v + y) , \quad \text{for } u + iv, x + iy \in V \oplus iV ,$$

(3) the Hermitian scalar product is defined as

$$(u + iv|w + ix)_{\mathbb{C}} := (u|w) + (v|x) + i(u|x) - i(v|w),$$
$$\text{for } u + iv, w + ix \in V \oplus iV .$$

■

Let us introduce a useful operator from $V \oplus iV$ to $V \oplus iV$.
The **complex conjugation** is defined as

$$J : u + iv \mapsto u - iv .$$

It is anti-linear and satisfies

$$(J(u + iv)|J(w + ix))_{\mathbb{C}} = (w + ix|u + iv)_{\mathbb{C}}$$

and

$$JJ = I .$$

An important property of J is the following: $s \in V \oplus iV$ satisfies $s \in V$ (i.e. $s = (u, 0) \in V \times V$) if and only if $Js = s$, whereas it satisfies $s \in iV$ (i.e. $s = (0, v) \in V \times V$) if and only if $Js = -s$. The proof is immediate from the definition.

A linear operator $A : V \oplus iV \to V \oplus iV$ is said to be **real** if $JA = AJ$.

Proposition 6.7 *Referring to Definition 6.6, let* $A \in \mathscr{L}(V \oplus iV|V \oplus iV)$. *The following facts hold.*

(a) *A is real if and only if there exists a uniquely determined* $\mathbb{R}$*-linear operator* $A_0 : V \to V$ *such that*

$$A(u + iv) = A_0 u + iA_0 v\,, \quad \textit{for all } u + iv \in V \oplus iV\,.$$

(b) *If* A *is real, then* $A^\dagger$ *is real as well.*

Proof Uniqueness in **(a)** is trivial. If $A(u + iv) = A_0 u + iA_0 v$ for all $u, v \in V$, then

$$JA(u + iv) = A_0 u - iA_0 v = A(u - iv) = AJ(u + iv)\,,$$

so A is real. Conversely, suppose that A is real. If $u \in V$, then $Ju = u$, hence

$$J(Au) = A(Ju) = Au\,,$$

so $Au \in V$. Therefore $A\restriction_V : V \to V$ is well defined. Define

$$A_0 := A\restriction_V\,.$$

Now, by complex linearity,

$$A(u + iv) = Au + iAv = A_0 u + iA_0 v\,,$$

which proves **(a)**.

Let us prove **(b)**. If A is real, then for all $u, v \in V \oplus iV$:

$$\begin{aligned}(A^\dagger Ju|v)_{\mathbb{C}} &= (Ju|Av)_{\mathbb{C}} = \overline{(u|JAv)_{\mathbb{C}}} = \overline{(u|AJv)_{\mathbb{C}}} \\ &= \overline{(A^\dagger u|Jv)_{\mathbb{C}}} = (JA^\dagger u|v)_{\mathbb{C}}\,,\end{aligned}$$

where we used the defining properties of J. Since this holds for all u, v, we conclude that

$$A^\dagger J = JA^\dagger\,.$$

Thus $A^\dagger$ is real. □

6.1.3 Square Roots

Definition 6.8 If V is a complex (respectively real) vector space equipped with a Hermitian (respectively real) scalar product $(\cdot|\cdot)$, and $A : V \to V$ is a positive operator (symmetric in the real case), then another positive operator $B : V \to V$ (Hermitian in the complex case, symmetric in the real case) such that

$$B^2 = A$$

is called a **square root of** A. ■

Notice that square roots, if they exist, are Hermitian by Lemma 6.5. The next theorem proves that the square root of a positive operator exists and is uniquely determined.

Theorem 6.9 *Let V be a finite-dimensional either complex or real vector space equipped with a Hermitian or, respectively, real scalar product $(\cdot|\cdot)$. Every positive operator $A : V \to V$ admits a unique square root, indicated by $\sqrt{A}$. $\sqrt{A}$ is Hermitian or, respectively, symmetric, and $\sqrt{A}$ is bijective if and only if A is bijective.*

Proof First consider the complex case. By Lemma 6.5, A is Hermitian. Using Proposition 6.3,

$$A = \sum_{\lambda\in\sigma(A)} \lambda P_\lambda \, .$$

Since $\lambda \geq 0$, we can define the linear operator

$$\sqrt{A} := \sum_{\lambda\in\sigma(A)} \sqrt{\lambda}\, P_\lambda \, .$$

By Proposition 6.3 we have

$$\begin{aligned}
\sqrt{A}\sqrt{A} &= \sum_{\lambda\in\sigma(A)} \sqrt{\lambda}\, P_\lambda \sum_{\mu\in\sigma(A)} \sqrt{\mu}\, P_\mu \\
&= \sum_{\lambda,\mu} \sqrt{\lambda\mu}\, P_\lambda P_\mu \\
&= \sum_{\lambda,\mu} \sqrt{\lambda\mu}\, \delta_{\lambda\mu} P_\lambda \\
&= \sum_{\lambda} \lambda P_\lambda = A \, .
\end{aligned}$$

Moreover, $\sqrt{A}$ is Hermitian by construction:

$$\sqrt{A}^\dagger = \sum_{\lambda\in\sigma(A)} \sqrt{\lambda}\, P_\lambda^\dagger = \sum_{\lambda\in\sigma(A)} \sqrt{\lambda}\, P_\lambda = \sqrt{A} \, .$$

Also, by construction,

$$\sigma(\sqrt{A}) = \{\sqrt{\lambda} \mid \lambda \in \sigma(A)\}\,,$$

and thus $\sqrt{A}$ is positive by Lemma 6.5. So $\sqrt{A}$ is a Hermitian square root of A. Notice that

$$0 \in \sigma(\sqrt{A}) \quad \text{if and only if} \quad 0 \in \sigma(A)\,,$$

so that $\sqrt{A}$ is bijective if and only if A is bijective.

Let us prove uniqueness. Suppose S is another positive square root of A. Then S is Hermitian, with $\sigma(S) \subset [0, +\infty)$, and it admits a unique spectral decomposition

$$S = \sum_{\nu \in \sigma(S)} \nu Q_\nu\,.$$

Define

$$B := \{\nu^2 \mid \nu \in \sigma(S)\}\,.$$

Since the map $\nu \mapsto \nu^2$ is injective on $[0, +\infty)$, it is simply proved that the map

$$B \ni \lambda \mapsto Q_{\sqrt{\lambda}}$$

satisfies:

(1)' $I = \sum_{\lambda \in B} Q_{\sqrt{\lambda}}$,
(2)' $Q_{\sqrt{\lambda}} Q_{\sqrt{\mu}} = Q_{\sqrt{\mu}} Q_{\sqrt{\lambda}} = 0$ for $\mu \neq \lambda$,
(3)' $A = S^2 = \sum_{\lambda \in B} \lambda Q_{\sqrt{\lambda}}$.

Therefore Proposition 6.3 entails that this spectral measure coincides with the spectral measure of A, so

$$Q_{\sqrt{\lambda}} = P_\lambda$$

for all $\lambda \in \sigma(A) = B$. In other words,

$$S = \sum_{\nu \in \sigma(S)} \nu Q_\nu = \sum_{\lambda \in \sigma(A)} \sqrt{\lambda}\, P_\lambda = \sqrt{A}\,.$$

Let us finally consider the case of a real space V. If $A \in \mathscr{L}(V|V)$ is positive and symmetric, define the operator

$$S : V \oplus iV \to V \oplus iV, \qquad S(u + iv) := Au + iAv\,.$$

Then S is a positive Hermitian operator on the complexification. By the first part of the proof, there is a unique positive Hermitian operator

$$\sqrt{S} \in \mathscr{L}(V \oplus iV | V \oplus iV)$$

such that

$$(\sqrt{S})^2 = S \, .$$

Since S is real, it commutes with J. Therefore, by Proposition 6.7, $S^\dagger = S$ is real as well, and hence the spectral projectors of S commute with J. It follows that $\sqrt{S}$ commutes with J, so $\sqrt{S}$ is real. By Proposition 6.7, there is a uniquely determined real-linear operator

$$R : V \to V$$

such that

$$\sqrt{S}(u + iv) = Ru + iRv \, .$$

Since $(\sqrt{S})^2 = S$, we get

$$R^2 = A \, .$$

Moreover, R is symmetric and positive because $\sqrt{S}$ is Hermitian and positive. Hence R is a square root of A, and we define

$$\sqrt{A} := R \, .$$

If A is bijective, then S is bijective and thus $\sqrt{S}$ is bijective; hence R is bijective as well. Conversely, if A is not bijective, choose $0 \neq u \in \ker A$. Then

$$0 = (u|Au) = (u|R^2u) = (Ru|Ru) \, ,$$

so $Ru = 0$, and therefore R is not injective.

To prove uniqueness, suppose that R' is another positive symmetric square root of A. Then the operator

$$B : u + iv \mapsto R'u + iR'v$$

is a positive Hermitian square root of S, and thus, by the uniqueness already proved in the complex case,

$$B = \sqrt{S} \, .$$

Therefore $R = R'$.

□

6.2 Polar Decomposition

The notions and results obtained above allow us to state and prove the polar decomposition theorem for operators in finite-dimensional vector spaces equipped with a scalar product.

Theorem 6.10 (Polar Decomposition of operators) *If $T \in \mathscr{L}(V|V)$ is a bijective operator, where V is a real (respectively complex) finite-dimensional vector space equipped with a real (respectively Hermitian) scalar product, then:*

(a) *there is a unique decomposition*

$$T = UP,$$

where U is orthogonal (respectively unitary) and P is bijective, symmetric (respectively Hermitian), and positive. In particular,

$$P = \sqrt{T^\dagger T}$$

and

$$U = T(\sqrt{T^\dagger T})^{-1}\,;$$

(b) *there is a unique decomposition*

$$T = P'U',$$

where U' is orthogonal (respectively unitary) and P' is bijective, symmetric (respectively Hermitian), and positive. In particular,

$$U' = U \quad \textit{and} \quad P' = UPU^\dagger\,.$$

Proof

(a) Consider $T \in \mathscr{L}(V|V)$ bijective. The operator $T^\dagger T$ is symmetric/self-adjoint, positive, and bijective. Define

$$P := \sqrt{T^\dagger T}\,,$$

which exists and is symmetric/self-adjoint, positive, and bijective by Theorem 6.9, and define

$$U := TP^{-1} .$$

Then U is orthogonal/unitary because

$$U^\dagger U = P^{-1}T^\dagger TP^{-1} = P^{-1}P^2P^{-1} = I ,$$

where we used $P^\dagger = P$. This proves that a polar decomposition of T exists, because by construction

$$UP = T .$$

Let us prove uniqueness. Suppose

$$T = U_1 P_1$$

is another such decomposition. Then

$$T^\dagger T = P_1 U_1^\dagger U_1 P_1 = P_1^2 .$$

Since also

$$T^\dagger T = P^2,$$

Theorem 6.9 implies that

$$P = P_1 .$$

Consequently,

$$U = TP^{-1} = TP_1^{-1} = U_1 .$$

This proves uniqueness in **(a)**.

(b) Define

$$P' := UPU^\dagger \quad \text{and} \quad U' := U .$$

Then P' is symmetric/self-adjoint, positive, and bijective, since U is orthogonal/unitary and P has the same properties. Moreover,

$$P'U' = UPU^\dagger U = UP = T .$$

So a decomposition of the required type exists. To prove uniqueness, suppose

$$T = QW$$

with Q positive, symmetric/self-adjoint, bijective, and W orthogonal/unitary. Then

$$TT^\dagger = QWW^\dagger Q = Q^2 .$$

On the other hand,

$$TT^\dagger = UPPU^\dagger = UP^2U^\dagger = (UPU^\dagger)^2 = (P')^2 .$$

By uniqueness of the positive square root,

$$Q = P' .$$

Hence

$$W = Q^{-1}T = (P')^{-1}T = U .$$

Therefore

$$Q = P' \quad \text{and} \quad W = U = U' ,$$

which proves uniqueness in **(b)**.

□

References

1. Lang, S.: Linear Algebra. Undergraduate Texts in Mathematics, 3rd edn. Springer, Berlin (1987)
2. Sernesi, E.: Geometria I and II. Bollati Boringhieri, Torino (1990)

Chapter 7
Special Relativity: A Geometric Presentation

This chapter develops Special Relativity as a geometric theory on Minkowski spacetime. The starting point is the affine structure of spacetime together with a Lorentzian metric of signature $(1, 3)$. From this structure one obtains the causal classification of vectors, the light cone, proper time, inertial reference frames, and the Lorentzian interpretation of temporal and spatial measurements.

The chapter deliberately adopts an axiomatic geometric viewpoint: rather than deriving the formalism from experiments, it uses the algebraic and metric tools developed in the previous chapters to formulate the theory intrinsically. The subsequent sections discuss kinematics, Lorentz and Poincaré transformations, time orientation, world lines, relativistic effects, four-momentum, massless particles, four-force, conservation laws, and the stress-energy tensor. In this way Special Relativity becomes the first major physical application of the tensorial language developed in the book.

7.1 Minkowski Spacetime

We henceforth adopt again the notation ∂_{x^a} to indicate the vector e_a of a basis of the space of translations V of an affine space $\mathbb{A}^n$, used to construct a Cartesian coordinate system $x^1, \ldots, x^n$ on $\mathbb{A}^n$ upon the choice of the origin $O \in \mathbb{A}^n$. Moreover, as before, dx^a denotes the corresponding element of the dual basis in V^*. Finally,

$$T_p\mathbb{A}^n := \{(p, u)\}_{u \in V} \quad \text{and} \quad T_p^*\mathbb{A}^n := \{(p, v^*)\}_{v^* \in V^*}$$

V. Moretti, *Geometric Methods in Mathematical Physics I*, La Matematica per il 3+2 184, https://doi.org/10.1007/978-3-032-31512-0_7

respectively indicate the vector spaces of vectors *applied to* p, and covectors *applied to* p, in the sense of Sect. 3.4. With these conventions

$$\partial_{x^i}|_p := (p, \partial_{x^i}) \quad \text{and} \quad dx^i|_p := (p, dx^i) .$$

We shall also use the shorthand notation

$$v_p := (p, v) \in T_p\mathbb{A}^n .$$

A *vector field* X on $\mathbb{A}^n$ is an assignment $\mathbb{A}^n \ni p \mapsto X_p \in T_p\mathbb{A}^n$. X is said to be *constant* if $X_p = X_q$ for all $p, q \in \mathbb{A}^n$. The vector field is said to be *of class* C^k if, once a basis of V is fixed (and thus, by translation, a basis of every $T_p\mathbb{A}^n$), the components of the applied vector X_p are C^k functions ($k = 0, 1, \ldots, \infty$) of a Cartesian coordinate system. If this fact is true for one Cartesian coordinate system, it is true for every other Cartesian coordinate system, since the transformations between Cartesian coordinate systems are C^∞ with C^∞ inverse.

7.1.1 General Minkowski Spacetime Structure

Let us recall some features of relativistic spacetime from a mathematical point of view. We adopt the notations and conventions for tensors on affine spaces introduced in Sect. 3.4.

Definition 7.1 A Minkowski spacetime $\mathbb{M}^4$ is a four-dimensional affine space whose real vector space of translations T^4 (identified with $\mathbb{R}^4$) is equipped with a pseudo-scalar product defined by a metric tensor $\mathbf{g}$ with signature $(1, 3)$, i.e. $(-1, +1, +1, +1)$. The following further definitions hold.

(i) The points of Minkowski spacetime are called **events**.
(ii) The Cartesian coordinate systems x^0, x^1, x^2, x^3 induced from the affine structure by arbitrarily fixing any pseudo-orthonormal basis $\mathbf{e}_0, \mathbf{e}_1, \mathbf{e}_2, \mathbf{e}_3$ in T^4 (with $\mathbf{g}(\mathbf{e}_0, \mathbf{e}_0) = -1$, $\mathbf{g}(\mathbf{e}_1, \mathbf{e}_1) = 1$, $\mathbf{g}(\mathbf{e}_2, \mathbf{e}_2) = 1$, $\mathbf{g}(\mathbf{e}_3, \mathbf{e}_3) = 1$) and any origin $O \in \mathbb{M}^4$, are called **Minkowskian coordinate systems** or **Minkowskian coordinate frames**.

■

Remarks 7.2 The elements of T^4 (as well as the vectors applied at every event) are often called *four-vectors* in the literature. ■

In practice, exploiting the affine structure and using standard notation for affine spaces, Minkowskian coordinates x^0, x^1, x^2, x^3 are defined in such a way that the map $\mathbb{M}^4 \ni p \mapsto (x^0(p), x^1(p), x^2(p), x^3(p)) \in \mathbb{R}^4$ satisfies

$$T^4 \ni p - O = x^a(p)\mathbf{e}_a$$

for every event $p \in \mathbb{M}^4$.

The pseudo-scalar product $(\cdot|\cdot)$, that is, the metric tensor $\mathbf{g}$ on T^4, has constant diagonal form in Minkowskian coordinates:

$$\mathbf{g} = -dx^0 \otimes dx^0 + dx^1 \otimes dx^1 + dx^2 \otimes dx^2 + dx^3 \otimes dx^3 \,. \tag{7.1}$$

Physically speaking, events are the minimal spacetime characterization of everything that occurs in the universe at any time. Modulo technicalities we shall examine shortly, Minkowskian coordinate frames define the class of coordinate systems of all *inertial observers* or *inertial reference systems*. Referring to the decomposition (7.1) of $\mathbf{g}$, the coordinates x^1, x^2, x^3 are thought of as "spatial coordinates", whereas the coordinate $x^0 = ct$ is a temporal coordinate. The number $c > 0$ is a constant with the dimension of a velocity whose value is that of the *speed of light* (about 3×10^5 km/s). In the following we assume a system of units such that $c = 1$.

The metric tensor $\mathbf{g}$ is used to perform measurements both in time and in space, as we clarify in the rest of this section. If $X_p \neq 0$ is a vector in $T_p\mathbb{M}^4$, it may represent either infinitesimal temporal displacements if $\mathbf{g}_p(X_p, X_p) \leq 0$, or infinitesimal spatial displacements if $\mathbf{g}_p(X_p, X_p) > 0$. In both cases $|\mathbf{g}_p(X_p, X_p)|$ has the physical meaning of the squared duration or squared length associated with X_p. Actually, the distinguished case $\mathbf{g}_p(X_p, X_p) = 0$ with $X_p \neq 0$ deserves a particular comment. These vectors represent an infinitesimal part of the history of a particle of light (we could call it a "photon", even if a photon is a quantum object).

Definition 7.3 For every event $p \in \mathbb{M}^4$, $T_p\mathbb{M}^4 \setminus \{0\}$ is decomposed into three pairwise disjoint subsets:

(i) the set of **spacelike** vectors, which satisfy $\mathbf{g}_p(X_p, X_p) > 0$;
(ii) the set of **timelike** vectors, which satisfy $\mathbf{g}_p(X_p, X_p) < 0$;
(iii) the set of **lightlike**, also called **null**, vectors, which satisfy $\mathbf{g}_p(X_p, X_p) = 0$.

The following further definitions hold.

(a) The union of the sets $\{0\}$ and of the timelike and lightlike vectors is a closed cone, $\overline{V_p}$, called the **closed light cone** at p. Its non-vanishing elements are called **causal** vectors.
(b) The interior V_p of $\overline{V_p}$ is called the **open light cone** at p (it therefore *does not* contain 0).
(c) The boundary ∂V_p is called the **light cone** at p (it therefore *does* contain 0). ■

Examples 7.4 (Timelike, Lightlike and Spacelike Vectors in $1 + 1$ Dimensions) Suppress the coordinates X^2, X^3 and work in units $c = 1$. In the plane with coordinates (X^0, X^1),

$$\mathbf{g}(X, X) = -(X^0)^2 + (X^1)^2 \,.$$

The following three vectors illustrate the classification:

$$T = (2, 1), \qquad L = (1, 1), \qquad S = (1, 2) \,.$$

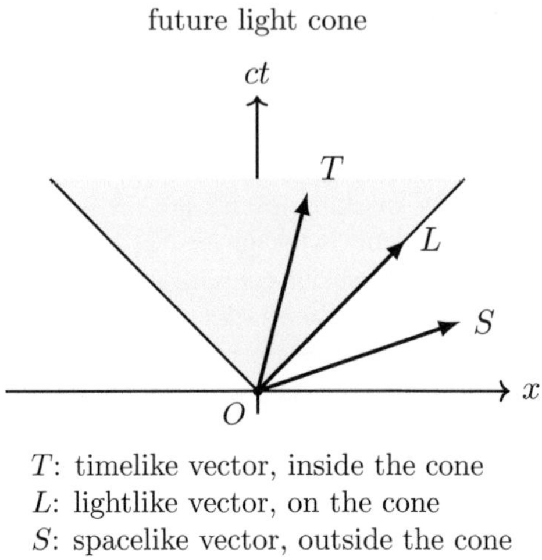

Fig. 7.1 Timelike, lightlike and spacelike vectors in a two-dimensional spacetime diagram

Indeed,

$$\mathbf{g}(T,T) = -4+1 = -3 < 0, \quad \mathbf{g}(L,L) = -1+1 = 0, \quad \mathbf{g}(S,S) = -1+4 = 3 > 0.$$

Therefore T is timelike, L is lightlike, and S is spacelike.

Figure 7.1 is only two-dimensional, but the same inequalities hold in 3 + 1 dimensions after replacing $(X^1)^2$ by $(X^1)^2 + (X^2)^2 + (X^3)^2$. ■

Exercises 7.5

1. Prove that, in Minkowski spacetime, if $\mathbf{g}(X,Y) = 0$ and X is timelike, then Y is spacelike. Show that $\mathbf{g}(X,Y) = 0$ may hold for X spacelike and Y of any type. (*Solution*. Fix a pseudo-orthonormal basis to decompose X and Y. Assume X timelike. We have

$$0 = \mathbf{g}(X,Y) = -X^0Y^0 + \sum_{k=1}^{3} X^kY^k .$$

However,

$$|X^0| > \sqrt{\sum_k (X^k)^2}$$

and

$$|X^0Y^0| = \left|\sum_{k=1}^{3} X^kY^k\right| \le \sqrt{\sum_k (X^k)^2}\sqrt{\sum_k (Y^k)^2} .$$

The only possibility is

$$|Y^0| < \sqrt{\sum_k (Y^k)^2} .$$

To prove the second statement, fix a pseudo-orthonormal basis and consider, e.g., the spacelike vector

$$X := \mathbf{e}_3 \equiv (0, 0, 0, 1) .$$

In that case $Y \equiv (1, 0, 0, 0)$ is timelike, $Y \equiv (0, 1, 0, 0)$ is spacelike, and $Y \equiv (1, 0, -1, 0)$ is lightlike, and $\mathbf{g}(X, Y) = 0$ holds in all three cases.)

2. Prove that, in Minkowski spacetime, if X is timelike, then there is a pseudo-orthonormal basis $\mathbf{e}_0, \mathbf{e}_1, \mathbf{e}_2, \mathbf{e}_3$ such that $\mathbf{e}_0$ is parallel to X.
   (*Hint*. Re-adapt, with a suitable choice of signs, the Gram-Schmidt orthogonalization procedure to a basis completing X, also taking the previous exercise into account.)

7.1.2 Time Orientation

Definition 7.6 Two continuous timelike vector fields T, T' on $\mathbb{M}^4$ are said to **have the same time orientation** if $\mathbf{g}_p(T_p, T'_p) < 0$ for every $p \in \mathbb{M}^4$. ■

To explain this fundamental definition, we start by noticing the following fact. Using a given system of Minkowskian coordinates and referring to the basis of $T_p\mathbb{M}^4$ associated with these coordinates, one sees that the open light cone is always pictured as the set

$$V_p = \{(X^0, X^1, X^2, X^3) \in \mathbb{R}^4 \setminus \{0\} \mid (X^0)^2 > (X^1)^2 + (X^2)^2 + (X^3)^2\} .$$

Thus, in those Minkowskian coordinates, V_p is made of two disjoint halves:

$$V_p^{(>)} := \{(X^0, X^1, X^2, X^3) \in V_p \mid X^0 > 0\} ,$$
$$V_p^{(<)} := \{(X^0, X^1, X^2, X^3) \in V_p \mid X^0 < 0\} . \tag{7.2}$$

Remarks 7.7 It could seem that this decomposition of V_p depends on the coordinate system used: the set $V_p^{(>)}$, for instance, could intersect both the sets $V_p^{(>)}$ and $V_p^{(<)}$ defined with respect to *another* Minkowskian coordinate system. Actually this is not the case for the following topological reason. $V_p^{(>)}$ and $V_p^{(<)}$ are open sets in $\mathbb{R}^4$ and thus in $T_p\mathbb{M}^4$, since the topology of the latter is homeomorphic to that of the former through the coordinate system itself (more precisely, through the map associating to each vector its components). Furthermore, they are disjoint connected sets. Hence they are the two connected components of the open light cone V_p. This fact does not depend on the particular Minkowskian coordinate system used. At most, changing Minkowskian coordinate system can swap $V_p^{(>)}$ and $V_p^{(<)}$. ■

Proposition 7.8 *The class $\mathscr{T}(\mathbb{M}^4)$ of continuous timelike vector fields on $\mathbb{M}^4$ satisfies the following.*

(a) *$\mathscr{T}(\mathbb{M}^4)$ is not empty.*
(b) *If $T, T' \in \mathscr{T}(\mathbb{M}^4)$, it holds either $\mathbf{g}(T_p, T'_p) < 0$ for all $p \in \mathbb{M}^4$, or $\mathbf{g}(T_p, T'_p) > 0$ for all $p \in \mathbb{M}^4$. More precisely, decomposing every V_p into the two disjoint halves (7.2),*
$\mathbf{g}(T_p, T'_p) < 0$ happens if and only if both T_p, T'_p belong to the same half of V_p,
whereas $\mathbf{g}(T_p, T'_p) > 0$ holds if and only if T_p, T'_p belong to different halves.
(c) *"to have the same time orientation" is an equivalence relation on $\mathscr{T}(\mathbb{M}^4)$, and it admits two equivalence classes only.*

Proof **(a)** $\mathscr{T}(\mathbb{M}^4)$ is not empty, since it contains the vector field ∂_{x^0} associated with any Minkowskian coordinate frame on $\mathbb{M}^4$. Let us prove **(b)**. Consider a continuous timelike vector field S. Using Minkowskian coordinates x^0, x^1, x^2, x^3 and the associated pseudo-orthonormal basis of elements $\mathbf{e}_{k,p} = \partial_{x^k}|_p \in T_p\mathbb{M}^4$, $k = 0, 1, 2, 3$, one has

$$(S_p^0)^2 > \sum_{i=1}^{3} (S_p^i)^2 \,. \tag{7.3}$$

Consider the two halves $V_p^{(>)}$ and $V_p^{(<)}$ introduced above. Since S is continuous and $\mathbb{M}^4$ is connected, S_p^0 cannot change sign as p varies in $\mathbb{M}^4$, because otherwise it would vanish somewhere, which is impossible in view of (7.3). Therefore either $S_p \in V_p^{(>)}$ for every $p \in \mathbb{M}^4$, that is

$$S_p^0 > \sqrt{\sum_{i=1}^{3} (S_p^i)^2}\,, \quad \text{for all } p \in \mathbb{M}^4\,, \tag{7.4}$$

or $S_p \in V_p^{(<)}$ for every $p \in \mathbb{M}^4$, that is

$$S_p^0 < -\sqrt{\sum_{i=1}^{3} (S_p^i)^2}\,, \quad \text{for all } p \in \mathbb{M}^4\,. \tag{7.5}$$

Now consider two continuous timelike vector fields T and T'. One has

$$\mathbf{g}_p(T_p, T'_p) = -T^0_p T'^0_p + \sum_{\alpha=1}^{3} T^\alpha_p T'^\alpha_p \,. \tag{7.6}$$

On the other hand, by the Cauchy-Schwarz inequality in $\mathbb{R}^3$,

$$\left|\sum_{\alpha=1}^{3} T^\alpha_p T'^\alpha_p\right| \leq \sqrt{\sum_{\alpha=1}^{3}(T^\alpha_p)^2}\sqrt{\sum_{\alpha=1}^{3}(T'^\alpha_p)^2} \,.$$

Taking into account that either (7.4) or (7.5) must hold for both T and T', we conclude that the sign of the right-hand side of (7.6) is the sign of $-T^0_p T'^0_p$. In other words, it holds either $\mathbf{g}_p(T_p, T'_p) < 0$ for all $p \in \mathbb{M}^4$, or $\mathbf{g}_p(T_p, T'_p) > 0$ for all $p \in \mathbb{M}^4$, and the first case happens if and only if both T and T' belong to the same half of V_p, whereas the second case occurs if and only if T and T' belong to different halves. This proves **(b)**.

Using **(b)**, we can prove **(c)**. Reflexivity follows immediately, since for every timelike vector field T one has $\mathbf{g}(T_p, T_p) < 0$ for all p. Symmetry is immediate from the symmetry of $\mathbf{g}$. Let us come to transitivity. Suppose that $\mathbf{g}(T_p, T'_p) < 0$ and $\mathbf{g}(T'_p, S_p) < 0$ for all $p \in \mathbb{M}^4$. Then, for every fixed p, both pairs (T_p, T'_p) and (T'_p, S_p) belong to the same half of V_p. Hence T_p and S_p belong to the same half of V_p as well, so that $\mathbf{g}(T_p, S_p) < 0$ for all $p \in \mathbb{M}^4$. This proves transitivity.

To conclude, notice that if T is a continuous timelike vector field on $\mathbb{M}^4$, then T and $-T$ belong to different equivalence classes, and if $\mathbf{g}(T, S) > 0$ then $\mathbf{g}(-T, S) < 0$. Therefore every continuous timelike vector field S belongs either to the equivalence class of T or to the equivalence class of $-T$. □

Definition 7.9 A pair $(\mathbb{M}^4, \mathcal{O})$, where $\mathcal{O}$ is one of the two equivalence classes of the relation "to have the same time orientation" in the set of continuous timelike vector fields on $\mathbb{M}^4$, is called a **time-oriented** Minkowski spacetime, and $\mathcal{O}$ is called a **time orientation** of $\mathbb{M}^4$. ■

There is an alternative way to fix a time orientation: from Proposition 7.8 we also conclude that:

Proposition 7.10 *The assignment of a time orientation $\mathcal{O}$ is equivalent to continuously selecting one of the two disjoint halves of V_p, for every $p \in \mathbb{M}^4$: namely, the one containing the timelike vectors which are restrictions to p of a continuous vector field in $\mathcal{O}$.*

Definition 7.11 Considering a time-oriented Minkowski spacetime $(\mathbb{M}^4, \mathcal{O})$ and an event $p \in \mathbb{M}^4$, the following definitions are valid:

(i) The half open cone at p containing the vectors which are restrictions to p of vector fields in $\mathcal{O}$ is denoted by V_p^+ and is called the **future open light cone** at p. Its elements are said to be **future-directed timelike vectors** at p.
(ii) ∂V_p^+ is called the **future light cone**, and the vectors in $\partial V_p^+ \setminus \{0\}$ are said to be **future-directed lightlike vectors** at p.
(iii) $\overline{V_p^+} := V_p^+ \cup \partial V_p^+$ is called the **closed future light cone**, and the vectors in $\overline{V_p^+} \setminus \{0\}$ are said to be **future-directed causal vectors**.

■

Remarks 7.12 We henceforth suppose to deal with a time-oriented Minkowski spacetime and indicate it simply by $\mathbb{M}^4$. ■

As a final technical proposition we have the following elementary though interesting result.

Proposition 7.13 *If A, B are causal vectors, then the* ***inverse Cauchy-Schwarz inequality*** *holds:*

$$|\mathbf{g}(A, B)| \geq \sqrt{-\mathbf{g}(A, A)}\sqrt{-\mathbf{g}(B, B)}\,.$$

If, in addition, the vectors are both future-directed or both past-directed, then

$$|\mathbf{g}(A, B)| = -\mathbf{g}(A, B)\,.$$

Proof It is sufficient to prove the thesis for A, B causal and future-directed. (If they are simply causal, the sign of $\mathbf{g}(A, B)$ may change, but the final result does not, because of the absolute value.) So we assume that A and B are future-directed. Then

$$-A_\mu B^\mu = A^0 B^0 - \vec{A} \cdot \vec{B} \geq A^0 B^0 - |\vec{A} \cdot \vec{B}| \geq A^0 B^0 - ||\vec{A}||||\vec{B}||$$

$$\geq \sqrt{\left((A^0)^2 - ||\vec{A}||^2\right)\left((B^0)^2 - ||\vec{B}||^2\right)} = \sqrt{-A_\mu A^\mu}\sqrt{-B_\mu B^\mu}\,.$$

The last inequality follows from the fact that $A^0, B^0 > 0$, and from the equivalent inequality

$$(A^0 B^0 - ||\vec{A}||||\vec{B}||)^2 \geq \left((A^0)^2 - ||\vec{A}||^2\right)\left((B^0)^2 - ||\vec{B}||^2\right),$$

whose proof is immediate, since it boils down to

$$(A^0||\vec{B}|| - B^0||\vec{A}||)^2 \geq 0\,.$$

□

7.2 Kinematics I: Basic Notions

We now pass to discuss the physical interpretation of some geometrical structures of Minkowski spacetime in terms of kinematical notions.

7.2.1 Curves as Histories or World Lines

If $I \subset \mathbb{R}$ is an open interval, a C^1 curve (Definition 3.14) $\gamma : I \to \mathbb{M}^4$ may represent the history of a particle of matter, also called its **world line**, provided its tangent vector (Definition 3.16) is causal and future-directed. A particle of light, in particular, has a world line represented by a curve with future-directed lightlike tangent vector.

Definition 7.14 A C^1 curve $\gamma : I \to \mathbb{M}^4$ is said to be **timelike**, **spacelike**, **causal**, or **lightlike** if all of its tangent vectors γ'_p, $p \in \gamma$, are respectively timelike, spacelike, causal, or lightlike.

A causal, timelike, or lightlike curve γ with future-directed tangent vectors is said to be a **future-directed** (resp. **causal**, **timelike**, **lightlike**) curve. ■

Notice that $\gamma'(\xi) \neq 0$ necessarily for $\xi \in I$ when γ is causal. However, in that case, $|\mathbf{g}(\gamma'(\xi), \gamma'(\xi))| \neq 0$ only if $\gamma'(\xi)$ is timelike.

Definition 7.15 If $\gamma = \gamma(\xi)$, $\xi \in I$ with $I \subset \mathbb{R}$ any interval of $\mathbb{R}$, is a C^1 future-directed timelike curve in $\mathbb{M}^4$, and $\xi_0 \in I$, then the function

$$\tau(\xi) := \int_{\xi_0}^{\xi} \sqrt{|\mathbf{g}(\gamma'(l), \gamma'(l))|}dl \tag{7.7}$$

is called the **proper time** of γ. The tangent vector obtained by reparametrizing the curve with the proper time, and denoted by $\dot{\gamma}$ (instead of γ'), is called the **four-velocity** of the curve. ■

Notice that the proper time is independent of the parametrization of γ, but it is defined up to the choice of the origin, that is, the point on γ where $\tau = 0$.

From the point of view of physics, when one considers timelike curves, proper time is the *time measured by an ideal clock at rest with the particle whose world line is* γ.

For timelike curves, proper time can be used as a natural parameter to describe the history γ of the particle because, directly from (7.7), one has

$$\frac{d\tau}{d\xi} > 0\,.$$

Notice also that, if $\dot{\gamma}$ is a four-velocity, then (7.7) implies that

$$\mathbf{g}(\dot{\gamma}, \dot{\gamma}) = -1 , \tag{7.8}$$

(where -1 is replaced by $-c^2$ if one restores the value of the speed of light c), that is, a four-velocity is *unit*. Indeed,

$$\frac{d\gamma}{d\tau} = \frac{d\gamma}{d\xi}\left(\frac{d\tau}{d\xi}\right)^{-1} = \frac{d\gamma}{d\xi}\left(\sqrt{|\mathbf{g}(\gamma'(\xi), \gamma'(\xi))|}\right)^{-1}$$

which produces (7.8), by noticing that $\mathbf{g}(\gamma'(\xi), \gamma'(\xi)) < 0$ along a timelike curve.

Remarks 7.16 An important fact is that future-directed causal curves can be parametrized by the x^0 coordinate of any Minkowskian system of coordinates. This is because, if the curve is represented as

$$I \ni \xi \mapsto (x^0(\xi), x^1(\xi), x^2(\xi), x^3(\xi))$$

in those coordinates, then

$$0 \geq \mathbf{g}(\gamma'(\xi), \gamma'(\xi)) = -\left(\frac{dx^0}{d\xi}\right)^2 + \sum_{k=1}^{3}\left(\frac{dx^k}{d\xi}\right)^2$$

implies $\frac{dx^0}{d\xi} \neq 0$ (the only other possibility would be $\gamma'(\xi) = 0$ somewhere, but this is excluded because γ' is causal), so that we can reparametrize the curve using x^0 instead of ξ. More precisely, if γ' and ∂_{x^0} have the same time orientation, i.e., if also ∂_{x^0} is future-directed, then

$$-\mathbf{g}(\gamma', \partial_{x^0}) = \frac{dx^0}{d\xi} > 0$$

so that, if x^0 increases, then $\gamma(\xi(x^0))$ evolves forward in time. ∎

From a physical point of view, future-directed causal curves which are straight lines (segments) with respect to the affine structure of $\mathbb{M}^4$ are thought of as the histories of particles not subjected to forces. *These are particles evolving with inertial motion.*

7.2.2 Minkowskian Reference Systems and Some Kinematical Effects

If x^0, x^1, x^2, x^3 are Minkowskian coordinates, the curves tangent to x^0 determine a constant vector field ∂_{x^0}, which is unit timelike, that is, it satisfies $\mathbf{g}(\partial_{x^0}, \partial_{x^0}) = -1$. It is constant in the sense that, by definition of Cartesian coordinates, it does not depend on the point in the affine space. A unit timelike constant vector field can always be viewed as the tangent vector to the coordinate x^0 of a Minkowskian coordinate frame. There are however *several* Minkowskian coordinate frames associated in this way with a given unit timelike constant vector field. They differ from each other only by the choice of the origin.

Definition 7.17 A constant timelike vector field $\mathscr{R}$ on $\mathbb{M}^4$ is said to be a **(Minkowskian) reference system** provided it is future-directed and unit timelike, i.e. $\mathbf{g}(\mathscr{R}, \mathscr{R}) = -1$. Furthermore the following definitions hold.

(a) Any Minkowskian coordinate system such that $\partial_{x^0} = \mathscr{R}$ is said to be **co-moving with** (or equivalently **adapted to**) $\mathscr{R}$.
(b) If $\gamma = \gamma(\xi)$, where $\xi \in \mathbb{R}$, is an integral curve of $\mathscr{R}$ (an affine straight line), the length function

$$t_{\mathscr{R}}(\gamma(\xi)) := \int_0^{\xi} \sqrt{|\mathbf{g}(\mathscr{R}_{\gamma(l)}, \mathscr{R}_{\gamma(l)})|}dl$$

defines a **time coordinate for $\mathscr{R}$ (along γ)**.

■

For a fixed $\mathscr{R}$, an adapted Minkowskian coordinate frame can be obtained by fixing a three-dimensional affine plane Σ orthogonal to $\mathscr{R}$ at some event O, and fixing a pseudo-orthonormal basis in $T_O\mathbb{M}^4$ such that $\mathbf{e}_0 = \mathscr{R}_O$ and $\mathbf{e}_1, \mathbf{e}_2, \mathbf{e}_3$ are parallel to Σ. One proves straightforwardly that the Minkowskian coordinate frame associated with $O, \mathbf{e}_0, \mathbf{e}_1, \mathbf{e}_2, \mathbf{e}_3$ is co-moving with $\mathscr{R}$ and, along the chosen integral curve γ of $\mathscr{R}$ through O, the coordinate x^0 coincides with $t_{\mathscr{R}}$ up to an additive constant in units where $c = 1$; if c is explicitly restored, $x^0 = ct_{\mathscr{R}}$ up to an additive constant. Notice that x^0 defines in this way a **global time coordinate** associated with $\mathscr{R}$, since the Cartesian coordinate x^0 associates a time coordinate value

$$t_{\mathscr{R}}(p) := x^0(p)$$

to every event $p \in \mathbb{M}^4$, not only to those along the initially chosen curve γ tangent to $\mathscr{R}$. Moreover, if a global time coordinate x^0 is defined for $\mathscr{R}$, every three-dimensional affine plane orthogonal to $\mathscr{R}$ contains only points with the same value of that time coordinate.

Definition 7.18 Let $\mathscr{R}$ be a reference system in $\mathbb{M}^4$ and let t be the physical time associated with a Minkowskian coordinate system co-moving with $\mathscr{R}$; equivalently,

$x^0 = ct$ when c is explicitly restored, and $x^0 = t$ in units where $c = 1$. Any three-dimensional affine plane $\Sigma_t^{(\mathscr{R})}$ orthogonal to $\mathscr{R}$, where $t \in \mathbb{R}$ is the value of that time coordinate of the points in the plane, is called the **rest space of $\mathscr{R}$ at time t.** ■

From the point of view of differential geometry, by construction, every rest space $\Sigma_t^{(\mathscr{R})}$ of $\mathscr{R}$ is an embedded three-dimensional submanifold of $\mathbb{M}^4$. Furthermore, the metric induced by $\mathbf{g}$ on $\Sigma_t^{(\mathscr{R})}$—i.e. the scalar product of vectors tangent to $\Sigma_t^{(\mathscr{R})}$—is positive definite. This follows immediately because that scalar product acts on the span of three orthonormal vectors $\mathbf{e}_1, \mathbf{e}_2, \mathbf{e}_3$ which are separately orthogonal to $\mathbf{e}_0 = \mathscr{R}$, and we know that

$$\mathbf{g}(\mathbf{e}_\alpha, \mathbf{e}_\beta) = \delta_{\alpha\beta} \quad \text{if } \alpha, \beta = 1, 2, 3.$$

From the point of view of physics, a Minkowski frame $\mathscr{R}$ is nothing but an *inertial reference system*. Furthermore:

(a) any global time coordinate of $\mathscr{R}$ constructed out of $\mathbf{g}$ is the *time measured by ideal clocks at rest with the reference system*;
(b) the metric induced by $\mathbf{g}$ on the rest spaces of $\mathscr{R}$ is the mathematical tool corresponding to *physical spatial measurements performed by rigid rulers at rest with the reference system.*

This mathematical picture is sufficiently developed to allow one to define the notion of *velocity of a particle* represented by a C^1 future-directed causal curve γ, *with respect to a reference system $\mathscr{R}$ at time t*. The procedure is straightforward. Consider the event

$$e := \Sigma_t^{(\mathscr{R})} \cap \gamma$$

where the curve intersects the rest space $\Sigma_t^{(\mathscr{R})}$ (each rest space intersects γ in exactly one point). Then $T_e\mathbb{M}^4$ is decomposed as the orthogonal direct sum

$$T_e\mathbb{M}^4 = \mathrm{Span}(\mathscr{R}_e) \oplus T_e\Sigma_t^{(\mathscr{R})}\,, \tag{7.9}$$

$\mathrm{Span}(\mathscr{R}_e)$ being the linear space spanned by the vector $\mathscr{R}_e$. As a consequence, the tangent vector γ'_e can be uniquely decomposed as

$$\gamma'_e = \delta x^0 \mathscr{R}_e + \delta X\,, \tag{7.10}$$

where $\delta X \in T_e\Sigma_t^{(\mathscr{R})}$ and $\delta x^0 \in \mathbb{R}$. The fact that γ'_e is causal prevents δx^0 from vanishing, and the fact that γ is future-directed implies that $\delta x^0 > 0$, so that it makes sense to give the following definition of velocity.

Definition 7.19 Let γ be a C^1 future-directed causal curve and let $\mathscr{R}$ be a reference system in $\mathbb{M}^4$. The **velocity of** γ with **respect to** $\mathscr{R}$ **at time** t is the vector $\mathbf{v}_{\gamma,t,\mathscr{R}} \in T_{\Sigma_t^{(\mathscr{R})}\cap\gamma}\Sigma_t^{(\mathscr{R})}$, given, when the speed of light is explicitly restored, by

$$\mathbf{v}_{\gamma,t,\mathscr{R}} := c\frac{\delta X}{\delta x^0} = \frac{\delta X}{\delta t},$$

and by $\delta X/\delta x^0$ in units where $c = 1$, referring to (7.9) and (7.10) with $e := \Sigma_t^{(\mathscr{R})} \cap \gamma$. ■

Notice that the ratio above does not depend on the parametrization of γ because, changing parametrization from ξ to ξ', the components of γ' would be multiplied by the same factor $\frac{d\xi}{d\xi'}$, and these factors would cancel when computing the ratio defining the velocity. Within this framework one has the following physically well-known result (where we explicitly write the value c of the speed of light).

Proposition 7.20 *Consider a future-directed causal curve γ and fix a reference system $\mathscr{R}$.*

(a) *The absolute value of $\mathbf{v}_{\gamma,t,\mathscr{R}}$ (the* speed *of γ at time t) is bounded above by c, and this value is attained only at those events along γ where the tangent vector of γ is lightlike.*

(b) *If γ is timelike and $\dot\gamma(t)$ is the four-velocity of γ at time t with respect to $\mathscr{R}$, then*

$$\dot\gamma(t) = \frac{c}{\sqrt{1-\mathbf{v}_{\gamma,t,\mathscr{R}}{}^2/c^2}}\mathscr{R} + \frac{\mathbf{v}_{\gamma,t,\mathscr{R}}}{\sqrt{1-\mathbf{v}_{\gamma,t,\mathscr{R}}{}^2/c^2}}. \tag{7.11}$$

(c) *If γ intersects the rest spaces $\Sigma_{t_1}^{(\mathscr{R})}$ and $\Sigma_{t_2}^{(\mathscr{R})}$ with $t_2 > t_1$ at proper times τ_1 and $\tau_2 > \tau_1$ respectively, the corresponding intervals of coordinate time and proper time satisfy*

$$\Delta\tau = \int_{t_1}^{t_2}\sqrt{1-\mathbf{v}_{\gamma,t,\mathscr{R}}{}^2/c^2}\,dt \tag{7.12}$$

and so $\Delta\tau < \Delta t$ unless $\mathbf{v}_{\gamma,t,\mathscr{R}} = 0$ throughout the considered interval of time, in which case $\Delta\tau = \Delta t$.

Remarks 7.21

(1) The absolute value $||\mathbf{v}_{\gamma,t,\mathscr{R}}||^2 = (\mathbf{v}_{\gamma,t,\mathscr{R}})^2$ is referred to the scalar product induced by $\mathbf{g}$ on the rest spaces of $\mathscr{R}$. As previously said, this is the physical metric tool used to perform spatial measurements.

(2) The value of $||\mathbf{v}_{\gamma,t,\mathscr{R}}||$ evaluated along curves with $\gamma'_e = \delta x^0\mathscr{R}_e + \delta X$ such that $\mathbf{g}(\gamma'_e, \gamma'_e) = 0$ is the same for *every* reference system. This value is denoted by c, as already said, and is sometimes put equal to 1 for the sake of simplicity.

It coincides, in particular, with the universal value of the speed of particles of light.

■

Proof of Proposition 7.20 Using Definition 7.19, decomposition (7.10), the orthogonality of δX and $\mathscr{R}$, and the fact that $\mathscr{R}$ is unit timelike, one has

$$0 \geq \mathbf{g}(\gamma', \gamma') = -(\delta x^0)^2 + \mathbf{g}(\delta X, \delta X) = (\delta x^0)^2 \left(-1 + \frac{||\mathbf{v}_{\gamma,t,\mathscr{R}}||^2}{c^2}\right).$$

The sign $=$ occurs only if γ' is lightlike. This immediately implies **(a)**. For **(b)** we write v in place of $\mathbf{v}_{\gamma,t,\mathscr{R}}$. Now explicitly assume that the curve is timelike and parametrized by proper time. By definition, if $\partial_{x^0} = \mathscr{R}$,

$$\dot{\gamma} = \dot{\gamma}^0 \partial_{x^0} + \sum_{\alpha=1}^{3} \dot{\gamma}^\alpha \partial_{x^\alpha}$$

where

$$-(\dot{\gamma}^0)^2 + \sum_{\alpha=1}^{3} (\dot{\gamma}^\alpha)^2 = -c^2,$$

and

$$v^\alpha = c\frac{\dot{\gamma}^\alpha}{\dot{\gamma}^0}.$$

Therefore

$$-1 + \sum_{\alpha=1}^{3} \left(\frac{v^\alpha}{c}\right)^2 = -\frac{c^2}{(\dot{\gamma}^0)^2},$$

and so

$$\dot{\gamma}^0 = \frac{c}{\sqrt{1 - v^2/c^2}} \quad \text{and} \quad \dot{\gamma}^\alpha = \frac{v^\alpha}{\sqrt{1 - v^2/c^2}}.$$

This yields (7.11).

Finally, **(c)** is a straightforward consequence of **(b)**, since the factor in front of $\mathscr{R}$ in (7.11) is

$$c\frac{dt}{d\tau} = \frac{c}{\sqrt{1 - \mathbf{v}_{\gamma,t,\mathscr{R}}^2/c^2}}.$$

Hence

$$\frac{d\tau}{dt} = \sqrt{1 - \mathbf{v}_{\gamma,t,\mathscr{R}}{}^2/c^2}\,,$$

which gives (7.12) after integration. □

We have a very interesting geometrical consequence.

Proposition 7.22 *The following facts are true in Minkowski spacetime.*

(a) *A timelike segment joining a pair of events in Minkowski spacetime is the unique* longest *causal curve joining these events.*
(b) *If a pair of events can be joined by a spacelike segment, then the infimum of the lengths of all C^1 curves joining those events is* 0.

Proof To prove **(a)** it is sufficient to adopt a Minkowskian coordinate system with ∂_{x^0} parallel to $\mathscr{R}$. With this choice, $\Delta\tau$ computed in (7.12) of Proposition 7.20 reduces to

$$\Delta\tau = \Delta t = t_2 - t_1$$

for the timelike segment joining the given events, because by construction those events lie on the same coordinate line tangent to ∂_{x^0}. If, keeping the same endpoints, we replace the segment by another causal curve γ joining them, we see that, since

$$\sqrt{1 - \mathbf{v}_{\gamma,t,\mathscr{R}}{}^2/c^2} \leq 1\,,$$

it necessarily holds

$$\Delta\tau \leq \Delta t\,,$$

which proves the assertion. The segment is the unique longest curve because, if $\Delta t = \Delta\tau$, then

$$\int_{t_1}^{t_2} \left(1 - \sqrt{1 - \mathbf{v}_{\gamma,t,\mathscr{R}}{}^2/c^2}\right) dt = 0\,.$$

Since the integrand is continuous and non-negative, this is possible only if the integrand vanishes identically. In other words,

$$\mathbf{v}_{\gamma,t,\mathscr{R}} = 0$$

along the whole curve. This implies that the curve is a straight timelike segment parallel to ∂_{x^0}, hence it coincides with the segment joining the two endpoints.

The proof of **(b)** is elementary. Let A and B be the two given events. Fix a Minkowskian coordinate system such that ∂_{x^0} is orthogonal to the spacelike segment joining them, and arrange the origin of the coordinates at the midpoint of the segment AB in such a way that AB lies along the axis x^1, with

$$x^1(A) = -\ell\,, \quad x^1(B) = \ell\,.$$

Consider a family of C^1 curves joining A and B which approximate arbitrarily well a broken continuous curve made of two lightlike segments AT and TB, where (assuming $c = 1$)

$$T \equiv (\ell, 0, 0, 0)\,.$$

The lengths of these lightlike segments are zero, proving the thesis. □

Examples 7.23

1. One recognizes in **(c)** of Proposition 7.20 the mathematical formulation of the celebrated relativistic phenomenon called *time dilation*. In particular, when the absolute value of $\mathbf{v}_{\gamma,t,\mathscr{R}}$ is constantly v, **(c)** specializes to

$$\Delta t = \frac{\Delta\tau}{\sqrt{1 - \frac{v^2}{c^2}}} \tag{7.13}$$

where $\Delta\tau$ is an interval of time measured—between the rest spaces $\Sigma_{t_1}^{(\mathscr{R})}$ and $\Sigma_{t_2}^{(\mathscr{R})}$—by a clock at rest with the particle moving with speed v in the reference system $\mathscr{R}$, whereas $\Delta t = t_2 - t_1 = \Delta x^0/c$ is the corresponding interval of time measured by a clock at rest in the reference system.

   This phenomenon is well known in particle physics. There are unstable particles called *pions* produced in Earth's atmosphere by the interaction of atmospheric particles and cosmic rays. If produced in a laboratory, these particles prove to have a certain known lifetime $\Delta\tau$. This lifetime is not enough to reach the ground from the regions of the upper atmosphere where they are produced, even if they moved with the speed of light. However these particles are detected at ground level. The explanation relies upon **(c)** of Proposition 7.20. On the one hand, $\Delta\tau$ is a feature of the particle measured at rest with the particle. On the other hand, the corresponding and larger interval of time Δt necessary to reach the ground, computed in (7.13) and referred to the rest frame of Earth, is larger than $\Delta\tau$ by enough to allow the pion with speed $v < c$ to reach the ground before decaying.[1]
2. The so-called *twin paradox* arises when one (wrongly) applies the time dilation phenomenon to the case of a pair of twin brothers, one of whom makes a journey into space in a high-speed rocket and returns home to find that the twin who

[1] Our planet cannot be considered an inertial reference system, but the corrections due to this fact can be proved to be negligible in this case.

remained at rest has aged more. The paradox consists in noticing that the point of view can be reversed by arguing that, from the reference frame of the moving brother, actually the other brother made a journey at high speed. Hence the younger twin should be the other one. The paradox is only apparent, because it is impossible for the twins to meet twice while both describing straight timelike worldlines in spacetime. At most one of them remained at rest in an inertial frame, and by Proposition 7.22 he is the older one at their second meeting. If none of them remained at rest in an inertial reference frame, the difference of age observed at their next meeting can be computed by integrating the lengths of their worldlines as in (7.12), and the result is simply a matter of computation.

3. The *Hafele-Keating* experiment was a famous test of special relativity. In October 1971, J.C. Hafele and R.E. Keating took four cesium-beam atomic clocks aboard commercial airliners. They flew twice around the world, first eastward, then westward, and compared the clocks with others that remained at the United States Naval Observatory. When reunited, the three sets of clocks were found to disagree with one another, and their differences were consistent with the predictions of special (and general) relativity.
4. Another interesting relativistic phenomenon is the so-called *length contraction*. Consider a rigid ruler with rest length L_0, measured in its rest frame, moving with constant velocity $\mathbf{v} = v\mathbf{e}_1$ with respect to the inertial reference system $\mathscr{R}$ (assuming that $\mathbf{e}_1$ is one of the spacelike vectors used to define a Minkowskian coordinate system co-moving with $\mathscr{R}$). What is the length L of the ruler measured in $\mathscr{R}$? Figure 7.2 illustrates the geometric situation. To derive the result, fix two events on the worldlines of the two endpoints of the ruler which are *simultaneous in* $\mathscr{R}$; their spatial distance in $\mathscr{R}$ is, by definition, L. Let (ct, x) be coordinates in $\mathscr{R}$ and (ct', x') coordinates in the rest frame $\mathscr{R}'$ of the ruler. Since the ruler is at rest in $\mathscr{R}'$, the spatial distance between its endpoints in $\mathscr{R}'$ is the constant value L_0. Therefore, for the two events chosen above, one has

$$\Delta x' = L_0 \, .$$

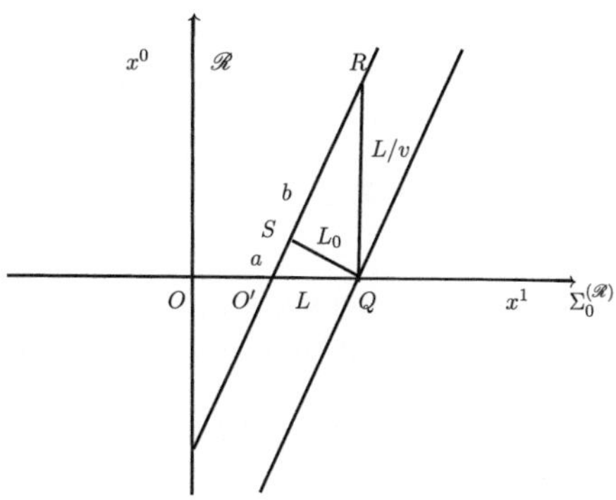

Fig. 7.2 Phenomenon of length contraction

However, those two events are not simultaneous in $\mathscr{R}'$. Since they are simultaneous in $\mathscr{R}$, one has

$$\Delta t = 0\,.$$

Using the Lorentz transformation between $\mathscr{R}'$ and $\mathscr{R}$ along the x^1 direction,

$$\Delta t = \gamma\left(\Delta t' + \frac{v}{c^2}\Delta x'\right), \qquad \Delta x = \gamma(\Delta x' + v\Delta t'),$$

where

$$\gamma = \frac{1}{\sqrt{1 - v^2/c^2}}\,,$$

the condition $\Delta t = 0$ gives

$$\Delta t' = -\frac{v}{c^2}\Delta x' = -\frac{vL_0}{c^2}\,.$$

Substituting into the second relation, we find

$$L = \Delta x = \gamma\left(L_0 - \frac{v^2}{c^2}L_0\right) = \gamma L_0\left(1 - \frac{v^2}{c^2}\right),$$

hence

$$L = L_0\sqrt{1 - \frac{v^2}{c^2}}\,. \tag{7.14}$$

Therefore $L < L_0$ as soon as $0 < v < c$. This is the well-known phenomenon of length contraction.

■

7.3 Kinematics II: Lorentz and Poincaré Groups

The following natural question arises: if $\mathscr{R}$ and $\mathscr{R}'$ are two Minkowskian reference systems equipped with co-moving Minkowskian coordinate systems x^0, x^1, x^2, x^3 and x'^0, x'^1, x'^2, x'^3 respectively, what is the most general relation between these different systems of coordinates?

From now on we exploit again the convention of summation over repeated indices. Since both coordinate frames are Minkowskian and, in turn, are Cartesian coordinate frames, the relation must be affine:

$$x'^a = \Lambda^a{}_b x^b + T^a \,, \tag{7.15}$$

the requirement of nonsingularity of the Jacobian determinant is equivalent to the nonsingularity of the matrix Λ of coefficients $\Lambda^\mu{}_\nu$. Finally, the requirement that in both coordinate systems $\mathbf{g}$ must have the form (7.1), i.e.

$$\mathbf{g} = \eta_{ab} dx^a \otimes dx^b = \eta_{ab} dx'^a \otimes dx'^b \tag{7.16}$$

where we have introduced the matrix $\eta = diag(-1, +1, +1, +1)$ of coefficients η_{ab}, leads to the requirement

$$\Lambda^t \eta \Lambda = \eta \,. \tag{7.17}$$

Notice that this relation implies the nonsingularity of matrices Λ because, taking the determinant of both sides, one gets:

$$(\det \Lambda)^2 \det \eta = \det \eta \,,$$

where we exploited $\det(\Lambda^t \eta \Lambda) = \det(\Lambda^t) \det(\eta) \det(\Lambda) = \det(\Lambda) \det(\eta) \det(\Lambda)$. Since $\det \eta = -1$, it must be $\det \Lambda = \pm 1$. Conversely, one sees that if x'^0, x'^1, x'^2, x'^3 is a Minkowskian coordinate frame and (7.15) holds with Λ satisfying (7.17), then x^0, x^1, x^2, x^3 is a Minkowskian coordinate frame too. Summarizing, one has the following straightforward result.

Proposition 7.24 *If x^0, x^1, x^2, x^3 and x'^0, x'^1, x'^2, x'^3 are Minkowskian coordinate systems on $\mathbb{M}^4$, the transformation rule between these coordinates has the form (7.15), where T^a, $a = 0, 1, 2, 3$, are suitable real constants and the matrix Λ satisfies (7.17).*

Conversely, if x'^0, x'^1, x'^2, x'^3 is a Minkowskian coordinate system and (7.15) holds for arbitrary real constants T^a, $a = 0, 1, 2, 3$, and an arbitrary matrix Λ satisfying (7.17), then x^0, x^1, x^2, x^3 is another Minkowskian coordinate system.

We stress that up to now nothing has been said about time orientation.

7.3.1 The Lorentz Group

The result proved above allows one to introduce the celebrated Lorentz group. In the following, $M(4, \mathbb{R})$ will denote the algebra of real 4×4 matrices and $GL(4, \mathbb{R})$ (see

Sect. 4.1) indicates the group of real 4×4 matrices with non-vanishing determinant. Consider the set of matrices

$$O(1,3) \quad := \{\Lambda \in M(4,\mathbb{R}) \mid \Lambda^t \eta \Lambda = \eta\}. \tag{7.18}$$

It is a subgroup of $GL(4,\mathbb{R})$. To prove this, it is sufficient to verify that it is closed with respect to matrix multiplication, and this is trivial from (7.18) using the fact that $\eta\eta = I$. The corresponding statement for inverses will be proved in Proposition 7.26(c). We are now in a position to give the following definition.

Definition 7.25 (Lorentz Group) The **Lorentz group** is the group of matrices, with group structure induced by that of $GL(4,\mathbb{R})$,

$$O(1,3) \quad := \{\Lambda \in M(4,\mathbb{R}) \mid \Lambda^t \eta \Lambda = \eta\}.$$

■

The next technical proposition will allow us to introduce some physically relevant subgroups of $O(1,3)$ later.

Proposition 7.26 *The Lorentz group enjoys the following properties.*

(a) η, $-I$, *and* $-\eta$ *belong to* $O(1,3)$.
(b) $\Lambda \in O(1,3)$ *if and only if* $\Lambda^t \in O(1,3)$.
(c) *If* $\Lambda \in O(1,3)$, *then* $\Lambda^{-1} = \eta\Lambda^t\eta$.
(d) *If* $\Lambda \in O(1,3)$, *then* $\det\Lambda = \pm 1$. *In particular, if* $\Lambda, \Lambda' \in O(1,3)$ *and* $\det\Lambda = \det\Lambda' = 1$, *then* $\det(\Lambda\Lambda') = 1$ *and* $\det(\Lambda^{-1}) = 1$. *As* $\det I = 1$ *as well, the set of matrices* $\Lambda \in O(1,3)$ *with* $\det\Lambda = 1$ *forms a subgroup of* $O(1,3)$ *denoted by* $SO(1,3)$ *and called the* ***special Lorentz group***.
(e) *If* $\Lambda \in O(1,3)$, *then either* ${\Lambda^0}_0 \geq 1$ *or* ${\Lambda^0}_0 \leq -1$. *In particular, if* $\Lambda, \Lambda' \in O(1,3)$ *and* ${\Lambda^0}_0 \geq 1$ *and* ${\Lambda'^0}_0 \geq 1$, *then* ${(\Lambda\Lambda')^0}_0 \geq 1$ *and* ${(\Lambda^{-1})^0}_0 \geq 1$. *As* ${I^0}_0 = 1$ *as well, the set of matrices* $\Lambda \in O(1,3)$ *with* ${\Lambda^0}_0 \geq 1$ *forms a subgroup of* $O(1,3)$ *indicated by* $O(1,3)\uparrow$ *and called the* ***orthochronous Lorentz group***.

Proof The proof of (a) is immediate from (7.18) also using $\eta\eta = I$. To prove (b), we start from $\Lambda^t\eta\Lambda = \eta$. Since Λ is nonsingular, Λ^{-1} exists and one has $\Lambda^t\eta\Lambda\Lambda^{-1} = \eta\Lambda^{-1}$, namely $\Lambda^t\eta = \eta\Lambda^{-1}$. Therefore, applying η on the right, $\Lambda^t = \eta\Lambda^{-1}\eta$. Finally, applying $\Lambda\eta$ on the left one gets

$$\Lambda\eta\Lambda^t = \eta\,,$$

so $\Lambda^t \in O(1,3)$ if $\Lambda \in O(1,3)$.

To show (c), we notice that $\eta\Lambda^t\eta \in O(1,3)$ because this set is closed with respect to the composition of matrices and η, $\Lambda^t \in O(1,3)$ by (a) and (b). Finally,

$$\Lambda(\eta\Lambda^t\eta) = (\Lambda\eta\Lambda^t)\eta = \eta\eta = I\,.$$

Since every $\Lambda \in O(1,3)$ is nonsingular as noticed below (7.17), we can conclude that $\eta\Lambda^t\eta = \Lambda^{-1}$ and also that $\Lambda^{-1} \in O(1,3)$ if $\Lambda \in O(1,3)$.

The first part of (d) has been proved previously. The remaining part is straightforward: $\det(\Lambda\Lambda') = (\det\Lambda)\cdot(\det\Lambda') = 1\cdot 1 = 1$ and $\det(\Lambda^{-1}) = (\det\Lambda)^{-1} = (1)^{-1} = 1$.

Let us conclude the proof by demonstrating (e), whose proof is a bit more involved. The constraint $(A^t\eta A)_{00} = \eta_{00}$ gives:

$$(A^0{}_0)^2 = 1 + \sum_{\alpha=1}^{3} A^\alpha{}_0 A^\alpha{}_0\,, \tag{7.19}$$

so that $A^0{}_0 \geq 1$ or $A^0{}_0 \leq -1$ if $A \in O(1,3)$. This proves the first statement of (e) if $\Lambda = A$. Let us pass to the second part. Suppose that $\Lambda, \Lambda' \in O(1,3)$ and $\Lambda^0{}_0 \geq 1$ and $\Lambda'^0{}_0 \geq 1$; we want to show that $(\Lambda\Lambda')^0{}_0 \geq 1$. Actually it is sufficient to show that $(\Lambda\Lambda')^0{}_0 > 0$ because of the first statement. We shall prove it.

We start from the identity

$$(\Lambda\Lambda')^0{}_0 = \Lambda^0{}_0\Lambda'^0{}_0 + \sum_{\alpha=1}^{3} \Lambda^0{}_\alpha\Lambda'^\alpha{}_0\,.$$

It can be rewritten as

$$(\Lambda\Lambda')^0{}_0 = (\Lambda^t)^0{}_0\Lambda'^0{}_0 + \sum_{\alpha=1}^{3} (\Lambda^t)^\alpha{}_0\Lambda'^\alpha{}_0 \tag{7.20}$$

Using Cauchy-Schwarz' inequality:

$$\left|\sum_{\alpha=1}^{3}(\Lambda^t)^\alpha{}_0\Lambda'^\alpha{}_0\right|^2 \leq \left(\sum_{\alpha=1}^{3}(\Lambda^t)^\alpha{}_0(\Lambda^t)^\alpha{}_0\right)\left(\sum_{\beta=1}^{3}\Lambda'^\beta{}_0\Lambda'^\beta{}_0\right) \leq \left((\Lambda^t)^0{}_0\Lambda'^0{}_0\right)^2,$$

where, in the last passage, we have used (7.19), which is valid both for Λ and Λ^t (since the Lorentz group is closed under transposition as already established), and furthermore $(\Lambda^t)^0{}_0 = \Lambda^0{}_0 > 0$ and $\Lambda'^0{}_0 > 0$ by hypothesis. We conclude from (7.20) that the sign of $(\Lambda\Lambda')^0{}_0$ must be the sign of $(\Lambda^t)^0{}_0\Lambda'^0{}_0$, i.e., strictly positive as wanted.

To prove the last statement, notice that, if $\Lambda^0{}_0 \geq 1$, then from (c),

$$(\Lambda^{-1})^0{}_0 = (\eta\Lambda^t\eta)^0{}_0 = \Lambda^0{}_0 \geq 1\,.$$

□

7.3.2 The Poincaré Group

Considering the complete transformation (7.15) we can introduce the celebrated Poincaré group, also called the inhomogeneous Lorentz group. To do it, we notice that the set

$$IO(1,3) := O(1,3) \times \mathbb{R}^4 , \tag{7.21}$$

is a group when equipped with the composition rule

$$(\Lambda, T) \circ (\Lambda', T') := (\Lambda\Lambda', T + \Lambda T') . \tag{7.22}$$

The proof of this fact is immediate.

Definition 7.27 The **Poincaré group** or **inhomogeneous Lorentz group** is the set

$$IO(1,3) := O(1,3) \times \mathbb{R}^4 ,$$

equipped with the composition rule

$$(\Lambda, T) \circ (\Lambda', T') := (\Lambda\Lambda', T + \Lambda T') .$$

■

We make only a few remarks.

Remarks 7.28 The composition rule (7.22) is nothing but that obtained by composing the two transformations of Minkowskian coordinates:

$$x_1^a = \Lambda^a{}_b x_2^b + T^a \quad \text{and } x_2^b = \Lambda'^b{}_c x_3^c + T'^b$$

obtaining

$$x_1^a = (\Lambda\Lambda')^a{}_c x_3^c + T^a + (\Lambda T')^a .$$

It is also evident that $IO(1,3)$ is nothing but the semi-direct product (see Sect. 4.1.2) of $O(1,3)$ and the group of spacetime displacements $\mathbb{R}^4$, namely $IO(1,3) = O(1,3) \times_\psi \mathbb{R}^4$. In this case, for every $\Lambda \in O(1,3)$, the group isomorphism $\psi_\Lambda : \mathbb{R}^4 \to \mathbb{R}^4$ is nothing but

$$\psi_\Lambda(T) := \Lambda T .$$

■

The subgroup $IO(1,3)\uparrow$ of $IO(1,3)$ whose elements (Λ, T) satisfy $\Lambda^0{}_0 \geq 0$ (equivalently $\Lambda^0{}_0 \geq 1$) is called the **orthochronous Poincaré group**. Its physical meaning should be clear: it is the group of Poincaré transformations connecting pairs of Minkowskian coordinate systems associated with corresponding pairs of *inertial reference systems*. This is due to the fact that the vectors $\mathscr{R}$ and $\mathscr{R}'$ of two inertial reference systems are both future oriented and thus $\mathbf{g}(\mathscr{R}, \mathscr{R}') < 0$. If x^0, x^1, x^2, x^3 and x'^0, x'^1, x'^2, x'^3 are Minkowskian coordinates respectively associated with these reference systems, we have:

$$0 > \mathbf{g}(\mathscr{R}', \mathscr{R}) = \mathbf{g}(\partial_{x'^0}, \partial_{x^0}) = \mathbf{g}(\partial_{x'^0}, \Lambda^a{}_0 \partial_{x'^a}) = \mathbf{g}(\partial_{x'^0}, \Lambda^0{}_0 \partial_{x'^0}) = -\Lambda^0{}_0 \,,$$

so that $\Lambda^0{}_0 \geq 0$ and, taking (e) of Proposition 7.26 into account, $\Lambda^0{}_0 \geq 1$, which is equivalent to saying that (Λ, T) belongs to the orthochronous subgroup. If we also impose the constraint $\det \Lambda = 1$, noticing that the intersection of subgroups is a subgroup, we obtain the so-called **orthochronous proper Poincaré group**, indicated by $ISO(1, 3)\uparrow$.

7.3.3 *Pure and Special Transformations*

From a kinematic point of view it makes sense to define the *four-velocity* of $\mathscr{R}$ (which does *not* depend on $\mathscr{R}'$) and the *velocity* of $\mathscr{R}$ *with respect to* $\mathscr{R}'$ (which *does* depend on $\mathscr{R}'$) when they are connected by a Poincaré transformation (Λ, T). The four-velocity and the velocity turn out to be constant in space and time. Indeed, let $\mathscr{R}$ and $\mathscr{R}'$ be Minkowskian reference systems with associated co-moving Minkowskian coordinate frames x^0, x^1, x^2, x^3 and x'^0, x'^1, x'^2, x'^3 respectively and suppose that (7.15) holds. Let $\gamma_{\mathscr{R}}$ represent the worldline of a material point *at rest* with $\mathscr{R}$, that is, $\gamma_{\mathscr{R}}$, projected to the rest space of $\mathscr{R}$, admits the trivial parametrization $x^\alpha(x^0) = x_0^\alpha$ constant for $\alpha = 1, 2, 3$. Since x^0 is the proper time of the history of any point at rest with $\mathscr{R}$, the four-velocity of that point is nothing but $\dot{\gamma}_{\mathscr{R}} = \mathscr{R}$ itself. However, it is interesting to write down the components of the four-velocity with respect to the other reference system in order to see how these components enter the coefficients of the Poincaré transformation connecting the two reference systems. From (7.15),

$$x'^a = \Lambda^a{}_b x^b + T^a$$

which implies

$$\partial_{x^b} = \Lambda^a{}_b \partial_{x'^a}$$

we obtain that, *referring to the coordinates* x'^0, x'^1, x'^2, x'^3, the components of the four-velocity

$$\dot{\gamma}_{\mathscr{R}} = \partial_{x^0}$$

are

$$\dot{\gamma}^0_{\mathscr{R}} = \Lambda^0{}_0\,, \quad \dot{\gamma}^\alpha_{\mathscr{R}} := \Lambda^\alpha{}_0\,, \quad \alpha = 1, 2, 3\,. \tag{7.23}$$

The result, due to Definition 7.19, immediately implies that also the velocity $\mathbf{v}^{(\mathscr{R}')}_{\mathscr{R}}$ of $\gamma_{\mathscr{R}}$ with respect to $\mathscr{R}'$ does not depend on the chosen point at rest with $\mathscr{R}$ and it is constant in $\mathscr{R}'$-time. Its components in coordinates x'^1, x'^2, x'^3 ($c = 1$) turn out to be:

$$\mathbf{v}^{(\mathscr{R}')\alpha}_{\mathscr{R}} = \frac{\dot{\gamma}^\alpha_{\mathscr{R}}}{\dot{\gamma}^0_{\mathscr{R}}} = \frac{\Lambda^\alpha{}_0}{\Lambda^0{}_0}\,, \quad \alpha = 1, 2, 3\,. \tag{7.24}$$

Conversely, it is interesting to wonder whether the information included in the components—with respect to some $\mathscr{R}'$—of a four-velocity $\dot{\gamma}_{\mathscr{R}}$ is enough to construct a Lorentz transformation connecting $\mathscr{R}'$ with an inertial system $\mathscr{R}$ having that four-velocity. Here a "four-velocity" is just understood as a future-oriented timelike vector normalized to -1 (or $-c^2$, depending on the units). The symbol $\dot{\gamma}_{\mathscr{R}}$ is just heuristic for the moment, since we do not yet know what $\mathscr{R}$ is.

However, the answer is positive. Consider a unit timelike future-oriented vector $\dot{\gamma}$ whose components are $\dot{\gamma}^0$ and $\vec{\dot{\gamma}}$ in Minkowskian coordinates x'^0, x'^1, x'^2, x'^3 of a Minkowskian reference frame $\mathscr{R}'$, then (where we assume $c \neq 1$)

$$\Lambda_{\dot{\gamma}} = \left[\begin{array}{c|c} \dot{\gamma}^0/c & \vec{\dot{\gamma}}^t/c \\ \hline \vec{\dot{\gamma}}/c & I + \frac{\vec{\dot{\gamma}}\vec{\dot{\gamma}}^t}{c^2(1+\dot{\gamma}^0/c)} \end{array}\right], \tag{7.25}$$

this is a *pure Lorentz transformation* connecting a Minkowskian coordinate frame x^0, x^1, x^2, x^3 with the Minkowskian coordinate frame x'^0, x'^1, x'^2, x'^3:

$$x'^a = (\Lambda_{\dot{\gamma}})^a{}_b x^b + T^a\,. \tag{7.26}$$

Indeed, by direct inspection, one immediately proves that (exercise)

(i) $(\Lambda_{\dot{\gamma}})^t \eta \Lambda_{\dot{\gamma}} = \eta$,
(ii) $(\Lambda_{\dot{\gamma}})^0{}_0 \geq 0$ and $\det \Lambda_{\dot{\gamma}} > 0$,
(iii) $\partial_{x^0} = \dot{\gamma} =: \mathscr{R}$.

As a consequence of (iii), the Minkowskian coordinates x^0, x^1, x^2, x^3 are really co-moving with the initial reference system $\mathscr{R} = \dot{\gamma}$ as expected. We observe that the components T^a play no role in the discussion; in fact, they can always be fixed to be $T^a = 0$ by changing the origin of the axes in $\mathscr{R}'$.

Lorentz transformations with the form (7.25), where $\dot{\gamma}$ is a unit future-oriented timelike vector represented in Minkowskian coordinates of a reference frame $\mathscr{R}'$, are called **pure Lorentz transformations**. When adding a spacetime translation as in (7.26), we obtain a **pure Poincaré transformation**.

A particular case occurs when $\vec{\gamma}$ has only one component. If $\vec{\gamma}$ has only one component along x'^3 and v is the only component of $\mathbf{v}_{\mathscr{R}}^{\mathscr{R}'}$, one immediately sees that (7.25) simplifies to the standard **special Lorentz transformation** along x^3:

$$\Lambda^{(3)} : \begin{cases} t' &= \frac{1}{\sqrt{1-v^2/c^2}}\left(t + \frac{vx^3}{c^2}\right) , \\ x'^1 &= x^1 , \\ x'^2 &= x^2 , \\ x'^3 &= \frac{1}{\sqrt{1-v^2/c^2}}\left(x^3 + vt\right) . \end{cases}$$

The product of two pure transformations is not, in general, a pure transformation, as is easy to check. However, in the special case of special transformations along a fixed axis we do have a subgroup of $SO(1,3)\uparrow$. The proof immediately arises from the fact that the above transformation can be rewritten

$$\Lambda^{(3)}_\chi : \begin{cases} ct' = & ct \cosh \chi + x^3 \sinh \chi , \\ x'^1 = & x^1 , \\ x'^2 = & x^2 , \\ x'^3 = & x^3 \cosh \chi + ct \sinh \chi , \end{cases}$$

where we have introduced the so-called **rapidity** χ:

$$\mathbb{R} \ni \chi := \tanh^{-1} \frac{v}{c} , \quad v \in (-c, c) . \tag{7.27}$$

Taking advantage of the addition formulas $\cosh(\chi' + \chi) = \cosh \chi \cosh \chi' + \sinh \chi \sinh \chi'$ and $\sinh(\chi' + \chi) = \sinh \chi \cosh \chi' + \cosh \chi \sinh \chi'$, one immediately proves that

$$\Lambda^{(3)}_\chi \Lambda^{(3)}_{\chi'} = \Lambda^{(3)}_{\chi+\chi'} \quad \chi, \chi' \in \mathbb{R}$$

which immediately proves that the set of special transformations along x^3 is a commutative subgroup of $SO(1,3)\uparrow$.

7.3.4 *Are Pure/Special Transformations and Spatial Rotations Sufficient to Reconstruct the Whole Lorentz Group?*

From physical arguments we expect that, starting from a pure transformation $\Lambda_{\vec{\gamma}}$ connecting two reference frames, it is always possible to reduce to this case by suitably orienting the spatial axes at rest with the two reference systems. Actually this fact can be proved mathematically (and we will come back again to this sort of result in the next section). To this end, we observe that $SO(1,3)\uparrow := SO(1,3) \cap O(1,3)\uparrow$ includes $SO(3)$ as a subgroup simply represented by the subgroup of matrices

$$\Omega_R = \left[\begin{array}{c|c} 1 & 0 \\ \hline 0 & R \end{array}\right], \quad R \in SO(3) \tag{7.28}$$

(if $R \in O(3)$, the corresponding matrix Ω_R is an element of $O(1,3)\uparrow$, which therefore contains $O(3)$ as a subgroup.) The composition $\Omega_R \Lambda_{\dot{\gamma}} \Omega_R^t$ yields

$$\Omega_R \Lambda_{\dot{\gamma}} \Omega_R^t = \left[\begin{array}{c|c} \dot{\gamma}^0/c & (R\vec{\dot{\gamma}})^t/c \\ \hline R\vec{\dot{\gamma}}/c & I + \frac{R\vec{\dot{\gamma}}(R\vec{\dot{\gamma}})^t}{c^2(1+\dot{\gamma}^0/c)} \end{array}\right], \tag{7.29}$$

It follows that, starting from a pure matrix of the form (7.29) and rotating the spatial axes of the reference system $\mathscr{R}'$ by a suitable rotation $R \in SO(3)$, we can always achieve a special Lorentz transformation along x^3, proving the assertion above.

A final issue is whether or not the product $\Omega_R \Lambda_{\dot{\gamma}}$ exhausts all possible elements of $SO(1,3)\uparrow$. The answer is positive and will be discussed in Chap. 8. We only state a minor version of the decomposition theorem, which will be established in full generality in Chap. 8, also taking the matrix Lie group structure into account.

Theorem 7.29 *If $\Lambda \in SO(1,3)\uparrow$, there exists a unique pair (Ω_R, P), where Ω_R is a spatial rotation of the form (7.28) and P is a pure Lorentz transformation of the form (7.25), such that*

$$\Lambda = \Omega_R P\,.$$

A nice immediate consequence of this theorem and of (7.29) is that, whatever the initial Poincaré transformation between a pair of Minkowskian reference frames $\mathscr{R}$ and $\mathscr{R}'$, we can always rearrange the used co-moving Minkowskian coordinate systems, changing the spacetime origin and rotating the spatial axes, in order to have a system of co-moving Minkowskian coordinate systems connected by a special Lorentz transformation along the axis x^3.

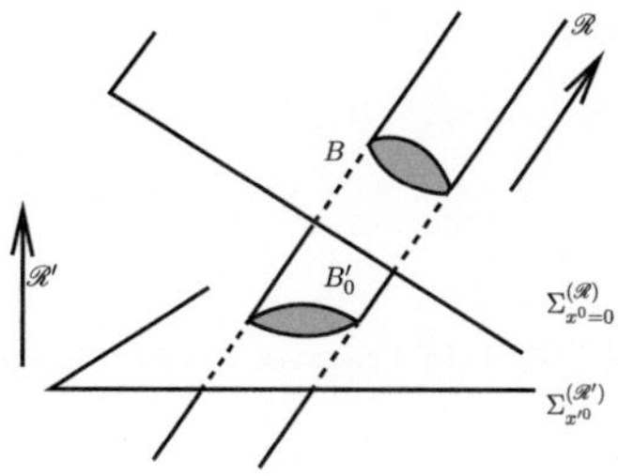

Fig. 7.3 Contraction of volumes

7.3.5 *Contraction of Volumes*

As an application of the explicit form of pure transformations provided in (7.25), we prove the formula of *contraction of volumes.*

Suppose that an extended body is at rest in a Minkowskian reference system $\mathscr{R}$, where it is defined by a region B of the rest space of $\mathscr{R}$. This region can be interpreted as a (Lebesgue) measurable set of $\mathbb{R}^3$ with respect to Minkowskian coordinates x^0, x^1, x^2, x^3 co-moving with $\mathscr{R}$. Let $Vol_{\mathscr{R}}(B)$ be the measure of that spatial region in the rest space of $\mathscr{R}$. Let us pass to the new reference system $\mathscr{R}'$. The points of $B'_{x'^0}$ are the intersections with the rest space of $\mathscr{R}'$ at time x'^0 of the worldlines of the points in B evolving parallel to $\mathscr{R}$. We are interested in the volume of $B'_{x'^0}$, denoted by $Vol_{\mathscr{R}'}(B'_{x'^0})$, measured in the other Minkowskian reference system $\mathscr{R}'$ equipped with co-moving Minkowskian coordinates x'^0, x'^1, x'^2, x'^3 (Fig. 7.3).

Remarks 7.30 It is important to stress that these volumes are independent of the choice of Minkowskian coordinate systems respectively co-moving with $\mathscr{R}$ and $\mathscr{R}'$, since the internal changes of Minkowskian coordinates just amount to spatial roto-translations, which are isometries and therefore preserve the Lebesgue measure. Similarly, a translation of the origin of the time axis just amounts to a corresponding translation in the index which labels the volumes B', without changing their values. ■

The relation between two different coordinate systems respectively co-moving with $\mathscr{R}$ and $\mathscr{R}'$ is implemented by some $(\Lambda, T) \in IO(1,3)\uparrow$. As discussed above, we can always rotate the axes in $\mathscr{R}'$ in order to have Λ of the form (7.25). Furthermore, changing the origin of the temporal and spatial axes of $\mathscr{R}'$, we can always fix $T = 0$. In this case we can take advantage of the expression of Λ in (7.25) in terms of the velocity $\mathbf{v}^{(\mathscr{R}')}_{\mathscr{R}}$ of $\mathscr{R}$ with respect to $\mathscr{R}'$. From (7.25) we have that

$$x'^0 = \dot{\gamma}^0_{\mathscr{R}} x^0 + \sum_{\alpha=1}^{3} \dot{\gamma}^\alpha_{\mathscr{R}} x^\alpha \tag{7.30}$$

and

$$x'^\beta = \dot{\gamma}^\beta_{\mathscr{R}} x^0 + x^\beta + \frac{\dot{\gamma}^\beta_{\mathscr{R}} \sum_{\alpha=1}^3 \dot{\gamma}^\alpha_{\mathscr{R}} x^\alpha}{1 + \dot{\gamma}^0_{\mathscr{R}}} .$$

In the rest space of $\mathscr{R}'$ and at time x'^0, the body B is therefore described by the set

$$B'_{x'^0} := \left\{ (x'^1, x'^2, x'^3) \in \mathbb{R}^3 \;\middle|\; x'^\beta = \dot{\gamma}^\beta_{\mathscr{R}} x^0 + x^\beta + \frac{\dot{\gamma}^\beta_{\mathscr{R}} \sum_{\alpha=1}^3 \dot{\gamma}^\alpha_{\mathscr{R}} x^\alpha}{1 + \dot{\gamma}^0_{\mathscr{R}}} ,\; (x^1, x^2, x^3) \in B \right\}$$

where x^0 is obtained by inverting (7.30) as

$$x^0 = \frac{x'^0 - \sum_{\alpha=1}^3 \dot{\gamma}^\alpha_{\mathscr{R}} x^\alpha}{\dot{\gamma}^0_{\mathscr{R}}}$$

and by inserting the result in the expression above for $B'_{x'^0}$. In summary,

$$B'_{x'^0} := \left\{ (x'^1, x'^2, x'^3) \in \mathbb{R}^3 \;\middle|\; x'^\beta = \frac{\dot{\gamma}^\beta_{\mathscr{R}}}{\dot{\gamma}^0_{\mathscr{R}}} x'^0 + x^\beta - \frac{\dot{\gamma}^\beta_{\mathscr{R}} \sum_{\alpha=1}^3 \dot{\gamma}^\alpha_{\mathscr{R}} x^\alpha}{\dot{\gamma}^0_{\mathscr{R}}(1 + \dot{\gamma}^0_{\mathscr{R}})} ,\; (x^1, x^2, x^3) \in B \right\} .$$

We conclude that, for $\alpha, \beta = 1, 2, 3$,

$$J^\beta_\alpha = \frac{\partial x'^\beta}{\partial x^\alpha} = \delta^\beta_\alpha - \frac{\dot{\gamma}^\beta_{\mathscr{R}} \dot{\gamma}^\alpha_{\mathscr{R}}}{\dot{\gamma}^0_{\mathscr{R}}(1 + \dot{\gamma}^0_{\mathscr{R}})} .$$

From standard results of measure theory, using also the fact that $\det J$ is a constant,

$$\begin{aligned} Vol_{\mathscr{R}'}(B'_{x'^0}) &= \int_{B'_{x'^0}} 1 \, dx'^1 dx'^2 dx'^3 = \int_B |\det J| \, dx^1 dx^2 dx^3 \\ &= |\det J| \int_B dx^1 dx^2 dx^3 = |\det J| Vol_{\mathscr{R}}(B) . \end{aligned}$$

The determinant of the matrix $J := [J_\alpha^\beta]_{\alpha,\beta=1,2,3}$ can be immediately computed by observing that the symmetric matrix of components

$$-\frac{\dot{\gamma}_{\mathscr{R}}^\beta \dot{\gamma}_{\mathscr{R}}^\alpha}{\dot{\gamma}_{\mathscr{R}}^0(1+\dot{\gamma}_{\mathscr{R}}^0)}$$

has an evident eigenvector $(\dot{\gamma}_{\mathscr{R}}^1, \dot{\gamma}_{\mathscr{R}}^2, \dot{\gamma}_{\mathscr{R}}^3)^t$ with eigenvalue

$$\lambda = -\frac{\sum_\alpha (\dot{\gamma}_{\mathscr{R}}^\alpha)^2}{\dot{\gamma}_{\mathscr{R}}^0(1+\dot{\gamma}_{\mathscr{R}}^0)} = \frac{1-(\dot{\gamma}_{\mathscr{R}}^0)^2}{\dot{\gamma}_{\mathscr{R}}^0(1+\dot{\gamma}_{\mathscr{R}}^0)} = \frac{1-\dot{\gamma}_{\mathscr{R}}^0}{\dot{\gamma}_{\mathscr{R}}^0}$$

and two vanishing eigenvalues corresponding to two mutually orthogonal vectors normal to $(\dot{\gamma}_{\mathscr{R}}^1, \dot{\gamma}_{\mathscr{R}}^2, \dot{\gamma}_{\mathscr{R}}^3)^t$ in $\mathbb{R}^3$. Hence, the 3×3 matrix J has the eigenvalues $1+\lambda$ and 1 with multiplicity 2. Restoring the value of the speed of light, we conclude that

$$\det J = \frac{c}{\dot{\gamma}_{\mathscr{R}}^0} = \sqrt{1 - \frac{(\mathbf{v}_{\mathscr{R}}^{(\mathscr{R}')})^2}{c^2}}\,, \tag{7.31}$$

so that we have established the classical formula of volume contraction,

$$Vol_{\mathscr{R}'}(B'_{x'^0}) = Vol_{\mathscr{R}}(B)\sqrt{1 - \frac{(\mathbf{v}_{\mathscr{R}}^{(\mathscr{R}')})^2}{c^2}}\,. \tag{7.32}$$

This formula is independent of the Minkowskian coordinate systems used in the two Minkowskian reference systems, since it only exploits and refers to intrinsic spatial metric quantities as discussed above and uses the relative velocity $\mathbf{v}_{\mathscr{R}}^{(\mathscr{R}')}$, which does not depend on any choice of coordinates but only on $\mathscr{R}$ and $\mathscr{R}'$.

7.4 Dynamics I: Material Points

We briefly discuss here some basic notions useful in relativistic dynamics, always adopting a geometric point of view.

Remarks 7.31 In this section we explicitly use the value of the speed of light $c \neq 1$. ■

7.4.1 Remarks on Relativistic Local Causality

Before entering the discussion of the formulation of relativistic dynamics, a few remarks are necessary about the structure of the arena where that dynamics is formulated. In relativistic theories, (macroscopic) interactions are described in terms of future-directed causal curves connecting the event p, which is the cause, with the event q, which is the effect, as we shall see shortly.

Definition 7.32 A pair of events p, q of (Minkowski) spacetime is said to be **causally connected** if $p = q$ or there is a future-directed causal curve joining them in one order or the other. Events which are not causally connected are said to be **spatially separated** or also **causally separated**.

Events which are not causally connected *cannot* enjoy causal relations. This requirement is usually named **locality** or **local causality** or **Einstein causality** or **Einstein locality**.

Minkowski spacetime possesses the light-cone structure and this fact permits one to state the definition above in an equivalent but easier-to-use form.

Proposition 7.33 *In Minkowski spacetime, the events p and q are causally connected if and only if q belongs to the future half $\overline{V_p^+}$ of the closed light-cone emanating from p (thus including its origin and its boundary), or* vice versa.

Proof If $q - p \in \overline{V^+}$, then $p = q$ or the segment from p to q is a future-directed causal curve. The same is valid swapping q and p. If $q - p \notin \overline{V^+}$ and $p - q \notin \overline{V^+}$, then we can arrange a reference frame $\mathscr{R}$ such that $q - p$ lies in a 3-rest space of $\mathscr{R}$ at time 0. A causal curve (if any) $x^a = x^a(s)$, $a = 0, 1, 2, 3$, with $s \in [a, b]$ joining p and q must satisfy $x^0(a) = x^0(b) = 0$ and thus there is a point $s_0 \in (a, b)$ where $dx^0/ds|_{s_0} = 0$. This is not possible because the curve is causal, so that for some choice of the parameter $dx^0/ds \neq 0$ everywhere. □

Some important remarks on these notions are in order.

(i) It is worth stressing that quantum phenomena may give rise to non-causal *correlations* between two spatially separated events as a consequence of the *Einstein-Podolsky-Rosen paradox* (see, e.g., Chapter 5 of [2]). These correlations have been experimentally observed and nowadays this type of phenomenology, based on the notion of quantum *entanglement*, is used in quantum information technology. The 2022 Nobel Prize in Physics was awarded to Aspect, Clauser, and Zeilinger for the definitive experimental proof of the entanglement phenomenon.

(ii) It is not difficult to prove that, if p and q are causally separated, there are three inertial reference systems $\mathscr{R}$, $\mathscr{R}'$, and $\mathscr{R}''$ equipped with global time coordinates, respectively t, t', and t'', such that $t(p) < t(q)$, $t'(p) = t'(q)$, and $t''(p) > t''(q)$. These reference systems can be constructed starting from a third reference system $\mathscr{R}_0$ such that the segment joining p and q belongs to a rest space of it. As a consequence, $t_0(p) = t_0(q)$ and thus $\mathscr{R}' = \mathscr{R}_0$. Next, placing $\mathscr{R}_0$ at the midpoint between p and q, we can tilt it towards p or q, defining in this way $\mathscr{R}$ and $\mathscr{R}''$. Notice that we have in particular found the fact that whether two spatially separated events happen simultaneously or not depends on the reference frame used to describe them.

(iii) Conversely, if p, q are causally related, their temporal order *cannot* be reversed by a suitable choice of the Minkowskian (= inertial) reference system. In fact, if $q \in \overline{V_p^+}$ (the other case is analogous), we have that the segment $\overrightarrow{pq}$ is either the zero vector or it is causal and future-directed and thus $\mathbf{g}(\mathscr{R}, \overrightarrow{pq}) \leq 0$, since also $\mathscr{R}$ is timelike and future-directed just by definition of inertial reference system. Therefore,

$$x^0(q) - x^0(p) = -\mathbf{g}(\mathscr{R}, \overrightarrow{pq}) \geq 0$$

for every global temporal coordinate x^0 of every inertial reference system $\mathscr{R}$.

(iv) The relativity of chronological order outside the light cone is in agreement with the previous remark that no causal relations between spatially separated events can be defined: here the time of the effect could precede the time of the cause for a suitable choice of the reference system. However, this possibility cannot be used to justify the impossibility of defining causal relations in view of the (potentially) wrong time ordering. This is because a closer scrutiny of the physical construction of Special Relativity shows that some conventional elements enter the game when the Einstein synchronization procedure is adopted, making the discussion on causality much more subtle.

(v) It should be clear that the notion of *rigid body* is untenable in relativity: with a very long rigid rod I could give rise to causal relations between causally separated events, simply by pushing one edge.

7.4.2 The Geometric Nature of the Relativity Principle in the Formulation of Relativistic Dynamics

One of the pillars Einstein exploited in founding Special Relativity is the so-called *Relativity Principle*, which can be stated by postulating that *all physical laws can be formulated with the same mathematical form in all inertial reference systems.*

If the physical laws concern the geometry of spacetime, then they can always be formulated with the *intrinsic* and *synthetic* language of geometry. In turn, we also know that the structure of (pseudo-)Euclidean spaces selects peculiar coordinate frames—the (pseudo-)orthonormal Cartesian systems of coordinates called Minkowskian coordinate systems—which are completely equivalent for the formulation of all geometric propositions written in coordinates and components: these statements have the same form in every Minkowskian coordinate system. This is a consequence of the fact that they are connected by means of isometries and the physical laws are invariant under the action of isometries.

This invariance-in-form property of the physical laws when represented in Minkowskian reference frames is often called *covariance* by physicists. However, covariance is more precisely related to the fact that the physical laws are written in terms of tensors and operations between tensors. Invariance of the physical laws (the Relativity Principle) is a more general property which is not necessarily made explicit through the use of tensors.

In summary, if the physical laws reflect the geometry of spacetime—i.e., if they have a geometric nature—they should take the same mathematical form in every Minkowskian system of coordinates when all intrinsic geometric objects are written in components.[2] As we have supposed that each Minkowskian system of coordinates defines (up to internal transformations) an inertial reference system, on assuming that the laws of physics are geometric propositions about the geometry of spacetime, we automatically have that the Relativity Principle is fulfilled. For this reason, in the rest of the section, we shall formulate the laws of relativistic dynamics using the intrinsic language of the geometry of Minkowski spacetime.

7.4.3 *Mass and Four-Momentum*

Special Relativity assumes that material points describing timelike worldlines are equipped with a (for the moment) strictly positive constant m called **mass** (or also **rest mass**). The idea is that, as Special Relativity should reduce to standard mechanics when the involved velocities are very small compared with c, then the Newtonian notion of mass should still have some corresponding tool in Special Relativity.[3] More precisely, the mass m of a material point coincides with its Newtonian mass since, by definition, it is measured in a Minkowskian reference system where the point has a very small velocity compared with c.

If $\dot{\gamma}(\tau)$ is the four-velocity of the considered material point, the future-directed timelike vector

$$P(\tau) := m\dot{\gamma}(\tau) \tag{7.33}$$

is called the **four-momentum**.

Remarks 7.34 Exactly as the four-velocity, the four-momentum is *intrinsically defined*, namely without referring to a reference system. That is completely different from the classical notions of velocity and momentum, which are defined only when a reference system is given. ■

---

[2] When the said laws are written in the language of tensor calculus, the fact that the form of the laws in components is identical in all Minkowskian systems of coordinates is known as *covariance of the physical laws*.

[3] This sort of argument is often called the "correspondence principle" and plays a crucial role in heuristically developing the theory.

From (7.11), we immediately have that, in the reference system $\mathscr{R}$ endowed with co-moving Minkowskian coordinates x^0, x^1, x^2, x^3, the components of P with respect to those coordinates read (restoring the constant c and observing that P is future-directed)

$$0 < P^0_{\mathscr{R}} = \frac{mc}{\sqrt{1 - \left(\frac{\mathbf{v}_{\mathscr{R}}}{c}\right)^2}} , \qquad P^\alpha_{\mathscr{R}} = \frac{m\mathbf{v}^\alpha_{\mathscr{R}}}{\sqrt{1 - \left(\frac{\mathbf{v}_{\mathscr{R}}}{c}\right)^2}} , \qquad \alpha = 1, 2, 3 , \tag{7.34}$$

where $\mathbf{v}_{\mathscr{R}}$ is the velocity of the material point with respect to $\mathscr{R}$ at the considered event. According to (7.34), if $\|\mathbf{v}_{\mathscr{R}}\| \ll c$, the spatial components of the four-momentum in a given Minkowskian reference system $\mathscr{R}$ become an approximation of the components of the classical momentum, since

$$\frac{m\mathbf{v}^\alpha_{\mathscr{R}}}{\sqrt{1 - \left(\frac{\mathbf{v}_{\mathscr{R}}}{c}\right)^2}} = m\mathbf{v}^\alpha_{\mathscr{R}} + O\left(\left(\frac{\mathbf{v}}{c}\right)^2\right) .$$

This fact provides a physical interpretation of the spatial components of the four-momentum: they are a generalization of the classical momentum. Instead, for the moment the meaning of P^0 is not clear in comparison with classical physics. From (7.34), however, we see that the mass coincides with the component P^0 of the four-momentum measured in an inertial reference frame at rest (possibly just *instantaneously*) with the particle.

7.4.4 Massless Particles

A useful pair of relations immediately arises from (7.34)

$$\mathbf{g}(P, P) = -m^2c^2 \quad \text{i.e., within the abstract index notation} \quad P_a P^a = -m^2c^2 , \tag{7.35}$$

and

$$P^0 = \sqrt{\sum_{\alpha=1}^{3} P^\alpha P^\alpha + m^2c^2} . \tag{7.36}$$

Equation (7.35) provides a definition of the mass m in case we assume P as a primitive notion. From this viewpoint, namely assuming P as primitive, it seems possible to extend the notion of four-momentum to particles satisfying (7.35) with $m = 0$. These particles have lightlike momentum and are a complete novelty, since they are not permitted in classical physics. It is worth stressing that experimental physics proved that they really exist.

Examples 7.35 *Photons* are an example of massless particles. In that case, in a semiclassical (non-quantum) view,

$$P^0 = \hbar\frac{\omega}{c}\,, \quad P^\alpha = \hbar k^\alpha\,, \quad \alpha = 1, 2, 3,$$

where ω is the angular frequency of the light wave associated with the photon and k^α the components of the wave vector in the considered reference system, $\hbar = \frac{h}{2\pi}$ with h the Planck constant (6.626×10^{-27} erg sec). Since

$$\sum_{\alpha=1}^{3}(k^\alpha)^2 = \left(\frac{\omega}{c}\right)^2$$

from electromagnetic theory, we have in this case $P_a P^a = 0$, so that *photons must be massless.* ■

Remarks 7.36 Since P is tangent to the worldlines of massless particles, proposition 7.20 implies that massless particles always move with the speed of light. In this case there is no rest frame for the particle, since P is lightlike. However (7.36) is still valid with obviously $m = 0$. ■

7.4.5 The Principle of Inertia

Recall that, if $\mathbb{A}^n$ is an affine space with space of translations V^n, then an **affine segment** is a map of the form

$$I \ni t \mapsto P + t\mathbf{v} \in \mathbb{A}^n =: P(t)\,, \tag{7.37}$$

where $I \subset \mathbb{R}$ is an interval, $P \in \mathbb{A}^n$ and $\mathbf{v} \in V^n$. The curve above can be reparametrized using another parameter s—defined on another interval I' and related with t through a C^1 map with C^1 inverse—obtaining a new map $P'(s) := P(t(s))$ with the same form as (7.37)

$$I' \ni s \mapsto P' + s\mathbf{v}' \in \mathbb{A}^n = P'(s)\,,$$

for suitable P' and $\mathbf{v}'$. This happens if and only if $s = at + b$ for $a \neq 0$ and $b \in \mathbb{R}$ arbitrary constants. All the parameters preserving the form (7.37) are called **affine parameters** of the affine segment.

In classical physics, an *isolated* material point is assumed to evolve with constant velocity in every inertial reference system, i.e., in every Minkowskian reference system. When assuming $m > 0$, from (7.11), we immediately have that histories with constant velocity in a given Minkowskian reference system are actually curves with constant four-velocity $\dot{\gamma}$ so that, in particular, they have constant velocity with respect to every other Minkowskian reference system. Hence, when parametrizing those curves using the proper time τ, we find

$$\gamma(\tau) = \gamma(\tau_0) + (\tau - \tau_0)\dot{\gamma}\,,$$

where $\dot{\gamma}$ is a constant timelike future-oriented vector. Those curves are nothing but *affine segments* in $\mathbb{M}^4$ and the proper time is an *affine parameter* for them. We can extend these assumptions to *isolated* massless particles evolving along lightlike worldlines, assuming that also in this case those lines are lightlike segments,

$$\gamma(s) = \gamma(s_0) + (s - s_0)\gamma'\,.$$

The difference is that now there is no preferred affine parameter describing them, as the proper time does for massive particles, since the integrand in (7.7) vanishes. The above affine parameter s can be changed with affine transformations $s' := as + b$ where $a > 0$, obtaining another physically equivalent (future-directed) affine parametrization. In summary, the principle of inertia in Minkowski spacetime states that

> **Principle of inertia.** *Isolated material points have worldlines which are causal future-oriented affine segments of* $\mathbb{M}^4$. *They are timelike, with the proper time as an affine parameter, or lightlike according to the value of their masses, strictly positive or vanishing respectively.*

Remarks 7.37 A substantial difference between this formulation of the inertia principle and the one of classical physics is that here the notion of reference frame plays no role, since the formulation is now geometrically intrinsic.

7.4.6 Four-Momentum Conservation Law

In Special Relativity, as in classical mechanics, it seems plausible to assume the conservation of the total four-momentum of a system of material points when these points form an isolated system but they interact with one another. This principle is harder to formulate than in classical mechanics in view of the difficult notion of total four-momentum of a system of interacting material points: *a priori*, the total four-momentum is computed by summing the momenta of the various particles forming the system on a given time slice of a Minkowskian reference system. However, the notion of "at a given time" depends on the reference system, and changing reference system we would sum the momenta of the same particles computed at different events. Conversely, the total four-momentum is expected to be an intrinsic, reference-independent notion. Dealing with material points, in a particular case the principle can nevertheless be formulated. We are referring to the case of mutual interactions *happening at isolated events*. In this case, the worldlines of the particles of the system define a graph whose lines are causal future-oriented affine segments

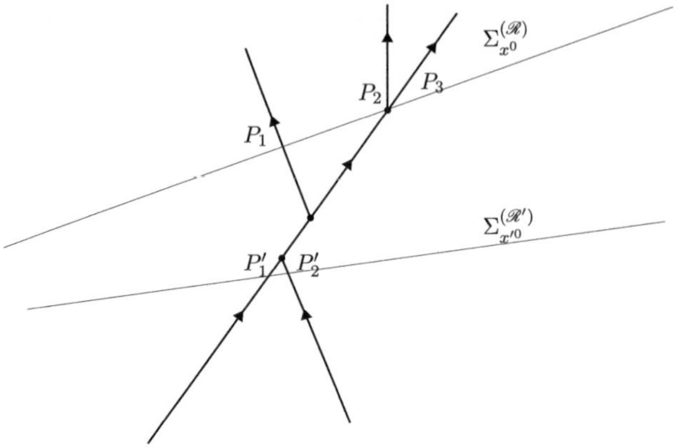

Fig. 7.4 Four-momentum conservation: $P'_1 + P'_2 = P_1 + P_2 + P_3$

of $\mathbb{M}^4$, according to the principle of inertia since the material points are isolated between the interactions they undergo, and the vertices are the very interaction events. The conservation of total four-momentum is stated as follows.

Four-momentum conservation. *For a system of material points whose mutual interactions are concentrated in isolated events (and no further interactions exist), the worldlines of the material points of the system form a graph whose lines are causal future-oriented segments and whose vertices are the interaction events. In this graph,*

(i) the four-momentum is constant along each line,
(ii) the sum of the four-momenta entering a vertex from the past is equal to the sum of the four-momenta exiting that vertex towards the future.

With this postulate an invariant notion of *total four-momentum* can be defined, at least when the number of interaction events and lines is finite. Indeed, every rest space $\Sigma^{(\mathscr{R})}_{x^0}$ of every Minkowskian reference system $\mathscr{R}$ is such that the sum of the four-momenta $P_1, P_2, \ldots, P_N$ of the worldlines crossing $\Sigma^{(\mathscr{R})}_{x^0}$ (notice that the number N generally depends on $\Sigma^{(\mathscr{R})}_{x^0}$) has a constant value $P = \sum_{i=1}^{N} P_i$ which is independent of *both* x^0 and $\mathscr{R}$. The proof of this fact uses an inductive argument. A crucial observation is that we cannot choose $\Sigma^{(\mathscr{R})}_{x^0}$ in such a way that it completely includes a segment of the graph; it may only cross each such line in a single event at most. That is because these segments are causal, whereas every segment which is embedded in $\Sigma^{(\mathscr{R})}_{x^0}$ is necessarily spacelike (Fig. 7.4).

Remarks 7.38

(1) $P = \sum_{i=1}^{N} P_i$ makes sense, even if the momenta P_i are located at different events, just because we are dealing with an *affine space*. However this sort of approach will become problematic once one relaxes the affine structure to that of a Lorentzian manifold in General Relativity.

(2) The requirement of conservation of the four-momenta as written in the postulate above can be assumed more generally valid when the worldlines entering a vertex and those exiting that vertex become affine segments in an arbitrarily small neighborhood of the vertex where the possible external interactions switch off. In this case, the four-momenta to be considered are computed exactly at the vertex along each worldline. In this more general situation, however, the notion of total four-momentum is more difficult to define.

■

Exercises 7.39

1. Prove that, even if mass is not conserved, no particles can be created out of the vacuum. (*Hint*. If the total final four-momentum P must be conserved, then its temporal component must be strictly positive because it is future-oriented.)
2. Prove that a photon cannot decay into a pair of massive particles. (*Hint*. Write down the law of conservation of the four-momentum in components in the reference system $\mathscr{R}_0$ that is parallel to the total conserved four-momentum P. Notice that P must be timelike because the four-momenta of the final particles are timelike.)

7.4.7 *The Interpretation of P^0 and the So-Called "Mass-Energy Equivalence"*

The above formulation of the conservation law of four-momentum permits processes where the number of particles is not conserved, whereas the total momentum is conserved. The masses of the involved particles generally change when crossing a vertex. The masses are however constant (possibly vanishing) along the lines due to (7.35). A natural issue in this context is whether or not the total mass of the material points is conserved during the interactions, as it happens in classical physics. In particular one may wonder whether, in processes where particles are created or destroyed—a particle breaks into many particles or many particles join into a single particle—the total mass remains constant. The answer to this question, on the one hand, suggests an interpretation of the component P^0 of the four-momentum; on the other hand, it leads to one of the great achievements of modern physics due to Einstein.

Considering the case of a particle decaying into two particles, if P is the four-momentum of the initial particle with mass M and P_1, P_2 are the momenta of the two final particles with masses m_1 and m_2 respectively, in the reference system of the initial particle, the conservation law of the four-momentum gives (we omit the index $\mathscr{R}$ for the sake of simplicity)

$$Mc = P_1^0 + P_2^0 \,, \quad 0 = P_1^\alpha + P_2^\alpha \,, \quad \alpha = 1, 2, 3 \,,$$

that is, regarding the first identity,

$$M = \frac{1}{c}P_1^0 + \frac{1}{c}P_2^0$$

which can be rearranged into

$$M = m_1 + m_2 + \left(\frac{1}{c}P_1^0 - m_1\right) + \left(\frac{1}{c}P_2^0 - m_2\right) . \tag{7.38}$$

We see that the mass is *not* conserved (more precisely, it is not additive), and the failure of additivity is represented by the terms in parentheses, which are however very small when the velocities of the two particles are small compared with c. In fact, from the former relation in (7.34)

$$\left(\frac{1}{c}P_1^0 - m_1\right)+\left(\frac{1}{c}P_2^0 - m_2\right) = \frac{1}{c^2}\left[\frac{1}{2}m_1\mathbf{v}_1^2 + \frac{1}{2}m_2\mathbf{v}_2^2 + O\left(\frac{\mathbf{v}_1^4}{c^2}\right) + O\left(\frac{\mathbf{v}_2^4}{c^2}\right)\right]. \tag{7.39}$$

We see that the initial mass M is equal to the sum of the final masses with a further contribution due to the classical kinetic energies of the two particles in the reference system of the initial decaying particle and further terms of higher order. Expansion (7.39) together with other theoretical results (see Sect. 7.4.9) leads one to define the **relativistic kinetic energy** in the reference system $\mathscr{R}$ of a particle of mass m and four-momentum P as

$$K_{\mathscr{R}} := cP^0_{\mathscr{R}} - mc^2 , \tag{7.40}$$

where P^0 is the temporal component of the four-momentum P in the reference system $\mathscr{R}$. On the other hand, it also leads one to rewrite (7.38) as

$$M = m_1 + m_2 + \frac{K_1}{c^2} + \frac{K_2}{c^2} . \tag{7.41}$$

The analogous result pops out when describing the time-reversed phenomenon where two particles merge into a unique particle: part of the mass M of the final particle, in addition to the masses m_1 and m_2, is due to the kinetic energies (in the reference system of the particle) of the two colliding particles.

Remarks 7.40 We stress that the discussed phenomena are really observed quite easily when dealing with elementary particles: the measured mass really ceases to be an additive quantity and it receives contributions from the kinetic energy of decay products according to (7.41) or similar equations, where all terms can be experimentally measured separately. ■

This phenomenon, where the mass of a particle seems to receive a contribution from the kinetic energies of constituents or of the products of decay, is corroborated from another completely theoretical point of view, if we assume that the mass of a system is c^{-1} times the value of the temporal component of its four-momentum P in its co-moving reference system $\mathscr{R}_0$, i.e., in a reference system defined by a normalized vector $\mathscr{R}_0$ parallel to P. (This is equivalent to saying that the three spatial components $P^\alpha_{\mathscr{R}_0}$ vanish in that reference system.) In fact, referring to the system of material points interacting only in isolated events that we discussed in the previous section, if $P = \sum_{i=1}^N P_i$ is the total four-momentum computed in the co-moving reference system $\mathscr{R}_0$ (whose associated rest space intersects N worldlines at the considered time), we have in that reference system

$$M := \frac{1}{c} P^0_{\mathscr{R}_0} = \sum_{i=1}^N \frac{1}{c} P^0_{\mathscr{R}_0\, i} = \sum_{i=1}^N m_i + \frac{K_{\mathscr{R}_0\, i}}{c^2}\,, \qquad P^\alpha = 0\,, \alpha = 1, 2, 3\,.$$

Above, m_i and $K_{\mathscr{R}_0\, i}$ are respectively the mass and the kinetic energy of the i-th material point in the co-moving reference system of the whole system at the considered time. The system is a "big" material point, at rest in the considered reference system, with a mass which is the sum of the masses of its constituents and a further contribution due to their kinetic energy in the said reference system. If such a system of molecules were confined in a box, we would expect that its mass increases if we warm up the box as a consequence of the fact that the temperature is nothing but a macroscopic expression of the (here only kinetic) energy of the molecules.

All that discussion, in particular the fact that $cP^0_{\mathscr{R}}$ is conserved in time and reduces to the kinetic energy in the classical limit up to an additive constant, leads to the following final postulate which implies the celebrated *mass-energy equivalence principle*.

Energy. *If $P^0_{\mathscr{R}}$ is the time component of the total momentum of a material point in a Minkowskian reference system $\mathscr{R}$, then*

$$E_{\mathscr{R}} := cP^0_{\mathscr{R}}$$

is the mechanical energy of the point in the said reference system.

That physical postulate has several implications we list below.

(i) The mechanical energy is conserved in every Minkowskian reference system when the hypotheses of the principle of conservation of four-momentum are satisfied.
(ii) If the mass of the system does not vanish, then it is

$$m = \frac{1}{c^2} E_{\mathscr{R}_0}\,,$$

where $\mathscr{R}_0$ is the reference system co-moving with the system, defined by the condition that $\mathscr{R}_0$ is parallel to P;

(iii) (**Mass-energy equivalence principle**) The mass of a system is not an additive (i.e. conserved) quantity and it contributes to the total mechanical energy $E_{\mathscr{R}}$ of the system, in a given reference system $\mathscr{R}$, by the amount

$$\Delta E_{\mathscr{R}} = mc^2 \, .$$

(iv) The kinetic energy,

$$K_{\mathscr{R}} = cP^0_{\mathscr{R}} - mc^2 \, ,$$

in a reference system $\mathscr{R}$ amounts to the difference between the total mechanical energy $E_{\mathscr{R}} = cP^0_{\mathscr{R}}$—which depends on the reference system—and the contribution due to the mass mc^2—which is independent of the reference system. In other words, the mechanical energy is the sum of the kinetic energy and the contribution due to the mass:

$$cP^0_{\mathscr{R}} = mc^2 + K_{\mathscr{R}} \, .$$

The postulate and the remarks above are extended to all physical systems, not necessarily material points. The reason is that, macroscopically speaking, every extended but spatially confined physical system can be described as a material point if dealing with spatial scales much larger than its typical size. From a general perspective, it is expected that the mass m of a body is nothing but the total amount of energy which can be localized inside the body, up to the conversion factor c^2, when it is measured in its rest frame, provided this reference frame can be defined. This energy, which appears to be a part of the total mechanical energy, actually accounts for all types of energies (e.g. chemical, thermodynamical, etc.) which can be localized in the region of the body. Physical transformations may transform the energy ascribed to the mass into different forms of energy and *vice versa*, in particular into kinetic energy. This possibility implies that the mass is not conserved.

7.4.8 The Notion of Four-Force

A theoretically interesting attempt to describe interactions which are not localized at events has been proposed in terms of a generalized version of *Newton's second law*. However, this approach is quite academic and the only concrete case where it is useful is when considering charged particles in interaction with the electromagnetic field. The idea, restricting ourselves to the case of a massive particle only, is to replace Newton's law with

$$F(\gamma(\tau), \dot{\gamma}(\tau)) = \frac{dP}{d\tau} \, , \quad P(\tau) = m(\tau)\dot{\gamma}(\tau) \, , \tag{7.42}$$

where $P = P(\tau)$ is the four-momentum at proper time τ of a particle with worldline $\gamma = \gamma(\tau)$ and the given **four-force** $F(p, S_p)$ defines a vector in $T_p\mathbb{M}^4$ for every $p \in \mathbb{M}^4$ and $S_p \in T_p\mathbb{M}^4$ with $\mathbf{g}(S_p, S_p) = -c^2$. It is worth stressing that the above system of differential equations includes the possibility of changing the mass m of the particle due to the interaction described by the four-force F. In fact, since

$$m^2 = -\frac{1}{c^2} P_a P^a$$

we also have

$$\frac{dm^2}{d\tau} = -\frac{2}{c^2} P_a \frac{dP^a}{d\tau} = -\frac{2}{c^2} P_a F^a = -\frac{2}{c^2} \mathbf{g}(P, F) .$$

Hence, the mass of the particle is constant in (proper) time if and only if F is always orthogonal to the four-momentum. A four-force satisfying

$$\mathbf{g}(S_p, F(p, S_p)) = 0 \quad \text{for all } p \in \mathbb{M}^4 \text{ and } S_p \in T_p\mathbb{M}^4 \text{ with } \mathbf{g}(S_p, S_p) = -c^2$$

is said to be **mechanical**. A discussion on the existence and uniqueness of the Cauchy problem for four-forces appears in [1].

A fundamental case of mechanical four-force is the Lorentz four-force acting on a charged particle with mass m, charge q (which is invariant, like the mass) and worldline γ, which has the relativistic expression

$$F^a(\gamma(\tau), \dot{\gamma}(\tau)) = q\, \dot{\gamma}_b(\tau)\, F^{ab}\,(\gamma(\tau)) \ , \tag{7.43}$$

Above, the antisymmetric tensor $\mathbb{M}^4 \ni p \mapsto F^{ab}(p)$ is the **electromagnetic tensor field**, which accounts for the electric and magnetic fields in every Minkowskian reference system:

$$[F^{ab}_{\mathscr{R}}]_{a,b=0,1,2,3} = \left[\begin{array}{cccc} 0 & E_x/c & E_y/c & E_z/c \\ \hline 0 & 0 & B_z/c & -B_y/c \\ & -B_z/c & 0 & B_x/c \\ -E_x/c & B_y/c & -B_x/c & 0 \end{array}\right] \tag{7.44}$$

where E^α and B^β are the components of the electric and magnetic field in the reference frame $\mathscr{R}$. A direct computation proves that

$$F^0 = \frac{q\mathbf{E}\cdot\mathbf{v}_{\mathscr{R}}}{\sqrt{1-\frac{\mathbf{v}^2_{\mathscr{R}}}{c^2}}} , \qquad F^\alpha = \frac{qE^\alpha}{\sqrt{1-\frac{\mathbf{v}^2_{\mathscr{R}}}{c^2}}} + \frac{q}{c}\left(\frac{\mathbf{v}_{\mathscr{R}}}{\sqrt{1-\frac{\mathbf{v}^2_{\mathscr{R}}}{c^2}}} \times \mathbf{B}\right)^\alpha \qquad \alpha = 1, 2, 3 ,$$

where the spatial components reduce to the standard classical formula when $\|\mathbf{v}_{\mathscr{R}}\| \ll c$, whereas the temporal component tends to the classical power of Lorentz' force.

Antisymmetry of F^{ab} is responsible for the conservation of m during the evolution of the particle:

$$q^{-1} F^a P_a = F^{ba} \dot{\gamma}_b P_a = m F^{ba} \dot{\gamma}_a \dot{\gamma}_b = m F^{ba} \dot{\gamma}_b \dot{\gamma}_a = -m F^{ba} \dot{\gamma}_b \dot{\gamma}_a = -m F^{ba} \dot{\gamma}_a \dot{\gamma}_b = 0\,.$$

7.4.9 The Relativistic vis viva Equation

A mechanical four-force satisfies the relativistic version of the relation connecting kinetic energy and power, known as the *vis viva theorem* in classical mechanics. Indeed, if a four-force acts on a material particle of (constant) mass $m > 0$, the mechanical-force constraint reads in components of a Minkowskian reference system $\mathscr{R}$, where $\dot{\gamma}(\tau) = \dot{\gamma}^0_{\mathscr{R}} \partial_{x^0} + \sum_{\alpha=1}^{3} \dot{\gamma}^\alpha_{\mathscr{R}} \partial_{x^\alpha}$ is the four-velocity of the particle,

$$\sum_{\alpha=1}^{3} \dot{\gamma}^\alpha_{\mathscr{R}} F^\alpha_{\mathscr{R}} = \dot{\gamma}^0_{\mathscr{R}} F^0_{\mathscr{R}}\,,$$

namely

$$\sum_{\alpha=1}^{3} \frac{v^\alpha_{\mathscr{R}}}{\sqrt{1-\left(\frac{\mathbf{v}_{\mathscr{R}}}{c}\right)^2}} F^\alpha_{\mathscr{R}} = \frac{c}{\sqrt{1-\left(\frac{\mathbf{v}_{\mathscr{R}}}{c}\right)^2}} F^0_{\mathscr{R}}$$

which implies

$$\sum_{\alpha=1}^{3} v^\alpha_{\mathscr{R}} F^\alpha_{\mathscr{R}} = c F^0_{\mathscr{R}} = \frac{d c P^0_{\mathscr{R}}}{d\tau} = \frac{d(c P^0_{\mathscr{R}} - mc^2)}{d\tau} = \frac{d K_{\mathscr{R}}}{d\tau}\,, \tag{7.45}$$

where we have used (7.40), noticing that m is constant. In summary, we have found that

$$\sum_{\alpha=1}^{3} v^\alpha_{\mathscr{R}} F^\alpha_{\mathscr{R}} = \frac{d K_{\mathscr{R}}}{d\tau}\,. \tag{7.46}$$

Since, for $x^0 = ct$,

$$\frac{d}{d\tau} = \frac{dt}{d\tau}\frac{d}{dt} = \frac{1}{\sqrt{1-\left(\frac{\mathbf{v}_{\mathscr{R}}}{c}\right)^2}}\frac{d}{dt} = \left(1 + O\left(\left(\frac{\mathbf{v}_{\mathscr{R}}}{c}\right)^2\right)\right)\frac{d}{dt}$$

Equation (7.46), for $\|\mathbf{v}_{\mathscr{R}}\| \ll c$, reduces to the standard result connecting the power of the total force acting on a material point and the derivative of the kinetic energy, known as the *vis viva equation*. On the other hand, this identity corroborates the idea that $K_{\mathscr{R}}$ defined in (7.40) can be interpreted as the kinetic energy in the relativistic realm.

Remarks 7.41 As a by-product of (7.45) above, we see that the temporal component $F^0_{\mathscr{R}}$ of a mechanical four-force, up to the factor $1/c$, always coincides with the power of the spatial component of the four-force according to the classical definition,

$$F^0_{\mathscr{R}} = \frac{1}{c} \sum_{\alpha=1}^{3} F^{\alpha}_{\mathscr{R}} v^{\alpha}_{\mathscr{R}}.$$

We stress that this result is valid for mechanical four-forces only. Also observe that in the classical regime $v \ll c$, $F^0_{\mathscr{R}}$ is therefore supposed to be negligible. This fact corresponds to the common classical evidence of the validity of the mass conservation law. ■

7.5 Dynamics II: The Stress-Energy Tensor for Macroscopic Systems

This section is devoted to the introduction of a fundamental tool in relativistic theories, the so-called *stress-energy tensor*. In field theory and particle physics, that mathematical object arises within the variational approach to dynamics as a by-product of *Noether's theorem*, when assuming that the Lagrangian density of the system is translationally invariant in space and time in Minkowski spacetime. Not all physical systems—especially macroscopic ones—admit a variational formulation and for that reason we adopt here a more general viewpoint to justify the introduction of that mathematical tool in the formalism.

7.5.1 The Non-interacting Gas of Particles

We now focus attention on continuous systems, and the most elementary one is a gas of massive non-interacting particles. This system is described by two ingredients:

(i) a C^1 assignment of timelike worldlines whose four-velocities define a C^1 timelike future-directed vector field $\mathbb{M}^4 \ni p \mapsto V(p) \in T_p\mathbb{M}^4$ in Minkowski spacetime, everywhere satisfying $V^a V_a = -c^2$. V is the **four-velocity field** of the continuous body we are studying;

(ii) a C^1 scalar field $\mu_0 : \mathbb{M}^4 \to [0, +\infty)$ representing the mass density in the local rest frames of the particles of the gas. This rest frame is defined at each event p by $\mathscr{R}_0 := \frac{1}{c}V(p)$. μ_0 is the **intrinsic mass density** of the continuous body we are studying.

Heuristically, we can think of $\mu_0(p)$ as the ratio

$$\mu_0(p) = \frac{\delta m_p}{\delta v_{0p}} ,$$

where δv_{0p} is the volume of a small portion of the gas in the rest space $\Sigma^{(\mathscr{R}_0)}_{x^0(p)}$ and δm_p is the total mass of the particles included in that volume: the particles in the said small volume can be considered at rest in $\mathscr{R}_0$. This heuristic (but strongly physically motivated!) view leads to the following requirement on the scalar field μ_0. If we fix a Minkowskian reference frame $\mathscr{R}$ ($\neq \mathscr{R}_0$ in general) and indicate by $\delta v^{(\mathscr{R})}_{0p}$ the volume occupied by the considered small portion of particles measured in $\mathscr{R}$, according to (7.32), the mass density $\mu_{\mathscr{R}}$ referred to the gas and measured at rest with $\mathscr{R}$ must be

$$\mu_{\mathscr{R}} = \mu_0 \frac{\delta v_{0p}}{\delta v^{(\mathscr{R})}_{0p}} = \mu_0 \frac{1}{\sqrt{1 - \frac{\|\mathbf{v}_p\|^2}{c^2}}} , \tag{7.47}$$

where $\mathbf{v}_p$ is the velocity associated with the four-velocity $V(p)$ in the rest space of $\mathscr{R}$. Observing that

$$\frac{1}{\sqrt{1 - \frac{\|\mathbf{v}_p\|^2}{c^2}}} = \frac{1}{c} V^0_{\mathscr{R}} = -\frac{1}{c}\mathbf{g}(V, \mathscr{R}),$$

we find

$$\mu_{\mathscr{R}}(p) := -\frac{\mu_0(p)}{c}\mathbf{g}(V, \mathscr{R}) . \tag{7.48}$$

We henceforth assume (7.48) as the *definition* of the **relative mass density**. For future convenience, we introduce the covariant unit vector,

$$n^{(\mathscr{R})} := -\mathbf{g}(\mathscr{R}, \ ) \qquad \text{namely} \quad n^{(\mathscr{R})}_a := -g_{ab}\mathscr{R}^b .$$

Hence the definition above can be rephrased as

$$\mu_{\mathscr{R}}(p) := \frac{\mu_0(p)}{c}\langle V(p), n^{(\mathscr{R})}\rangle . \tag{7.49}$$

Remarks 7.42 Notice that, introducing a Minkowskian system of coordinates $x^0 = ct, x^1, x^2, x^3$ co-moving with $\mathscr{R}$, we immediately have that

$$\mathscr{R} = \partial_{x^0}\,, \qquad n^{(\mathscr{R})} = dx^0\,. \tag{7.50}$$

We shall take advantage of those identities shortly. ■

7.5.2 *Mass Conservation Law in Local Form for the Non-interacting Gas*

Consider a spatial portion of continuum $B_0 \in \Sigma_0^{(\mathscr{R})}$ at time $t = 0$ in a Minkowskian system of coordinates $x^0 = ct, x^1, x^2, x^3$ co-moving with an inertial reference system $\mathscr{R}$. B_0 is assumed here to be a generic sufficiently regular measurable set of $\mathbb{R}^3$, for instance a coordinate 3-ball. If V is C^1, its evolution through the *local flow*[4] ϕ_t of V gives rise to a measurable regular set $B_t := \phi_t(B_0) \in \Sigma_t^{(\mathscr{R})}$ as well, since this set is the image of B_0 through a C^1 diffeomorphism. On the physical side, *since we are assuming that the particles do not interact and evolve freely*, we must admit that the mass is conserved. Hence,

$$\int_{\Sigma_t^{(\mathscr{R})}\cap B_t} \mu_{\mathscr{R}}(ct, x^1, x^2, x^3)dx^1dx^2dx^3 = \int_{\Sigma_0^{(\mathscr{R})}\cap B_0} \mu_{\mathscr{R}}(0, x^1, x^2, x^3)dx^1dx^2dx^3\,, \tag{7.51}$$

for every choice of B_0. We intend to write that requirement in local form. To this end observe that (7.49) permits one to rephrase the identity above as, for $p \equiv (ct, x^1, x^2, x^3)$ and $p_0 \equiv (0, x^1, x^2, x^3)$,

$$\int_{\Sigma_t^{(\mathscr{R})}\cap B_t} \mu_0(p)\langle V(p), n^{(\mathscr{R})}\rangle dx^1dx^2dx^3$$
$$+ \int_{\Sigma_0^{(\mathscr{R})}\cap B_0} \mu_0(p_0)\langle V(p_0), -n^{(\mathscr{R})}\rangle dx^1dx^2dx^3 = 0\,.$$

In turn, this identity has a geometric interpretation with respect to the Euclidean geometry of the space $\mathbb{R}^4$ where the coordinates of $\mathscr{R}$ range. In fact

(1) the components of $n_i^{(\mathscr{R})} = \delta_i^0$ form the $\mathbb{R}^4$ vector normal to the surfaces $x^0 = 0$ and $x^0 = ct$, and this vector is normalized *with respect to the standard Euclidean metric of* $\mathbb{R}^4$;

[4] We assume that the flow of V is complete (and this happens in particular if the C^1 vector field V has bounded components in Cartesian coordinates of $\mathbb{M}^4$) or that the interval $[0, t]$ and B_0 are sufficiently small to guarantee that $B_t \in \Sigma_t^{(\mathscr{R})}$ exists.

(2) the pairing $\langle V(p), n^{(\mathscr{R})}\rangle$ is nothing but the standard $\mathbb{R}^4$ scalar product of the vector $(V^0_{\mathscr{R}}(p), V^1_{\mathscr{R}}(p), V^2_{\mathscr{R}}(p), V^3_{\mathscr{R}}(p))^t$ and the vector $(n_0^{(\mathscr{R})}, n_1^{(\mathscr{R})}, n_2^{(\mathscr{R})}, n_3^{(\mathscr{R})})^t$, and the same holds for the other integrand;

(3) we can add to the two integrals above a vanishing integral computed on the lateral boundary L of a tube $S \subset \mathbb{R}^4$ made of the integral lines of V and joining the two faces $\Sigma_0^{(\mathscr{R})} \cap B_0$ and $\Sigma_t^{(\mathscr{R})} \cap B_t$:

$$\int_L \mu_0(q)\langle V(q), n_L^{(\mathscr{R})}\rangle d\nu_{\mathscr{R}}(q)\,,$$

$\nu_{\mathscr{R}}$ denoting the natural measure on the 3-surfaces in $\mathbb{R}^4$. This boundary integral simply vanishes because the unit vector $n_L^{(\mathscr{R})}$ (viewed as a vector in $\mathbb{R}^4$) normal to L is consequently normal to $(V^0_{\mathscr{R}}, V^1_{\mathscr{R}}, V^2_{\mathscr{R}}, V^3_{\mathscr{R}})^t$ and produces the result 0 when multiplied with $(V^0_{\mathscr{R}}, V^1_{\mathscr{R}}, V^2_{\mathscr{R}}, V^3_{\mathscr{R}})^t$ in the integrand;

(4) $n^{(\mathscr{R})}$ on $\Sigma_t^{(\mathscr{R})} \cap B_t$ and $-n^{(\mathscr{R})}$ on $\Sigma_0^{(\mathscr{R})} \cap B_0$ define, in components, outward vectors on the boundary of S contained in the tube with bases B_0 and B_t.

Summing up, we have that, in coordinates, (7.51) is equivalent to

$$\int_{+\partial S} \mu_0(q)\langle V(q), n^{(\mathscr{R})}(q)\rangle d\nu_{\mathscr{R}}(q) = 0\,. \tag{7.52}$$

The divergence theorem in $\mathbb{R}^4$ implies that

$$\int_S \nabla_a[\mu_0(q)V(q)^a]dx^0dx^1dx^2dx^3 = 0\,.$$

Evidently, since the Lebesgue measure of $\mathbb{R}^4$ is translationally invariant, the found result is also valid if the temporal interval is $[t_1, t_1 + t]$ instead of $[0, t]$. Exploiting the arbitrariness of S and the fact that the integrand is continuous, we have that the validity of (7.51) implies the **local form of the mass conservation law**

$$\nabla_a(\mu_0 V^a) = 0 \quad \text{everywhere in } \mathbb{M}^4. \tag{7.53}$$

In fact, if this identity were false at some event, then in a neighborhood of that event $\nabla_a(\mu_0 V^a)$ would be strictly positive or strictly negative. Choosing S as a sufficiently small tube—with bases B_0 and B_t—constructed around the said event, we would have that (7.51) is false simply by going backward along our reasoning. Instead, we have supposed that the identity holds for every such tube.

The local form of the mass conservation law implies back the integral form (7.51) of the same physical law again by going backward along our reasoning.

Remarks 7.43

(1) A crucial observation is that *the identity (7.53) does not depend on the Minkowskian system of coordinates because these coordinates are Cartesian with respect to the affine structure of* $\mathbb{M}^4$ *and we can apply Definition 3.19.*

(2) It is worth stressing that (7.53) is generally false for a gas of interacting particles where particles are destroyed or created.

■

7.5.3 *Four-Momentum Conservation in Local Form for the Non-interacting Gas*

Let us pass to discuss the conservation of the four-momentum associated with our gas of non-interacting particles. If $\mathscr{R}$ is a given inertial reference frame, the natural definition of **four-momentum density** in $\mathscr{R}$ is

$$\mathcal{P}_{\mathscr{R}} := \mu_{\mathscr{R}} V .$$

Notice that the definition depends on the reference frame just because we want eventually to integrate $\mathcal{P}_{\mathscr{R}}$ over the rest space of that reference frame. Yet, for that reason the relevant mass density is here $\mu_{\mathscr{R}}$ and not μ_0. Taking advantage of (7.49), we can rearrange this expression to

$$\mathcal{P}_{\mathscr{R}} = \frac{\mu_0(p)}{c} V \langle V(p), n^{(\mathscr{R})} \rangle . \tag{7.54}$$

This expression is more interesting since it can be interpreted in the following way, which will prove to be quite universal. First, we introduce the so-called **stress-energy tensor** field for the gas of non-interacting particles,

$$T := \mu_0 V \otimes V \qquad \text{namely} \quad T^{ab} := \mu_0 V^a V^b , \tag{7.55}$$

and from now on we use the following notation: if $\omega : \mathbb{M}^4 \ni p \mapsto \omega(p) \in T_p^*\mathbb{M}$ is a C^1 covariant vector field, then $T(\omega)$ is the contravariant vector field obtained by contraction of ω with the left index of T (we omit to specify the event and the following identity is supposed to hold pointwise in $\mathbb{M}^4$)

$$T(\omega)^j := T^{jk}\omega_k . \tag{7.56}$$

With this definition we immediately have that, referring to Minkowskian coordinates $x^0 = ct, x^1, x^2, x^3$ co-moving with the inertial reference system $\mathscr{R}$,

$$\mathcal{P}^a_{\mathscr{R}} = \langle T(dx^a), n_{\mathscr{R}} \rangle , \tag{7.57}$$

so that we can define the **components of the total momentum** at time t with respect to Minkowskian coordinates $x^0 = ct, x^1, x^2, x^3$ co-moving with $\mathscr{R}$ as

$$P^a_{\mathscr{R}}(t) := \int_{\Sigma_t^{(\mathscr{R})}} \langle T(dx^a), n_{\mathscr{R}} \rangle dx^1 dx^2 dx^3 \,, \quad a = 0, 1, 2, 3 \,. \tag{7.58}$$

Remarks 7.44 We stress that up to now we do *not* know whether these components, which are defined for every fixed Minkowskian coordinate system, really define a contravariant vector when varying the coordinate systems. ■

To conclude we prove that the above quantities are in fact conserved, i.e., constant, in time. This result arises *only* from a local equation satisfied by T which, again, does not depend on the reference frame. Later we also prove that the said local equation implies that the components $P^a_{\mathscr{R}}$ actually define a vector P that does *not* depend on $\mathscr{R}$.

Theorem 7.45 *Consider a gas of non-interacting particles with C^1 proper mass density μ_0 and C^1 field of four-velocities V. If the local mass conservation law (7.53) is valid,*

$$\nabla_a(\mu_0 V^a) = 0$$

and the particles have inertial motion (see below), so that

$$V^b \nabla_b V^a = 0 \,, \tag{7.59}$$

then the stress-energy tensor (7.55) satisfies

$$\nabla_a T^{ab} = 0 \quad \textit{everywhere in } \mathbb{M}^4. \tag{7.60}$$

Proof It is just a matter of computation:

$$\nabla_a T^{ab} = \nabla_a \left(\mu_0 V^a V^b \right) = (\nabla_a(\mu_0 V^a))V^b + \mu_0 V^a \nabla_a V^b = 0 + 0 \,.$$

□

Remarks 7.46

(1) Equation (7.59) is the so-called **geodesic equation**, which describes the motion of free-falling particles in a general spacetime. The fact that (7.60) is true in view of that equation (and (7.53)) is quite relevant, since it allows us to generalize the statement of the theorem to General Relativity. In Minkowski spacetime, the meaning of (7.59) is the following. In Minkowskian coordinates, if $x^b = x^b(\tau)$ is the worldline integral curve of V,[5] we have

$$0 = V^b(p)\nabla_b V^a(p) = \frac{dx^b}{d\tau}\frac{\partial V^a}{\partial x^b} = \frac{d}{d\tau}V^a(x(\tau)) = \frac{d}{d\tau}\frac{d}{d\tau}x^a(\tau) = \frac{d^2x^a}{d\tau^2},$$

so that

$$x^a(\tau) = x^a(0) + V^a(x^0(0), x^1(0), x^2(0), x^3(0))\tau$$

and, as expected, the motion of the particles of the continuous body is given by affine segments according to the principle of inertia. *Vice versa*, (7.59) is satisfied if the worldlines of the particles are affine segments.

(2) As before, according to Definition 3.19, the local conservation equation (7.60) is intrinsic and does not depend on the used Minkowskian coordinate system.

■

We conclude with this pair of very nice consequences of the sole equation (7.60) and of definition (7.58), which are completely *independent* of the specific form (7.55) of the stress-energy tensor. From now on, a **tube** $U \subset \mathbb{M}^4$ is a connected open set with C^1 boundary whose intersection with every rest space $\Sigma^{(\mathscr{R})}_{x^0}$ defines an open connected set $U^{(\mathscr{R})}_{x^0}$ with C^1 boundary, which is bounded in the $\mathbb{R}^3$ Euclidean topology of the coordinates. As a trivial example of U, think of an infinite cylinder defined in Minkowskian coordinates of an inertial reference frame with constant basis

$$\{(x^1, x^2, x^3) \in \mathbb{R}^3 \mid (x^1)^2 + (x^2)^2 + (x^3)^2 < R^2\}$$

for some real $R > 0$.

Theorem 7.47 *Let $\mathbb{M}^4 \ni p \mapsto T(p) \in T_p\mathbb{M}^4 \otimes T_p\mathbb{M}^4$ be a C^1 tensor field and assume that the components $P^a_{\mathscr{R}}(t)$ are defined as in (7.58) for every Minkowskian system of coordinates $x^0 = ct, x^1, x^2, x^3$ co-moving with every inertial reference frame $\mathscr{R}$. If T smoothly vanishes before getting to the boundary ∂U of a tube U, then*

(a) *$P^a_{\mathscr{R}}(t) = P^a_{\mathscr{R}}(0)$ $(a = 0, 1, 2, 3)$.*
(b) *The four functions $P^a_{\mathscr{R}}$ define pointwise a vector P which does not depend on the choice of inertial reference system $\mathscr{R}$ and of the co-moving system of Minkowskian coordinates.*

---

[5] We can interpret the parameter τ of the integral lines as the proper time because of the constraint $V^aV_a = \eta_{ab}\frac{dx^a}{d\tau}\frac{dx^b}{d\tau} = -c^2$.

Proof

(a) Referring to the $\mathbb{R}^4$ space constructed out of the Minkowskian coordinates $x^0 = ct, x^1, x^2, x^3$ of $\mathscr{R}$, define

$$S := U \cap \{(x^0 = c\tau, x^1, x^2, x^3) \in \mathbb{R}^4 \mid 0 \leq \tau < t\}\,.$$

Working within that $\mathbb{R}^4$ space, with the same argument we used to prove (7.52) but proceeding backward, the identity $\nabla_a T^{ab} = 0$ implies in particular

$$0 = \int_U \nabla_a T^{ab} dx^0 dx^1 dx^2 dx^3 = \int_{+\partial S} n_a^{(S\cap U)} T^{ab} dv$$

$$= \int_{\Sigma_t^{(\mathscr{R})} \cap U_t^{(\mathscr{R})}} n_a^{(\mathscr{R})} T^{ab} dx^1 dx^2 dx^3 + \int_{\Sigma_0^{(\mathscr{R})} \cap U_0^{(\mathscr{R})}} (-n^{(\mathscr{R})})_a T^{ab} dx^1 dx^2 dx^3\,,$$

where we have used the fact that T^{ab} smoothly vanishes before reaching the lateral boundary of S, so we have omitted the surface integral arising from that part of the boundary of S. The found identity can be rearranged to

$$P^a_{\mathscr{R}}(t) = \int_{\Sigma_t^{(\mathscr{R})}} n_a^{(\mathscr{R})} T^{ab} dx^1 dx^2 dx^3 = \int_{\Sigma_0^{(\mathscr{R})}} n_a^{(\mathscr{R})} T^{ab} dx^1 dx^2 dx^3 = P^a_{\mathscr{R}}(0)$$

that is the thesis in (a).

(b) Consider another reference frame $\mathscr{R}'$ together with $\mathscr{R}$ and a rest space of the former $\Sigma_{t'}^{\mathscr{R}'}$ such that the portion S of U between the surfaces $\Sigma_0^{(\mathscr{R})}$ and $\Sigma_{t'}^{(\mathscr{R}')}$ stays far from $\Sigma_0^{(\mathscr{R})} \cap \Sigma_{t'}^{(\mathscr{R}')}$. Notice that now S is a bounded solid with non-parallel bases (see Fig. 7.5). Endow both reference frames with respective

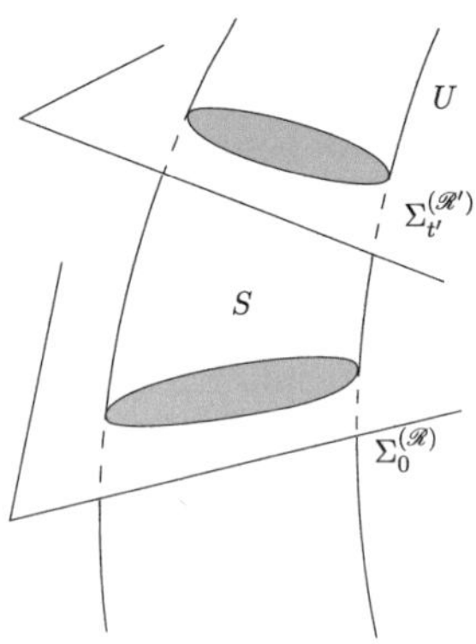

Fig. 7.5 Proof of (b) in Theorem 7.47

systems of co-moving Minkowskian coordinates and suppose that, for every constant covariant vector $\omega \in (T^4)^*$,

$$\int_{\Sigma_0^{(\mathscr{R})}} \langle T(\omega), n^{(\mathscr{R})} \rangle dx^1 dx^2 dx^3 = \int_{\Sigma_{t'}^{(\mathscr{R}')}} \langle T(\omega), n^{(\mathscr{R}')} \rangle dx'^1 dx'^2 dx'^3 . \tag{7.61}$$

Under this hypothesis, the linear maps

$$(T^4)^* \ni \omega \mapsto \int_{\Sigma_0^{(\mathscr{R})}} \langle T(\omega), n^{(\mathscr{R})} \rangle dx^1 dx^2 dx^3$$

and

$$(T^4)^* \ni \omega \mapsto \int_{\Sigma_{t'}^{(\mathscr{R}')}} \langle T(\omega), n^{(\mathscr{R}')} \rangle dx'^1 dx'^2 dx'^3$$

coincide, thus they define the same vector P such that $\langle P, \omega \rangle$ coincides with the common value of the right-hand sides above. By construction, the components of P in every $\mathscr{R}$ and co-moving Minkowskian coordinate system are just the constant right-hand side of (7.58) at every time t, by part (a). To conclude the proof, it is therefore sufficient to establish (7.61). To this end, we work in the $\mathbb{R}^4$ of the coordinates x^0, x^1, x^2, x^3. They are related to the coordinates x'^0, x'^1, x'^2, x'^3 through a Poincaré transformation

$$x'^a = \Lambda^a{}_b x^b + C^a .$$

The three-dimensional surface $\Sigma_{t'}^{(\mathscr{R}')}$ is therefore described as the plane $x'^0 = ct'$ in $\mathbb{R}^4$:

$$\Sigma_{t'}^{(\mathscr{R}')} \equiv \{(x^0, x^1, x^2, x^3) \in \mathbb{R}^4 \mid \Lambda^0{}_b x^b - ct' - C^0 = 0\} .$$

We shall suppose that the intersection of $\Sigma_0^{(\mathscr{R})}$ and $\Sigma_{t'}^{(\mathscr{R}')}$ does not pass through the solid S. That is possible, since S is bounded, just by taking $t' > 0$ sufficiently large. The generic case is encompassed when taking statement (a) into account (replacing $\mathscr{R}$ by $\mathscr{R}'$).

Since ω is constant and thus $\nabla_a(T(\omega)^a) = \nabla_a(T^{ab}\omega_b) = (\nabla_a T^{ab})\omega_b = 0$, the $\mathbb{R}^4$ divergence theorem implies, with the same argument as above,

$$\int_{\Sigma_0^{(\mathscr{R})}} \langle T(\omega), n^{(\mathscr{R})} \rangle dx^1 dx^2 dx^3 = \int_{\Sigma_{t'}^{(\mathscr{R}')}} T(\omega)^a N_a d\nu , \tag{7.62}$$

where ν is the natural measure on $\Sigma_{t'}^{(\mathscr{R}')}$ induced by that of $\mathbb{R}^4$ and (N_0, N_1, N_2, N_3) is the normal vector to $\Sigma_{t'}^{(\mathscr{R}')}$ in $\mathbb{R}^4$ with unit *Euclidean norm* in $\mathbb{R}^4$. From the definition of $\Sigma_{t'}^{(\mathscr{R}')}$ it holds

$$N^a = \frac{\Lambda^0{}_a}{\sqrt{\sum_{b=0}^{3}(\Lambda^0{}_b)^2}}, \quad a = 0, 1, 2, 3, \tag{7.63}$$

As a consequence, if we parametrize $\Sigma_{t'}^{(\mathscr{R}')}$ with the coordinates x^1, x^2, x^3 through the canonical projection $\pi : \mathbb{R}^4 \ni (x^0, x^1, x^2, x^3) \mapsto (x^1, x^2, x^3)$, the volume $Vol(G)$ of a subset $G \subset \Sigma_{t'}^{(\mathscr{R}')}$ with respect to the $\mathbb{R}^4$ measure induced on $\Sigma_{t'}^{(\mathscr{R}')}$ and that of its projection $\pi(G)$ are related by the relation, where the dot denotes the standard scalar product in $\mathbb{R}^4$,

$$\frac{Vol(\pi(G))}{Vol(G)} = N \cdot n_{\mathscr{R}} = \sum_{a=0}^{3} \frac{\Lambda^0{}_a}{\sqrt{\sum_{b=0}^{3}(\Lambda^0{}_b)^2}} \delta^0_a = \frac{\Lambda^0{}_0}{\sqrt{\sum_{b=0}^{3}(\Lambda^0{}_b)^2}} .$$

In other words, the induced measure on $\Sigma_{t'}^{(\mathscr{R}')}$ can be written

$$dv = \frac{\sqrt{\sum_{b=0}^{3}(\Lambda^0{}_b)^2}}{\Lambda^0{}_0} dx^1 dx^2 dx^3 . \tag{7.64}$$

Collecting (7.63) and (7.64), the right-hand side of (7.62) has consequently the form

$$\int_{\Sigma_{t'}^{(\mathscr{R}')}} T(\omega)^a N_a dv = \int_{\Sigma_{t'}^{(\mathscr{R}')}} \frac{\Lambda^0{}_a T(\omega)^a}{\sqrt{\sum_{b=0}^{3}(\Lambda^0{}_b)^2}} \frac{\sqrt{\sum_{b=0}^{3}(\Lambda^0{}_b)^2}}{\Lambda^0{}_0} dx^1 dx^2 dx^3$$

$$= \int_{\Sigma_{t'}^{(\mathscr{R}')}} \frac{T'(\omega)^0}{\Lambda^0_0} dx^1 dx^2 dx^3 .$$

Now suppose that the two Minkowskian coordinate systems are connected by a pure transformation $x'^a = (\Lambda_{\dot\gamma})^a{}_b x^b$. Here $\dot\gamma$ is the four-velocity of every point at rest with $\mathscr{R}$. Its components in $\mathscr{R}'$ define the pure transformation $\Lambda_{\dot\gamma}$. As before, we parametrize the spatial volumes in $\Sigma_{t'}^{(\mathscr{R}')}$ in terms of the coordinates x^1, x^2, x^3 co-moving with $\mathscr{R}$ taking advantage of the canonical projection $\pi : \Sigma_{t'}^{(\mathscr{R}')} \to \Sigma_0^{(\mathscr{R})}$. The mathematical interplay between the two spatial volumes is the same as for the contraction of volumes: there is a spatial volume at rest with $\mathscr{R}$ evolving with the integral lines of $\dot\gamma$ and we want to evaluate the corresponding volume in $\mathscr{R}'$ by crossing the worldlines emanating from the former volume with the rest spaces $\Sigma_{t'}^{(\mathscr{R}')}$ of $\mathscr{R}'$.[6] From the contraction-of-volumes formula (7.32) we know that

$$\det\left[\frac{\partial x'^{\alpha}}{\partial x^{\beta}}\right]_{\alpha,\beta=1,2,3} = \sqrt{1-\frac{v^2}{c^2}} = \frac{1}{\dot{\gamma}^0/c} = \frac{1}{(\Lambda_{\dot{\gamma}})^0{}_0}\,.$$

If the Poincaré transformation connecting $\mathscr{R}$ and $\mathscr{R}'$

$$x'^a = \Lambda^a{}_b x^b + C^a$$

is not pure, leaving fixed the reference frames but changing the co-moving coordinates, we can always reduce to a pure *Lorentz* transformation by a translation of the origin of the four axes in $\mathscr{R}'$ and a spatial rotation (7.28) in the same space. Let us discuss this point with some details. First, we stress that these changes of coordinates are isometries, so that they do not change the measure on $\Sigma_{t'}^{(\mathscr{R}')}$, which is the Lebesgue one and thus invariant under isometries. Similarly, a translation along t' does not affect the measure on the rest spaces of $\mathscr{R}'$. The relevant measure on $\Sigma_{t'}^{(\mathscr{R}')}$ is always written as $dx'^1dx'^2dx'^3$ independently of the choice of the Minkowskian coordinates co-moving with $\mathscr{R}'$. With a four-translation in $\mathscr{R}'$, without changing the measure on $\Sigma_{t'}^{(\mathscr{R}')}$, we pass from $x'^a = \Lambda^a{}_b x^b + C^a$ to $x'^a = \Lambda^a{}_b x^b$. Next, due to the decomposition theorem of the Lorentz group, $\Omega_{R^{-1}}\Lambda$ is pure if $R \in SO(3)$ is suitably chosen. As said above, the measure on $\Sigma_t'^{(\mathscr{R}')}$ is not affected by this change of coordinates. Furthermore, the product with a spatial rotation does not affect the term $\Lambda^0{}_0$, as one immediately proves by direct inspection:

$$(\Omega_{R^{-1}}\Lambda)^0{}_0 = \Lambda^0{}_0 \quad .$$

We conclude that

$$\det\left[\frac{\partial x'^{\alpha}}{\partial x^{\beta}}\right]_{\alpha,\beta=1,2,3} = \frac{1}{\Lambda^0{}_0}$$

is valid in general when $x'^a = \Lambda^a{}_b x^b + C^a$ holds (with $\Lambda \in O(1,3)\uparrow$), and this result gives the transformation law between the natural measures on the considered rest spaces:

$$dx'^1dx'^2dx'^3 = \frac{1}{\Lambda^0_0}dx^1dx^2dx^3\,.$$

[6] We observe that the interplay of the two spatial volumes has nothing to do with the particles of the continuum we are studying, whose worldlines are described by V, and the reasoning above is of pure Lorentzian geometric nature.

So that

$$\int_{\Sigma_{t'}^{(\mathscr{R}')}} T(\omega)^a N_a dv = \int_{\Sigma_{t'}^{(\mathscr{R}')}} \frac{T'(\omega)^0}{\Lambda^0_0} dx^1 dx^2 dx^3 = \int_{\Sigma_{t'}^{(\mathscr{R}')}} T'(\omega)^0 dx'^1 dx'^2 dx'^3$$

$$= \int_{\Sigma_{t'}^{(\mathscr{R}')}} \langle T'(\omega), n^{(\mathscr{R}')} \rangle dx'^1 dx'^2 dx'^3 .$$

Inserting this result into (7.62) we obtain (7.61), concluding the proof.

□

Remarks 7.48

(1) The proof also shows that the hypothesis concerning the existence of the tube U can be relaxed by assuming a sufficiently rapid decay of T at spatial infinity of an inertial reference frame. We do not pursue this generalization here.

(2) Introducing the *divergence theorem in covariant form* with respect to the *Levi-Civita connection*, the proof of (7.61) would be trivial.

(3) What is really fundamental in the proof of Theorem 7.47 is the divergence theorem in $\mathbb{R}^4$, which is exploited just in view of the validity of the equation $\nabla_a T(\omega)^a = 0$, where $T(\omega)^a = \omega_b T^{ab}$ and $\omega = dx^k$ (referred to some Minkowskian coordinate system x^0, x^1, x^2, x^3). Actually, $\nabla_a T(\omega)^a = 0$ is more generally valid if T satisfies (7.60) and ω is a constant covariant field or, more weakly, if

(i) T is *symmetric* and

(ii) ω satisfies the identity, known as the **Killing condition**,

$$\nabla_a \omega_b + \nabla_b \omega_a = 0 . \tag{7.65}$$

Indeed, under this pair of assumptions,

$$\nabla_a \left(\omega_b T^{ba}\right) = (\nabla_a \omega_b) T^{ba} + \omega_b \nabla_a T^{ba} = \frac{1}{2}(\nabla_a \omega_b)(T^{ba} + T^{ab}) + \omega_b \nabla_a T^{ab}$$

$$= \frac{1}{2}(\nabla_a \omega_b + \nabla_b \omega_a) T^{ba} + \omega_b \nabla_a T^{ab} = 0 .$$

When $\nabla_a T(\omega)^a = 0$, the quantities

$$Q_{\mathscr{R}}[\omega] = \int_{\Sigma_t^{(\mathscr{R})}} \langle T(\omega), n^{(\mathscr{R})} \rangle dx^1 dx^2 dx^3$$

are conserved in time (they do not depend on t) and $Q_{\mathscr{R}}[\omega] = Q_{\mathscr{R}'}[\omega]$ for every choice of the reference frames $\mathscr{R}, \mathscr{R}'$. The proof is the same as that of Theorem 7.47.

Condition (7.65) can be stated more generally on manifolds and it determines the continuous isometries of a manifold obtained by constructing the flow of the contravariant version of its solutions ω. In that context, conserved quantities $Q[\omega]$ similarly arise. In this sense, the known relation between symmetries and conserved quantities can be recast in General Relativity.

(4) Within the hypotheses of (3), if $\mathbb{M}^4 \ni e \mapsto T_e(\omega) \in V$ is a C^1 timelike future-directed vector field, we can consider a tube between $\Sigma_t^{(\mathscr{R})}$ and $\Sigma_{t'}^{(\mathscr{R}')}$ with one basis $B \subset \Sigma_t^{(\mathscr{R})}$ and the other

$$B' \subset \Sigma_{t'}^{(\mathscr{R}')} ,$$

where $B' = \{\phi_s(e)\,|\, e \in B, s \in I\} \cap \Sigma_{t'}^{(\mathscr{R}')}$ and ϕ is the flow of the contravariant vector field $T(\omega)$, where $I \subset \mathbb{R}$ is a sufficiently large interval. If everything goes as expected when choosing B sufficiently regular and small, so that $\phi_s(B)$ is well defined on some interval $I \ni s$ and B' sufficiently close to B is well defined, a local version of the conservation of Q can be established:

$$\int_{B'} \langle T(\omega), n^{(\mathscr{R}')}\rangle dx'^1 dx'^2 dx'^3 = \int_B \langle T(\omega), n^{(\mathscr{R})}\rangle dx^1 dx^2 dx^3$$

with the same argument used in proving the local conservation of the mass, starting from $\nabla_a T^{ab} = 0$ and also taking advantage of some constructions and arguments exploited in the proof of (b) in Theorem 7.47. As a matter of fact, the validity of the integral identity above for a suitable large family of sets B is equivalent to the law $\nabla_a T^{ab} = 0$. The issue with the generality of the above integral identity is that $T(\omega)$ is not timelike for all physical systems. Quite often that vector field is causal and vanishes somewhere, or it is spacelike. In general, the validity of the integral equation above should be checked case by case.

(5) Sometimes the stress-energy tensor of a *single particle* is defined in terms of a *Dirac measure* (or *Dirac distribution*). Here is the sketch of this formalism. Fix a reference frame $\mathscr{R}$ and co-moving Minkowskian coordinates $ct = x^0, x^1, x^2, x^3$. Let us suppose that the worldline γ is described, in coordinates, by $x^a = x^a(\tau)$. As $\frac{dct}{d\tau} > 0$ because the worldline is timelike and future-directed, we can reparametrize the worldline using the coordinate t as new parameter. If the particle has mass $m > 0$, the mass density on $\Sigma_t^{\mathscr{R}}$ can be defined as the Dirac measure $\delta(\vec{x} - \vec{x}(t))$, where $\vec{x} = (x^1, x^2, x^3)^t$ is a generic point on $\Sigma_t^{(\mathscr{R})} \equiv \mathbb{R}^3$. Notice that $\vec{x}(t)$ is assigned at each time t. The measure is therefore concentrated on $\vec{x}(t)$. The stress-energy tensor can be defined, generalizing (7.55), as

$$T^{ab}(t, \vec{x}) = V^a(t) V^b(t) m J(t, \vec{x}) \delta(\vec{x} - \vec{x}(t))$$

where the four-velocity V is constructed out of the worldline γ and the Jacobian $J(t, \vec{x})$ is necessary to interpret the Dirac-delta density as a measure on the *instantaneous* rest space of the particle as in (7.47):

$$J(t, \vec{x}) = \sqrt{1 - \frac{1}{c^2}\left(\frac{d\vec{x}}{dt}\right)^2} .$$

In summary,

$$T(t, \vec{x}) = m\sqrt{1 - \frac{1}{c^2}\left(\frac{d\vec{x}}{dt}\right)^2}\, \delta(\vec{x} - \vec{x}(t))\, V(t) \otimes V(t) .$$

The object above, though obtained quite heuristically, possesses good mathematical properties in the appropriate theoretical context. For instance, viewed as a tensor-valued distribution on $\mathbb{M}^4$, it takes the same form independently of the choice of the reference frame (and co-moving coordinates). Furthermore, if viewed as a compact-support tensor-valued distribution on $\Sigma_t^{(\mathscr{R})}$, the four-momentum of the particle is obtained (exercise!) as the integral

$$mV^a(t) = \int_{\Sigma_t^{(\mathscr{R})}} T^{ab}(t, \vec{x}) n_b^{(\mathscr{R})} dx^1 dx^2 dx^3 .$$

■

Exercises 7.49

1. Prove that, assuming $T^{ab} = \mu_0 V^a V^b$, with V everywhere timelike and future-directed and $\mu_0 \geq 0$, then the conservation equation $\nabla_a T^{ab} = 0$ implies both inertial motion $V^a \nabla_a V^b = 0$ (where $\mu_0 \neq 0$) and mass conservation $\nabla_a(\mu_0 V^a) = 0$.
   Solution. $V_b \nabla_a T^{ab} = 0$ can be expanded to

$$0 = V_b(\nabla_a T^{ab}) = (V_b \nabla_a V^b)\mu_0 V^a + V^b V_b \nabla_a(\mu_0 V^a) ,$$

   where

$$V_b \nabla_a V^b = \frac{1}{2}\nabla_a(V_b V^b) = -\frac{1}{2}\nabla_a c^2 = 0 ,$$

   and thus it remains

$$-c^2 \nabla_a(\mu_0 V^a) = 0 .$$

   Taking this identity into account, by expanding $\nabla_a(\mu_0 V^a V^b) = 0$, one finds $\mu_0 V^a \nabla_a V^b = 0$. Hence $V^a \nabla_a V^b = 0$ where $\mu_0 \neq 0$.

2. Prove that, assuming $T^{ab} = \mu_0 V^a V^b$, with V everywhere timelike and future-directed and $\mu_0 \geq 0$, then $\nabla_a T^{ab} = f^b$, where $V_b f^b = 0$, can be interpreted as the equation of motion for the particles of the non-interacting gas under the action of the density of *mechanical* four-force f, and the density is measured at rest with the particles:

$$\mu_0 \frac{d}{d\tau}\left(\frac{dx^b}{d\tau}\right) = f^b , \tag{7.66}$$

where $x^b = x^b(\tau)$ is a worldline of a particle of the continuous body in Minkowskian coordinates. Prove also that the mass conservation law $\nabla_a(\mu_0 V^a) = 0$ is still valid and that μ_0 is constant along the said worldlines.
Solution. $V_b \nabla_a T^{ab} = V_b f^b = 0$ can be expanded to

$$0 = (V_b \nabla_a V^b)\mu_0 V^a + V^b V_b \nabla_a(\mu_0 V^a) ,$$

where

$$V_b \nabla_a V^b = \frac{1}{2}\nabla_a(V_b V^b) = -\frac{1}{2}\nabla_a c^2 = 0 ,$$

and thus it remains

$$-c^2 \nabla_a(\mu_0 V^a) = 0 ,$$

so that mass is conserved. Using this result in $\nabla_a T^{ab} = f^b$, we obtain $\mu_0 V^a \nabla_a V^b = f^b$. To conclude, using Minkowskian coordinates as in (1) of Comments 7.46 to describe those worldlines, the equation $\mu_0 V^a \nabla_a V^b = f^b$ obtained above can be rephrased as $\mu_0 \frac{d^2 x^b}{d\tau^2} = f^b$.

3. Prove that Eq. (7.66) has the heuristic meaning of the equation of motion of an "infinitesimal" portion of continuum in agreement with Eq. (7.42).
Solution. Equation (7.66) has the heuristic meaning of the equation of motion of a portion of continuous body of mass δm referred to a co-moving volume δv_0 such that $\mu_0 = \frac{\delta m}{\delta v_0}$. Inserting this ratio into (7.66) we have

$$\delta v_0 \mu_0 \frac{d}{d\tau}\left(\frac{dx^b}{d\tau}\right) = f^b \delta v_0 ,$$

that is, since the mass of the volume does not change along its worldline,

$$\frac{d}{d\tau}\delta P^b = f^b \delta v_0 , \quad \text{where} \quad \delta P^b = \delta m \frac{dx^b}{d\tau} ,$$

this equation is nothing but (7.42) for the considered portion of continuous body. We can pass from the proper time to the global time t of the used Minkowski frame $\mathscr{R}$:

$$\frac{d}{dt}\delta P^b = f^b \frac{\delta v_0}{V^0} .$$

Since the volume of the portion of continuum we are considering is $\delta v = \frac{\delta v_0}{V^0}$ when referred to the rest space of our reference frame $\mathscr{R}$, the previous equation becomes

$$\frac{d}{dt}\delta P^b = f^b \delta v .$$

7.5.4 The Macroscopic Stress-Energy Tensor in the General Case

The pair of theorems proved above suggests the idea that the energy and momentum content of every isolated continuous system should be described by means of a $(2, 0)$-type tensor field T satisfying the local equation (7.60)

$$\nabla_a T^{ab} = 0 \quad \text{everywhere in } \mathbb{M}^4 ,$$

and such that the total momentum of the system P should be defined by (7.58)

$$P^b_{\mathscr{R}}(t) := \int_{\Sigma_t^{(\mathscr{R})}} \langle T(dx^b), n^{(\mathscr{R})} \rangle dx^1 dx^2 dx^3 , \quad b = 0, 1, 2, 3 ,$$

where we have fixed a reference system $\mathscr{R}$ and co-moving Minkowskian coordinates. As established before, these choices do not affect the definition of the vector P, which does not depend on $\mathscr{R}$, on the used co-moving coordinates, and is also constant in time due to the validity of (7.60) as proved in Theorem 7.47. These properties of P are true provided the stress-energy tensor vanishes outside compact sets in the rest spaces of reference frames (or vanishes sufficiently rapidly at infinity on those 3-spaces). Conversely, $\nabla_a T^{ab} = 0$ is a local equation and can be considered valid even if P cannot be defined.

Remarks 7.50 With this interpretation T^{00} and $\frac{1}{c}T^{0\alpha}$ respectively have the meaning of spatial energy density and the spatial density of the component $\alpha = 1, 2, 3$ of the momentum in a reference frame $\mathscr{R}$ with co-moving coordinates x^0, x^1, x^2, x^3. ■

The most direct way to construct the stress-energy tensor of a physical system in Special Relativity passes through variational approaches and the various formulations of Noether's theorem when this formalism is suitable for the considered system. Typically this happens for a field to be quantized and often describing elementary particles. This approach concerns (quantum) *microscopic* systems. However, the existence of the stress-energy tensor for *macroscopic* systems can be assumed as a direct generalization of the *stress tensor* (see Examples 3.20) of classical mechanics together with the general classical laws of continuum mechanics, as we are going to illustrate.

The fundamental equations of *classical* continuum mechanics in classical mechanics are the following ones, valid in a *classical* inertial reference frame and referring to orthonormal Cartesian coordinates x^1, x^2, x^3 in the rest space of the reference system and to the absolute time t:

$$\frac{\partial \mu}{\partial t} + \sum_{\beta=1}^{3} \frac{\partial}{\partial x^\beta} \mu v^\beta = 0\,, \tag{7.67}$$

$$\mu \left(\frac{\partial v^\alpha}{\partial t} + \sum_{\beta=1}^{3} v^\beta \frac{\partial}{\partial x^\beta} v^\alpha \right) = \sum_{\beta=1}^{3} \frac{\partial}{\partial x^\beta} \sigma^{\alpha\beta} + f^\alpha\,, \quad \alpha = 1, 2, 3\,. \tag{7.68}$$

Above, v^α is a component of the velocity field vector of the continuum, μ is the mass density, and σ is Cauchy's stress tensor describing the *internal* stresses of the system. Furthermore f^α is the α-th component of the spatial density of *external* forces. Equation (7.67) is the (classical) *conservation law of mass*. Equation (7.68) is *Newton's second law* for every particle of the continuous body, taking into account both internal stresses and external forces. The (classical) conservation law of angular momentum is equivalent to the fact that σ is symmetric, when the previous two equations are taken into account:

$$\sigma^{\alpha\beta} = \sigma^{\beta\alpha}\,. \tag{7.69}$$

When one specifies the type of continuous body, further information must be supplied regarding the relation between the stress tensor, the field of velocities, and other geometrical structures of the continuous body such as the *deformation tensor*. However, the above three general equations are completely sufficient for an elementary heuristic formulation.

In order to compare (7.67) and (7.68) with the corresponding relativistic equations, we are going to rewrite the second equation in another form. The final system is however equivalent to (7.67) and (7.68).

Taking advantage of the trivial identity

$$\mu \frac{\partial v^\alpha}{\partial t} = \frac{\partial}{\partial t} \mu v^\alpha - v^\alpha \frac{\partial \mu}{\partial t}\,,$$

and expanding the derivative $\frac{\partial \mu}{\partial t}$ in terms of the spatial derivatives using (7.67), we have

$$\mu \frac{\partial v^\alpha}{\partial t} = \frac{\partial}{\partial t}\mu v^\alpha + v^\alpha \sum_{\beta=1}^{3} \frac{\partial}{\partial x^\beta}\mu v^\beta \,.$$

Inserting this result into the left-hand side of (7.68), we find

$$\frac{\partial}{\partial t}\mu v^\alpha + v^\alpha \sum_{\beta=1}^{3} \frac{\partial}{\partial x^\beta}\mu v^\beta + \sum_{\beta=1}^{3} \mu v^\beta \frac{\partial}{\partial x^\beta} v^\alpha = \sum_{\beta=1}^{3} \frac{\partial}{\partial x^\beta}\sigma^{\alpha\beta} + f^\alpha \,.$$

That is,

$$\frac{\partial}{\partial t}\mu v^\beta + \sum_{\alpha=1}^{3} \frac{\partial}{\partial x^\alpha}\left(\mu v^\alpha v^\beta - \sigma^{\alpha\beta}\right) = f^\beta \,.$$

In summary, since the passages above are reversible (when keeping both equations), the set of equations (7.67) and (7.68) can be equivalently rearranged as follows:

$$\frac{\partial \mu}{\partial t} + \sum_{\alpha=1}^{3} \frac{\partial}{\partial x^\alpha}\mu v^\alpha = 0 \,, \tag{7.70}$$

$$\frac{\partial}{\partial t}\mu v^\beta + \sum_{\alpha=1}^{3} \frac{\partial}{\partial x^\alpha}\left(\mu v^\alpha v^\beta - \sigma^{\alpha\beta}\right) = f^\beta \,, \quad \beta = 1, 2, 3 \,. \tag{7.71}$$

Let us finally assume that the continuous body is nothing but the *classical limit* of a relativistic system admitting a stress-energy tensor T satisfying a natural generalization of (7.60),

$$\nabla_a T^{ab} = f^b \,, \tag{7.72}$$

where the vector field f on $\mathbb{M}^4$ is interpreted as the *density of external four-force* and T^{00} and $\frac{1}{c}T^{0\beta}$ respectively represent the energy density and the momentum densities. From this perspective, (7.72) can be interpreted as a direct relativistic generalization of the set of Eqs. (7.67) and (7.68). To show it, fix a Minkowskian reference frame $\mathscr{R}$ equipped with co-moving Minkowskian coordinates $x^0 = ct$, x^1, x^2, x^3, identified with the classical ones, and expand (7.72) in components:

$$\frac{\partial}{\partial t}\frac{1}{c^2}T^{00} + \sum_{\alpha=1}^{3} \frac{\partial}{\partial x^\alpha}\frac{1}{c}T^{\alpha 0} = \frac{1}{c}f^0 \,, \tag{7.73}$$

$$\frac{\partial}{\partial t}\frac{1}{c}T^{0\beta} + \sum_{\alpha=1}^{3} \frac{\partial}{\partial x^\alpha} T^{\alpha\beta} = f^\beta\,, \quad \beta = 1, 2, 3\,. \tag{7.74}$$

When we say that the relativistic system admits a classical limit we just mean that, within the regime of small velocity v^α (with respect to c), T^{00}/c^2 and $T^{\beta 0}/c$ become the corresponding classical quantities

$$T^{00}/c^2 \simeq \mu\,, \quad T^{\beta 0}/c \simeq \mu v^\beta$$

and $\frac{1}{c} f^0$ is negligible (see Comment 7.41), that is equivalent to postulating the validity of the (classical!) mass conservation law. In that case, Eq. (7.73) becomes (7.70). Similarly, (7.74) tends to become (7.71) provided one assumes, in that regime,

$$T^{\alpha\beta} \simeq \mu v^\alpha v^\beta - \sigma^{\alpha\beta}\,, \quad T^{0\beta}/c \simeq T^{\beta 0}/c \simeq \mu v^\beta\,. \tag{7.75}$$

According to this interpretation, let us study the special choice of a reference frame exactly *at rest* with a particle of the continuum at a given event $e \in \mathbb{M}^4$. Here (7.75) boils down to some *exact* relations valid exactly at e (we will omit to write this explicitly):

$$T^{\alpha\beta} = -\sigma^{\alpha\beta}\,, \quad T^{0\beta} = T^{\beta 0} = 0\,, \quad \alpha, \beta = 1, 2, 3 \tag{7.76}$$

Since $\sigma^{\alpha\beta} = \sigma^{\beta\alpha}$, in particular $T^{ab} = T^{ba}$ is valid in the said reference frame, i.e. in the considered basis of $T_e\mathbb{M}^4 \otimes T_e\mathbb{M}^4$. However, as we know, the symmetry property of a tensor does not depend on the chosen basis, so that we are committed to assuming that *the stress-energy tensor is a symmetric tensor*, at least when dealing with macroscopic continuous media. Symmetry also permits one to present the conservation law of relativistic angular momentum in an easy formulation (see Exercises 7.54 below).

Remarks 7.51 In the very general case, the symmetry of the stress-energy tensor is a delicate issue, especially when dealing with variational approaches for systems which do not admit a classical limit in terms of continuous bodies. The stress-energy tensor obtained in that way is not symmetric in general and it must be symmetrized, preserving the total conserved quantities, obtaining the so-called *Belinfante-Rosenfeld stress-energy tensor*. ■

Examples 7.52 Referring to the gas of non-interacting particles we have $\sigma^{\alpha\beta} = 0$ and, indeed, $T^{\alpha\beta} = \mu_0 V^\alpha V^\beta = 0$ if $\alpha, \beta = 1, 2, 3$ in the rest frame of a particle of the continuum, since V has only the temporal component. ■

This discussion suggests that the notion of stress-energy tensor is a very promising tool to describe the content of momentum, energy, but also internal stresses for a relativistic extended continuous system. We expect that T^{ab} can also be defined for systems which are not continuous systems in the classical sense: (quantum) fields, in particular. In that case the classical limit (7.70) and (7.71) does not make sense. We can give the following general interpretation of the terms which decompose T^{ab} in a given reference frame $\mathscr{R}$.

(i) $T^{00}_{\mathscr{R}}$ is the *density of energy* of the system referred to the rest space $\Sigma_t^{(\mathscr{R})}$,
(ii) $cT^{\alpha 0}_{\mathscr{R}}$ is the α-component of the *internal flux of energy* of the system referred to the rest space $\Sigma_t^{(\mathscr{R})}$,
(iii) $\frac{1}{c}T^{0\beta}_{\mathscr{R}}$ is the *density of the spatial momentum* of the system along the β axis referred to the rest space $\Sigma_t^{(\mathscr{R})}$,
(iv) $T^{\alpha\beta}_{\mathscr{R}}$ is the α-component of the *internal flux of spatial momentum* along the β axis of the system referred to the rest space $\Sigma_t^{(\mathscr{R})}$.

Remarks 7.53 It is important to notice that the internal flux of spatial momentum is not a purely kinematic object as it includes the internal stresses in the case of a macroscopic continuous body. Indeed, in the classical limit it becomes

$$T^{\alpha\beta} \simeq \mu v^\alpha v^\beta - \sigma^{\alpha\beta} .$$

■

The interpretation above is a consequence of the following integral representation of $\nabla_a T^{ab} = 0$. Fix a spatial bounded geometrical volume B—an open set whose closure is compact with orientable regular boundary—at rest with an inertial reference frame $\mathscr{R}$. Working in Minkowskian coordinates $x^0 = ct, x^1, x^2, x^3$, let us integrate the equation above in B. We have

$$\int_B \nabla_0 T^{0a} dx^1 dx^2 dx^3 = -\sum_{\alpha=1}^{3} \int_B \nabla_\alpha T^{\alpha a} dx^1 dx^2 dx^3 .$$

Since $\overline{B}$ is compact, if we assume that the integrands above are continuous in all coordinates, we can extract the temporal derivative (for instance through a suitable use of Lebesgue's dominated convergence theorem and Lagrange's theorem) and use the divergence theorem on the right-hand side, obtaining

$$\frac{d}{dt}\int_B \frac{1}{c} T^{0a} dx^1 dx^2 dx^3 = -\oint_{+\partial B} \sum_{\alpha=1}^{3} T^{\alpha a} n_\alpha dv ,$$

where n_α denotes the α-th component of the outward normal vector to ∂B. It is now evident that the variation per unit time of the total quantity in the volume B (energy or spatial momentum, up to constant factors) amounts to the flux of the corresponding quantity passing through the boundary ∂B of the volume. All this corroborates the interpretation proposed above for the integrands in the right-hand side.

7.5.5 Interacting Macroscopic Continuous Systems

When we add a density of external four-force, the physical interpretation of T^{ab} and its conservation law still holds, but the components of the four-force act as *external sources* of energy and momentum. To explain this extension of the formalism, consider the case of two continuous systems, where each subsystem has its own stress-energy tensor T, T'. Here, the total stress-energy tensor T_{tot} is the sum of those terms, possibly with a third part describing the interaction between them,

$$T_{tot} = T + T' + T_{int} .$$

If the overall system is isolated, then T_{tot} satisfies (7.60), but T, T', T_{int} do not separately. Equivalently, the *density of external force* acting on the first and the second system are respectively

$$f^b = -\nabla_a(T'^{ab} + T^{ab}_{int}) \quad \text{and} \quad f'^b = -\nabla_a(T^{ab} + T^{ab}_{int}) .$$

Notice that $f^b \neq -f'^b$ unless $\nabla_a T^{ab}_{int} = 0$.

Exercises 7.54

1. Consider a Minkowskian coordinate system co-moving with an inertial reference frame $\mathscr{R}$. Assume that both $\nabla_a T^{ab} = 0$ and $T^{ab} = T^{ba}$ and define

$$M^{ab}_{\mathscr{R}}(t) := \frac{1}{c}\int_{\Sigma_t^{(\mathscr{R})}} \left(x^a T^{0b} - x^b T^{0a}\right) dx^1 dx^2 dx^3 .$$

Above, the integrand, and more strongly the functions $\mathbb{R}^4 \ni x \mapsto T^{ab}(x) \in \mathbb{R}$, are C^1 and have spatial compact support in a neighborhood of the considered instant t. Prove that

$$M^{ab}_{\mathscr{R}}(t) = M^{ab}_{\mathscr{R}}(0) .$$

2. Provide a physical interpretation of the quantities

$$L^{(\mathscr{R})}_\gamma := \frac{1}{2}\sum_{\alpha,\beta=1}^{3} \epsilon_{\gamma\alpha\beta} M^{\alpha\beta}_{\mathscr{R}} \quad \text{where } \gamma = 1, 2, 3.$$

(*Hint.* $L^{(\mathscr{R})}_\gamma$ is the γ-th component of the total angular momentum in $\mathscr{R}$ computed with respect to the origin of the coordinates.)

3. Prove that if $\mathscr{R}$ and $\mathscr{R}'$ are connected by the Poincaré transformation $x'^a = \Lambda^a{}_b x^b + C^a$ and $M^{ab}_{\mathscr{R}}$ and $M^{ab}_{\mathscr{R}'}$ are defined as above, then

$$M^{ab}_{\mathscr{R}'} = \Lambda^a{}_c \Lambda^b{}_d M^{cd}_{\mathscr{R}} + C^a P^b_{\mathscr{R}'} - C^b P^a_{\mathscr{R}'} .$$

4. Specializing $M^{ab}_{\mathscr{R}}(t) = M^{ab}_{\mathscr{R}}(0)$ to the case $a = 0, b = \beta$, prove that

$$\frac{1}{c}\int_{\Sigma_t^{(\mathscr{R})}} T^{0\beta}(t)dx^1dx^2dx^3 = \frac{d}{dt}\int_{\Sigma_t^{(\mathscr{R})}} x^\beta \frac{T^{00}}{c^2}dx^1dx^2dx^3$$

and provide a physical interpretation of this equation.
(*Hint*. The argument of the temporal derivative in the right-hand side is Mx^β_G, where x^β_G is the β-th coordinate of the *center of mass* of the physical system in the reference frame $\mathscr{R}$, where the *total mass* is

$$M = \int_{\Sigma_t^{(\mathscr{R})}} \frac{T^{00}}{c^2}dx^1dx^2dx^3 ,$$

and the left-hand side is the β-th component of the total momentum.)

7.5.6 *The Stress-Energy Tensor of the Ideal Fluid and of the Electromagnetic Field*

In classical mechanics, the simplest example of continuum is an *ideal fluid*, which is defined by the requirement that its stress tensor is completely isotropic: in orthonormal Cartesian coordinates of an inertial reference frame

$$\sigma^{\alpha\beta} = -p\delta^{\alpha\beta} , \quad \alpha, \beta = 1, 2, 3 , \tag{7.77}$$

where $p \geq 0$ is the **pressure** of the fluid at the considered time and place in space (see Sect. 4.2.3, where we adopted a different notation). The pressure p is related to the other properties of the fluid as the mass density μ_0 and the velocity field v through some *constitutive relations* which depend on the type of fluid. To relativistically generalize this physical system, we observe that in the instantaneous rest frame $\mathscr{R}_e$ of a particle of fluid at the event $e \in \mathbb{M}^4$, according to the discussion in the previous section,

$$(T_0)^{\alpha\beta} = p\delta^{\alpha\beta} , \quad (T_0)^{00} = c^2\mu_0 , \quad (T_0)^{0\alpha} = (T_0)^{\alpha 0} = 0 , \quad \alpha, \beta = 1, 2, 3 .$$

Hence, in a generic inertial reference frame $\mathscr{R}$ and co-moving Minkowskian coordinates x^0, x^1, x^2, x^3 we therefore have at the event e

$$T^{ab} = \Lambda^a{}_c \Lambda^b{}_d (T_0)^{cd} ,$$

where Λ is the Lorentz part of the orthochronous Poincaré transformation connecting a Minkowskian coordinate frame at rest with $\mathscr{R}_e$ with the Minkowskian coordinates co-moving with $\mathscr{R}$. If the worldlines of the particles of the fluid have components V^a in $\mathscr{R}$, we can always arrange a co-moving Minkowskian coordinate system such that the transformation between this reference frame and the one instantaneously at rest with a particle is a pure Lorentz transformation $\Lambda = \Lambda_V$. From (7.25),

$$\Lambda_V = \left[\begin{array}{c|c} V^0/c & \vec{V}^t/c \\ \hline \vec{V}/c & I + \frac{\vec{V}\vec{V}^t}{c^2(1+V^0/c)} \end{array}\right] .$$

In fact, this matrix transforms the vector $(c, 0, 0, 0)^t \in \mathbb{R}^4$, *representing the components of the four-velocity of a particle in its rest frame*, to the vector of $\mathbb{R}^4$ whose components are those of V. By direct computation, the components of T in $\mathscr{R}$ are

$$T^{ab} = \mu_0 V^a V^b + p\left(g^{ab} + \frac{V^a V^b}{c^2}\right) . \tag{7.78}$$

Notice that choosing a different co-moving Minkowskian coordinate system in $\mathscr{R}$ is completely equivalent (due to the decomposition theorem of the Lorentz group) to composing the right-hand side with a spatial rotation Ω_R and including a spacetime translation, which however has no effect on tensor components. These operations do not change the structure of the found tensor: in all reference frames and co-moving Minkowskian coordinates T always admits the form (7.78) in components. This form is therefore intrinsic. If $\tilde{g}$ denotes the metric tensor with both indices raised, then

$$T = \mu_0 V \otimes V + p\left(\tilde{g} + \frac{1}{c^2} V \otimes V\right) ,$$

where $\mathbb{M}^4 \ni e \mapsto V(e) \in T_e\mathbb{M}^4$ is the vector field of four-velocities of the continuum and p and μ_0 are scalar functions on $\mathbb{M}^4$ with the said physical meaning.

This type of stress-energy tensor is called the **stress-energy tensor of the ideal fluid**. In (7.78), μ_0 and p are respectively the mass density and the pressure measured at rest with the particles of the fluid and V is its four-velocity field.

Remarks 7.55 The form (7.78) of the stress-energy tensor is also exploited in General Relativity and cosmological applications: there the fluid is made of galaxies. In that context, the classical constraint $p \geq 0$ is relaxed, and generally speaking, negative pressures are admitted. ■

There are continuous systems which do not admit a four-velocity vector field. These systems *cannot* be interpreted as continuous systems of particles and they are in particular *fields*, as the *electromagnetic field*. However, these physical systems also admit their own stress-energy tensor, usually produced by Noether's theorem when the system admits a variational description. Usually some further manipulations are necessary to assure that the final tensor is symmetric. The *symmetric stress-energy tensor of the electromagnetic field* reads

$$T_{EM}^{ab} = F^{ac}F^{b}{}_{c} - \frac{1}{4}g^{ab}F_{cd}F^{cd} . \tag{7.79}$$

Above F^{ab} is the electromagnetic tensor introduced in (7.44). The interpretation of the densities T_{EM}^{00} and $T_{EM}^{0\alpha}$ as the density of energy and the α-th component of the density of momentum (up to multiplicative constants) is extended to this stress-energy tensor and it is in agreement with experimental facts. $T^{\alpha 0} = T^{0\alpha}$ is in fact the corresponding component of the *Poynting vector*,

$$T^{0\alpha} = \frac{1}{c^2}(\vec{E} \times \vec{B})^{\alpha} ,$$

which describes the density of momentum of the electromagnetic field. $T_{EM}^{00} \geq 0$ is satisfied, as requested by the interpretation of T_{EM}^{00} as a density of energy up to positive factors, because T_{EM}^{00} has the classical expression

$$T_{EM}^{00} = \frac{1}{c^2}\left(\vec{E}^2 + \vec{B}^2\right) .$$

Finally $T_{EM\,a}^{a} = 0$. When quantizing the electromagnetic field, this condition is related to the fact that photons have zero mass.

Exercises 7.56

1. Maxwell equations are written in terms of the electromagnetic tensor F^{ab}

$$\nabla_a F^{ab} = -J^b , \qquad \epsilon^{abcd}\nabla_b F_{cd} = 0 ,$$

where J is the four-current density $J^a = \rho_0 V^a$, where ρ_0 is the charge density computed at rest with the particles of a continuous charged body. Prove that

$$\nabla_a T_{EM}^{ab} = -F^{bc}J_c .$$

So that, in particular, the conservation equation $\nabla_a T^{ab}_{EM} = 0$ is valid in the absence of charges. Observe that, according to (7.43), $F^{bc} J_c$ is the Lorentz-force density acting on a particle of continuum.

2. Consider an electrically charged gas of non-interacting particles with $\rho_0 = \kappa\mu_0$ (for a constant κ with physical dimensions [charge]/[mass]) subjected to the Lorentz-force density

$$f^a = \rho_0 F^{ab} V_b$$

due to the electromagnetic field. Prove that

$$\nabla_a \left(T^{ab}_{EM} + T^{ab}\right) = 0$$

where $T^{ab} = \mu_0 V^a V^b$.

Solution. $\nabla_a T^{ab}_{EM} = J_c F^{cb} = \kappa\mu_0 V_c F^{cb} = -f^b$ and $\nabla_a T^{ab} = f^b$ according to (2) of Exercises 7.49.

References

1. Moretti, V.: Analytical Mechanics: Classical, Lagrangian and Hamiltonian Mechanics, Stability Theory, Special Relativity. Springer (2023)
2. Moretti, V.: Fundamental Mathematical Structures of Quantum Theory Spectral Theory: Foundational Issues, Symmetries, Algebraic Formulation, Quantum Measurement Theory, 2nd revised and enlarged edition. Springer (2026)

Chapter 8
Lorentz Group Structure

This chapter studies the internal structure of the Lorentz group and its physically relevant subgroups. The aim is to connect the algebraic description of Lorentz transformations with the geometric distinction between spatial rotations, boosts, time orientation, and connected components. This analysis clarifies which transformations relate physically admissible inertial frames and how the proper orthochronous Lorentz group is singled out.

A central theme is the relation between the decomposition of Lorentz transformations used in physics and the polar decomposition theorem proved earlier. The chapter therefore combines matrix Lie group methods, metric arguments, and finite-dimensional operator theory. The final result is a precise decomposition of elements of $SO(1,3)\uparrow$ into rotations and boosts, a fact that will be used in the subsequent discussion of $SL(2,\mathbb{C})$ and spinorial representations.

Remarks 8.1 We henceforth assume that the reader is familiar with the basic notions of matrix Lie groups [1]. ■

8.1 Lie Group Structure, Distinguished Subgroups, and Connected Components of the Lorentz Group

We have a first elementary but important result concerning the Lorentz group and its Lie group structure.

Proposition 8.2 *The Lorentz group $O(1,3)$ is a Lie group which is a Lie subgroup of $GL(4,\mathbb{R})$. Similarly, the Poincaré group $IO(1,3)$ is a Lie group which can be viewed as a Lie subgroup of $GL(5,\mathbb{R})$.*

Proof From the general theory of Lie groups [1] we know that, to show that $O(1,3)$ is a Lie subgroup of the Lie group $GL(4,\mathbb{R})$, it is sufficient to prove that the former is a topologically closed algebraic subgroup of the latter. The fact that $O(1,3)$

V. Moretti, *Geometric Methods in Mathematical Physics I*, La Matematica per il 3+2 184, https://doi.org/10.1007/978-3-032-31512-0_8

is a closed subset of $GL(4, \mathbb{R})$, where the latter is equipped with the topology (and the differentiable structure) induced by $\mathbb{R}^{16}$, follows from $\Lambda^t \eta \Lambda = \eta$, since matrix multiplication and transposition are continuous operations. Concerning the second statement, it is possible to provide $IO(1, 3)$ with the structure of a Lie group which is also a subgroup of $GL(5, \mathbb{R})$ and which includes $O(1, 3)$ and $\mathbb{R}^4$ as Lie subgroups, as follows. One starts with the injective map

$$IO(1,3) \ni (\Lambda, T) \mapsto \left[\begin{array}{c|c} 1 & 0 \\ \hline T & \Lambda \end{array}\right] \in GL(5, \mathbb{R}), \tag{8.1}$$

and one verifies that the map is in fact an injective group homomorphism. The matrix group on the right-hand side of (8.1) defines, in fact, a closed subset of $\mathbb{R}^{25}$ and thus of $GL(5, \mathbb{R})$. As a consequence, this matrix group is a Lie group which is a Lie subgroup of $GL(5, \mathbb{R})$. □

Consider two Minkowskian frames $\mathscr{R}$ and $\mathscr{R}'$ and let x^0, x^1, x^2, x^3, x'^0, x'^1, x'^2, x'^3 be two co-moving Minkowskian coordinate frames respectively. We know that $\partial_{x^0} = \mathscr{R}$ and $\partial_{x'^0} = \mathscr{R}'$, and we also know that $\mathscr{R}$ and $\mathscr{R}'$ must have the same time orientation, since they are both future directed. We conclude that it must be $g(\partial_{x^0}, \partial_{x'^0}) < 0$. Only transformations (7.15) satisfying that constraint may make physical sense. From (7.15) one has $\partial_{x^0} = \Lambda^\mu{}_0 \partial_{x'^\mu}$, so that the requirement $g(\partial_{x^0}, \partial_{x'^0}) < 0$ is equivalent to $\Lambda^0{}_0 > 0$, which is, in turn, equivalent to $\Lambda^0{}_0 \geq 1$ because of the first statement in (e) of Proposition 7.26. One expects that the Poincaré or Lorentz transformations fulfilling this constraint form a subgroup. Indeed this is the case, and the group is called the *orthochronous subgroup*: it embodies all physically sensible coordinate transformations between inertial frames in Special Relativity. The following proposition states that result, introducing also two other relevant subgroups. The proof of the following proposition is immediate from (e) and (d) of Proposition 7.26.

Proposition 8.3 *The subsets of $IO(1, 3)$ and $O(1, 3)$ defined by*

$$IO(1,3)\uparrow := \{(\Lambda, T) \in IO(1,3) \,|\, \Lambda^0{}_0 \geq 1\}, \quad O(1,3)\uparrow := \{\Lambda \in O(1,3) \,|\, \Lambda^0{}_0 \geq 1\}, \tag{8.2}$$

and called respectively the **orthochronous Poincaré group** *and the* **orthochronous Lorentz group**, *are (Lie) subgroups of $IO(1, 3)$ and $O(1, 3)$ respectively.*

The subsets of $IO(1, 3)$ and $O(1, 3)$ defined by

$$\begin{aligned} ISO(1,3) &:= \{(\Lambda, T) \in IO(1,3) \mid \det \Lambda = 1\}, \\ SO(1,3) &:= \{\Lambda \in O(1,3) \mid \det \Lambda = 1\}, \end{aligned} \tag{8.3}$$

and called respectively the **proper Poincaré group** *and the* **proper Lorentz group**, *are (Lie) subgroups of $IO(1, 3)$ and $O(1, 3)$ respectively.*

We remark that the condition $\Lambda^0{}_0 \geq 1$ can be replaced by the equivalent constraint $\Lambda^0{}_0 > 0$, whereas the condition $\det \Lambda = 1$ can be replaced by the equivalent constraint $\det \Lambda > 0$. Since the intersection of a pair of (Lie) groups is a (Lie) group, we can give the following final definition.

Definition 8.4 The (Lie) subgroups of $IO(1,3)$ and $O(1,3)$ defined by

$$ISO(1,3)\uparrow := IO(1,3)\uparrow \cap ISO(1,3)\,, \quad SO(1,3)\uparrow := O(1,3)\uparrow \cap SO(1,3) \tag{8.4}$$

are called respectively the **orthochronous proper Poincaré group** and the **orthochronous proper Lorentz group**. ■

To conclude the short survey of properties of the Lorentz group initiated in the previous chapter, we state and partially prove (the complete proof needs a result we shall achieve later) the following proposition about the connected components of the Lorentz group. These are obtained by starting from $SO(1,3)\uparrow$ and transforming it under the left action of elements of the discrete subgroup of $O(1,3)$: $\{I, \mathscr{P}, \mathscr{T}, \mathscr{P}\mathscr{T}\}$, where $\mathscr{T} := \eta$ and $\mathscr{P} := -\eta$ (so that $\mathscr{P}\mathscr{T} = \mathscr{T}\mathscr{P} = -I$, $\mathscr{P}\mathscr{P} = \mathscr{T}\mathscr{T} = I$). In this context $\mathscr{T}$ is called the *time reversal* operator—since it changes the time orientation of causal vectors—and $\mathscr{P}$ is also said to be the *(parity) inversion* operator—since it corresponds to spatial inversion in the rest space.

Proposition 8.5 $O(1,3)$ *admits four connected components which are respectively, with obvious notation,* $SO(1,3)\uparrow$, $\mathscr{P}SO(1,3)\uparrow$, $\mathscr{T}SO(1,3)\uparrow$, $\mathscr{P}\mathscr{T}SO(1,3)\uparrow$. *Only the first is a subgroup.*

Proof By construction: (1) if $\Lambda \in \mathscr{P}SO(1,3)\uparrow$, both $\det \Lambda = -1$ and ${\Lambda^0}_0 \geq 1$, (2) if $\Lambda \in \mathscr{T}SO(1,3)\uparrow$, both ${\Lambda^0}_0 \leq -1$ and $\det \Lambda = -1$, (3) if $\Lambda \in \mathscr{P}\mathscr{T}SO(1,3)\uparrow$, both ${\Lambda^0}_0 \leq -1$ and $\det \Lambda = 1$.

Thus the last statement is an immediate consequence of the fact that, as I satisfies $\det I = 1$ and $(I)^0_0 = 1$, it cannot belong to the three sets by construction: $\mathscr{P}SO(1,3)\uparrow$, $\mathscr{T}SO(1,3)\uparrow$, $\mathscr{P}\mathscr{T}SO(1,3)\uparrow$. Assume that $SO(1,3)\uparrow$ is connected. We shall prove this later. Since the maps $O(1,3) \ni \Lambda \mapsto \mathscr{T}\Lambda$ and $O(1,3) \ni \Lambda \mapsto \mathscr{P}\Lambda$ are continuous, they transform connected sets into connected sets. As a consequence, $\mathscr{P}SO(1,3)\uparrow$, $\mathscr{T}SO(1,3)\uparrow$, $\mathscr{P}\mathscr{T}SO(1,3)\uparrow$ are connected sets. To conclude, it is sufficient to prove that the considered sets are pairwise disconnected. To this end it is sufficient to exhibit continuous real-valued functions defined on $O(1,3)$ which cannot vanish but which change sign when passing from one set to another.[1] By construction two functions are sufficient: $O(1,3) \ni \Lambda \mapsto \det \Lambda$ and $O(1,3) \ni \Lambda \mapsto {\Lambda^0}_0$. □

[1] Indeed, assuming that X is a topological space and $f : X \to \mathbb{R}$ a continuous function, if X is connected then $f(X)$ is connected. Therefore, if $a, b \in f(X)$ with $a < 0$ and $b > 0$, then $f(X)$ is a connected subset of $\mathbb{R}$ including a, b. The connected subsets of $\mathbb{R}$ are the intervals, so that $f(X)$ has to contain all reals between $a < 0$ and $b > 0$, in particular $0 \in f(X)$. If f cannot vanish, X cannot be connected.

8.2 Spatial Rotations and Boosts

The final aim of this chapter is to state and prove the decomposition theorem for the group $SO(1,3)\uparrow$. To this end, two ingredients have to be introduced: spatial rotations and boosts, which are distinguished types of Lorentz transformations in $SO(1,3)\uparrow$.

8.2.1 Spatial Rotations

Consider a reference system $\mathscr{R}$. What is the relation between two Minkowskian coordinate frames x^0, x^1, x^2, x^3 and x'^0, x'^1, x'^2, x'^3 both co-moving with $\mathscr{R}$? These transformations are called **internal** to $\mathscr{R}$. The answer is quite simple. The class of all internal Poincaré transformations is completely obtained by imposing the further constraint $\partial_{x^0} = \partial_{x'^0} (= \mathscr{R})$ on Eqs. (7.15) and assuming that $\Lambda \in O(1,3)\uparrow$.

Since $\partial_{x^0} = \Lambda^\mu{}_0 \partial_{x'^\mu}$, the constraint is equivalent to imposing $\Lambda^0{}_0 = 1$ and $\Lambda^i{}_0 = 0$ for $i = 1, 2, 3$. The Lorentz condition $\Lambda^t \eta \Lambda = \eta$ implies in particular that:

$$(\Lambda^0{}_0)^2 = 1 + \sum_{i=1}^{3} \Lambda^0{}_i \Lambda^0{}_i \ ,$$

and thus, since $\Lambda^0{}_0 = 1$, we find that $\Lambda^0{}_i = 0$. Summing up, internal Lorentz transformations must have the already known form

$$\Omega_R = \left[\begin{array}{c|c} 1 & 0 \\ \hline 0 & R \end{array}\right] , \tag{8.5}$$

By direct inspection one finds that, in this case, $\Omega^t \eta \Omega = \eta$ reduces to

$$R^t R = I \ , \tag{8.6}$$

This is nothing but the equation determining the orthogonal group $O(3)$. Conversely, starting from any matrix $R \in O(3)$ and thus satisfying (8.6), and defining Ω_R as in (8.5), it is immediate to verify that $\Omega_R \in O(1,3)\uparrow$ and

$$x'^\mu = (\Omega_R)^\mu{}_\nu x^\nu + T^\mu \ ,$$

with $T \in \mathbb{R}^4$ fixed arbitrarily, is an internal Poincaré transformation. It is immediate to show also that Ω_R with the form (8.5) belongs to $SO(1,3)\uparrow$ if and only if $R \in SO(3)$, the group of special rotations made of rotations of $O(3)$ with unit determinant.

Remarks 8.6 It is worthwhile noticing, from a kinematic point of view, that the velocity of $\mathscr{R}$ with respect to $\mathscr{R}'$ seen in Sect. 7.3.3 is invariant under changes of co-moving Minkowskian coordinates when the coordinate transformations are internal to $\mathscr{R}$ and $\mathscr{R}'$. We leave the simple proof of this fact to the reader. ■

Definition 8.7 The Lorentz transformations Ω_R defined in (8.5) with $R \in O(3)$ are called **spatial rotations**. If $R \in SO(3)$, Ω_R is called a **spatial proper rotation**. ■

Since the translational part is trivial, from now on we will focus on the Lorentz part of the Poincaré group only.

8.2.2 Lie Algebra Analysis

Focusing on the Lorentz group, we wish to extract the *non-internal* part of a Lorentz transformation, i.e. what remains after one has taken spatial rotations into account. This goal will be achieved after a preliminary analysis of the Lie algebra of $SO(1,3)\uparrow$ and the corresponding exponentiated operators. As $SO(1,3)\uparrow$ is a matrix Lie group which is a Lie subgroup of $GL(4,\mathbb{R})$, its Lie algebra can be obtained as a Lie algebra of matrices in $M(4,\mathbb{R})$ with the commutator $[\cdot,\cdot]$ given by the usual matrix commutator. The topology and the differential structure on $SO(1,3)\uparrow$ are those induced by $\mathbb{R}^{16}$.

As the maps $O(1,3) \ni \Lambda \mapsto \det\Lambda$ and $O(1,3) \ni \Lambda \mapsto \Lambda^0{}_0$ are continuous and $\det I = 1$ and $(I)^0{}_0 > 0$, every $\Lambda \in O(1,3)$ sufficiently close to I must belong to $SO(1,3)\uparrow$ and vice versa. Hence the Lie algebra of $O(1,3)$, $o(1,3)$, coincides with that of its Lie subgroup $SO(1,3)\uparrow$ because it is completely determined by the behavior of the group in an arbitrarily small neighborhood of the identity.

Proposition 8.8 *The Lie algebra of $SO(1,3)\uparrow$ admits a vector basis made of the following 6 matrices called* **boost generators** K_1, K_2, K_3 *and* **spatial rotation generators** S_1, S_2, S_3*:*

$$K_1 = \left[\begin{array}{c|ccc} 0 & 1 & 0 & 0 \\ \hline 1 & 0 & 0 & 0 \\ 0 & 0 & 0 & 0 \\ 0 & 0 & 0 & 0 \end{array}\right], \quad K_2 = \left[\begin{array}{c|ccc} 0 & 0 & 1 & 0 \\ \hline 0 & 0 & 0 & 0 \\ 1 & 0 & 0 & 0 \\ 0 & 0 & 0 & 0 \end{array}\right], \quad K_3 = \left[\begin{array}{c|ccc} 0 & 0 & 0 & 1 \\ \hline 0 & 0 & 0 & 0 \\ 0 & 0 & 0 & 0 \\ 1 & 0 & 0 & 0 \end{array}\right]. \tag{8.7}$$

$$S_i = \left[\begin{array}{c|ccc} 0 & 0 & 0 & 0 \\ \hline 0 & & & \\ 0 & & T_i & \\ 0 & & & \end{array}\right] \quad \textit{with } T_1 = \begin{bmatrix} 0 & 0 & 0 \\ 0 & 0 & -1 \\ 0 & 1 & 0 \end{bmatrix}, \quad T_2 = \begin{bmatrix} 0 & 0 & 1 \\ 0 & 0 & 0 \\ -1 & 0 & 0 \end{bmatrix}, \quad T_3 = \begin{bmatrix} 0 & -1 & 0 \\ 1 & 0 & 0 \\ 0 & 0 & 0 \end{bmatrix}. \tag{8.8}$$

These generators enjoy the following commutation relations, which, as a matter of fact, determine the structure tensor of $o(1, 3)$*:*

$$[S_i, K_j] = \sum_{k=1}^{3} \epsilon_{ijk} K_k\,, \quad [S_i, S_j] = \sum_{k=1}^{3} \epsilon_{ijk} S_k\,, \quad [K_i, K_j] = -\sum_{k=1}^{3} \epsilon_{ijk} S_k\,. \tag{8.9}$$

Above ϵ_{ijk} *is the usual completely antisymmetric Ricci symbol with* $\epsilon_{123} = 1$.

Proof If a matrix $N \in M(4, \mathbb{R})$ is in $o(1, 3)$, the generated one-parameter subgroup $\{e^{uN}\}_{u\in\mathbb{R}}$ in $GL(4, \mathbb{R})$ satisfies $(e^{uN})^t \eta e^{uN} = \eta$ for u in a neighborhood of 0, that is, $e^{uN^t} \eta e^{uN} = \eta$ for the same values of u. Taking the derivative at $u = 0$ one gets the necessary condition

$$N^t \eta + \eta N = 0\,. \tag{8.10}$$

These equations are also sufficient. Indeed, from standard properties of the exponential map of matrices, one has

$$\frac{d}{du}\left(e^{uN^t} \eta e^{uN} - \eta\right) = e^{uN^t}\left(N^t \eta + \eta N\right) e^{uN}\,.$$

Thus, the validity of (8.10) implies that $e^{uN^t} \eta e^{uN} - \eta$ is constant. For $u = 0$ one recognizes that the constant is 0 and so $(e^{uN})^t \eta e^{uN} = \eta$ is valid (for every $u \in \mathbb{R}$). Equation (8.10) supplies 10 linearly independent conditions, so that it determines a subspace of $M(4, \mathbb{R})$ with dimension 6. The 6 matrices $S_i, K_j \in M(4, \mathbb{R})$ are linearly independent and satisfy (8.10), so they are a basis for $o(1, 3)$. The relations (8.9) can be checked by direct inspection. □

From now on **K**, **S**, **T** respectively denote the formal vector with components K_1, K_2, K_3, the formal vector with components S_1, S_2, S_3, and the formal vector with components T_1, T_2, T_3. $\mathbb{S}^2$ will indicate the sphere of three-dimensional unit vectors.

Generators S_1, S_2, S_3 produce proper spatial rotations as stated in the following proposition.

Proposition 8.9 *The following facts about proper rotations hold.*

(a) Every proper spatial rotation has the form $\Omega_R = e^{\theta \mathbf{n}\cdot\mathbf{S}}$*—or equivalently* $R = e^{\theta \mathbf{n}\cdot\mathbf{T}}$ *for every* $R \in SO(3)$*—with suitable* $\theta \in \mathbb{R}$ *and* $\mathbf{n} \in \mathbb{S}^2$ *depending on* R.

(b) *Every matrix* $e^{\theta \mathbf{n}\cdot\mathbf{S}}$ *with* $\theta \in \mathbb{R}$ *and* $\mathbf{n} \in \mathbb{S}^2$ *is a proper rotation* Ω_R*, and is associated with* $R = e^{\theta \mathbf{n}\cdot\mathbf{T}} \in SO(3)$.

(c) *The following equivalent identities hold true, for every* $U \in SO(3)$, $\mathbf{n} \in \mathbb{S}^2$, $\theta \in \mathbb{R}$*:*

$$\Omega_U e^{\theta \mathbf{n}\cdot S} \Omega_U{}^t = e^{\theta (U\mathbf{n})\cdot S}, \quad U e^{\theta \mathbf{n}\cdot T} U^t = e^{\theta (U\mathbf{n})\cdot T}. \tag{8.11}$$

The latter holds, more generally, also if $U \in SL(3, \mathbb{C})$.

(d) *The Lie group of spatial proper rotations* $SO(3)$ *is connected, but not simply connected, its fundamental group being* $\pi_1(SO(3)) = \mathbb{Z}_2$.

Proof (a) and (c). Since, from the given definitions,

$$e^{\theta \mathbf{n}\cdot S} = \left[\begin{array}{c|ccc} 1 & 0 & 0 & 0 \\ \hline 0 & & & \\ 0 & & e^{\theta \mathbf{n}\cdot\mathbf{T}} & \\ 0 & & & \end{array}\right], \tag{8.12}$$

it follows immediately that $\Omega_R = e^{\theta \mathbf{n}\cdot\mathbf{S}}$ is completely equivalent to $R = e^{\theta \mathbf{n}\cdot\mathbf{T}}$, so we deal with the latter.

If $R \in SO(3)$, the induced operator in $\mathbb{C}^3$ is unitary and thus it admits a basis of eigenvectors with eigenvalues λ_i with $|\lambda_i| = 1$, $i = 1, 2, 3$. As the characteristic polynomial of R is real, an eigenvalue must be real, the remaining pair of eigenvalues being either real or complex conjugates. Since $\det R = \lambda_1\lambda_2\lambda_3 = 1$, 1 is one of the eigenvalues. If another eigenvalue coincides with 1 all three eigenvalues must do so and $R = I$. In this case every non-vanishing real vector is an eigenvector of R. Otherwise, the eigenspace of $\lambda = 1$ must be one-dimensional and thus, as R is real, it must contain a real eigenvector. We conclude that, in every case, R has a real normalized eigenvector $\mathbf{n}$ with eigenvalue 1. Consider an orthonormal basis $\mathbf{n}_1$, $\mathbf{n}_2$, $\mathbf{n}_3 := \mathbf{n}$, related to the initial one by means of $R' \in SO(3)$, and represent R in the new basis. Imposing the requirement that $\mathbf{n}_3$ is an eigenvector with eigenvalue 1 as well as that the represented transformation belongs to $SO(3)$, one can easily prove that, in such a basis, R is represented by the matrix $e^{\theta T_3}$ for some $\theta \in [0, 2\pi]$. In other words, coming back to the initial basis, $R = R' e^{\theta T_3} R'^t$ for some $R' \in SO(3)$. Now notice that $(T_i)_{jk} = -\epsilon_{ijk}$. This fact entails that $\sum_{i,j,k} U_{pi} U_{qj} U_{rk} \epsilon_{ijk} = \epsilon_{pqr}$ for all $U \in SL(3, \mathbb{C})$. That identity can be re-written as $\mathbf{n} \cdot U\mathbf{T}U^t = (U\mathbf{n}) \cdot \mathbf{T}$ for every $U \in SL(3, \mathbb{C})$. Consequently, if $U \in SO(3)$ in particular, it also holds $U e^{\theta \mathbf{n}\cdot\mathbf{T}} U^t = e^{\theta (U\mathbf{n})\cdot\mathbf{T}}$. (This proves the latter in (8.11); the former is a trivial consequence of the given definitions). Therefore, the identity found above for any $R \in SO(3)$, $R = R' e^{\theta T_3} R'^t$ with $R' \in SO(3)$, can equivalently be written as $R = e^{\theta \mathbf{n}\cdot\mathbf{T}}$ for some versor $\mathbf{n} = R'\mathbf{e}_3$.

(b) Finally, every matrix $e^{\theta \mathbf{n}\cdot\mathbf{T}}$ belongs to $SO(3)$ because $(e^{\theta \mathbf{n}\cdot\mathbf{T}})^t = e^{\theta \mathbf{n}\cdot\mathbf{T}^t} = e^{-\theta \mathbf{n}\cdot\mathbf{T}} = (e^{\theta \mathbf{n}\cdot\mathbf{T}})^{-1}$ and $\det e^{\theta \mathbf{n}\cdot\mathbf{T}} = e^{\theta \mathrm{tr}(\mathbf{n}\cdot\mathbf{T})} = e^0 = 1$.

(d) We sketch here only the idea of the proof. First consider the subgroup made of matrices $e^{\theta \mathbf{e}_3 \cdot \mathbf{T}}$ with $\theta \in \mathbb{R}$. By the explicit form of these matrices, i.e.

$$e^{\theta T_3} = \begin{bmatrix} \cos\theta & -\sin\theta & 0 \\ \sin\theta & \cos\theta & 0 \\ 0 & 0 & 1 \end{bmatrix}, \tag{8.13}$$

one sees that all these matrices are determined by pairs $(\theta, \mathbf{n}) \in [0, \pi] \times \{-\mathbf{e}_3, \mathbf{e}_3\}$ with the only exception given by the pairs $(\pi, -\mathbf{e}_3)$ and $(\pi, \mathbf{e}_3)$ which produce the same rotation. This result generalizes to the subgroup $\{e^{\theta \mathbf{n}\cdot\mathbf{T}}\}_{\theta\in\mathbb{R}}$ for every fixed $\mathbf{n} \in \mathbb{S}^2$, in view of (c). From now on we restrict θ to range in $[0, \pi]$. By direct inspection, reducing to the case $\mathbf{n} = \mathbf{e}_3$ using (c) again, one sees that $e^{\theta \mathbf{n}\cdot\mathbf{T}} = e^{\theta' \mathbf{n}'\cdot\mathbf{T}}$ is possible only if $\mathbf{n} = \mathbf{n}'$ and $\theta = \theta'$, with the only exception $\theta = \theta' = \pi$, where also $\mathbf{n} = -\mathbf{n}'$ is allowed as a consequence of the analysis above. The proof of this fact is an immediate consequence of the fact that $\mathbf{n}'$ is an eigenvector (with eigenvalue 1) of $e^{\theta' \mathbf{n}'\cdot\mathbf{T}}$, so that, if $e^{\theta \mathbf{e}_3\cdot\mathbf{T}} = e^{\theta' \mathbf{n}'\cdot\mathbf{T}}$, $\mathbf{n}'$ must be an eigenvector of the matrix (8.13) (with eigenvalue 1). As a conclusion we obtain that $SO(3)$ is biunivocally determined by the points of a set B constructed as follows. B is the ball in $\mathbb{R}^3$ with radius π, where each two points on the surface of the ball which belong to the same diameter are identified (in other words, the pairs $(\pi, \mathbf{n})$ and $(\pi, -\mathbf{n})$ determine the same element of $SO(3)$ as we said above). A closer scrutiny proves that this one-to-one correspondence is actually a homeomorphism when B is endowed with the natural topology induced by $\mathbb{R}^3$ and the said identifications. The group of continuous closed paths in B is $\mathbb{Z}_2$ as one may simply prove. □

8.2.3 Boosts

From the analysis performed above, we conclude that what remains of a Lorentz transformation in $SO(1, 3)\uparrow$, once one has taken spatial rotations into account, are transformations obtained by exponentiating the generators $\mathbf{K}$.

Definition 8.10 The elements of $SO(1, 3)\uparrow$ with the form $\Lambda = e^{\chi \mathbf{m}\cdot\mathbf{K}}$, with $\chi \in \mathbb{R}$ and $\mathbf{m} \in \mathbb{S}^2$, are called **boosts** or **pure transformations**. ■

Let us investigate the basic properties of boosts. These are given by the following proposition.

Proposition 8.11 *Boosts enjoy the following properties.*

(a) All matrices $\Lambda = e^{\chi \mathbf{m}\cdot\mathbf{K}}$ with arbitrary $\chi \in \mathbb{R}$ and $\mathbf{m} \in \mathbb{S}^2$ belong to $SO(1, 3)\uparrow$ and thus are boosts.

(b) For every pair $\mathbf{m} \in \mathbb{S}^2$, $\chi \in \mathbb{R}$, and every $R \in SO(3)$ one has

$$\Omega_R e^{\chi \mathbf{m}\cdot\mathbf{K}} \Omega_R{}^t = e^{\chi (R\mathbf{m})\cdot\mathbf{K}}. \tag{8.14}$$

(c) For $\mathbf{n} \in \mathbb{S}^2$*, the explicit form of* $e^{\chi \mathbf{n}\cdot\mathbf{K}}$ *reads:*

$$e^{\chi \mathbf{n}\cdot\mathbf{K}} = \left[\begin{array}{c|c} \cosh\chi & (\sinh\chi)\mathbf{n}^t \\ \hline (\sinh\chi)\mathbf{n} & I - (1-\cosh\chi)\mathbf{n}\mathbf{n}^t \end{array}\right], \tag{8.15}$$

(d) Every boost is symmetric and strictly positive definite.

(e) For every fixed $\mathbf{m} \in \mathbb{S}^2$*,* $\{e^{\chi \mathbf{m}\cdot\mathbf{K}}\}_{\chi\in\mathbb{R}}$ *is a subgroup of* $SO(1,3)$*. However, the set of all boost transformations is not a subgroup of* $SO(1,3)$*, and it is homeomorphic to* $\mathbb{R}^3$ *when equipped with the topology induced by* $SO(1,3)\uparrow$*.*

Proof

(a) This was proved in the proof of Proposition 8.8, taking into account that $N := \chi\mathbf{m}\cdot\mathbf{K} \in o(1,3)$.

(b) Fix $\mathbf{n} \in \mathbb{S}^2$ and $j = 1, 2, 3$. Now, for $i = 1, 2, 3$, define the functions

$$f_j(\theta) := e^{\theta\mathbf{n}\cdot S} K_j e^{-\theta\mathbf{n}\cdot S}, \quad g_j(\theta) := \sum_{k=1}^{3} \left(e^{\theta\mathbf{n}\cdot T}\right)_{jk} K_k .$$

Taking the first derivative in θ and using both (8.9) and the explicit form of the matrices T_h, one finds that the smooth functions f_k and the smooth functions g_k satisfy the same system of first-order differential equations written in normal form: for $j = 1, 2, 3$,

$$\frac{df_j}{d\theta} = \sum_{k=1}^{3} n_k \epsilon_{kjp} f_p(\theta), \quad \frac{dg_j}{d\theta} = \sum_{k=1}^{3} n_k \epsilon_{kjp} g_p(\theta) .$$

Since $f_j(0) = g_j(0)$ for $j = 1, 2, 3$, we conclude that these functions coincide for every $\theta \in \mathbb{R}$:

$$e^{\theta\mathbf{n}\cdot S} K_j e^{-\theta\mathbf{n}\cdot S} = \sum_{k=1}^{3} \left(e^{\theta\mathbf{n}\cdot T}\right)_{jk} K_k .$$

Now, by exponentiation, we get (8.14) exploiting (c) of Proposition 8.9.

(c) First consider the case $\mathbf{n} = \mathbf{e}_3$. In this case, directly from Taylor expansion,

$$e^{\chi \mathbf{e}_3\cdot\mathbf{K}} = \left[\begin{array}{c|ccc} \cosh\chi & 0 & 0 & \sinh\chi \\ \hline 0 & 1 & 0 & 0 \\ 0 & 0 & 1 & 0 \\ \sinh\chi & 0 & 0 & \cosh\chi \end{array}\right] = \left[\begin{array}{c|c} \cosh\chi & (\sinh\chi)\mathbf{e}_3^t \\ \hline (\sinh\chi)\mathbf{e}_3 & I - (1-\cosh\chi)\mathbf{e}_3\mathbf{e}_3^t \end{array}\right], \tag{8.16}$$

If $\mathbf{n} \in \mathbb{S}^2$ there is $R \in SO(3)$ such that $\mathbf{n} = R\mathbf{e}_3$. Using this R in (b) with $\mathbf{m} = \mathbf{e}_3$ one gets (8.15) with $\mathbf{n}$.

(d) Symmetry is evident from (8.15). Using (b), it is sufficient to prove positivity in the case $\mathbf{m} = \mathbf{e}_3$. In this case strict positivity can be checked by direct inspection using (8.16).

(e) The first statement is trivial, since $\{e^{\chi\mathbf{m}\cdot\mathbf{K}}\}_{\chi\in\mathbb{R}}$ is a one-parameter subgroup of $SO(1,3)\uparrow$. By direct inspection using (8.15) for $\mathbf{n} = \mathbf{e}_1$ and $\mathbf{n} = \mathbf{e}_3$ one verifies that the product of these boosts cannot be represented as in (8.15). Concerning the last statement, one proves that $e^{\mathbf{u}\cdot\mathbf{K}} = e^{\mathbf{u}'\cdot\mathbf{K}}$ implies $\mathbf{u} = \mathbf{u}'$, reducing to the simpler case $e^{\chi\mathbf{e}_3\cdot\mathbf{K}} = e^{\mathbf{u}'\cdot\mathbf{K}}$ and making use of (b). This case can be examined by taking advantage of (8.16) and (8.15), yielding $\chi\mathbf{e}_3 = \mathbf{u}'$ straightforwardly. Summarizing, $\mathbb{R}^3 \ni \mathbf{u} \mapsto e^{\mathbf{u}\cdot\mathbf{K}}$ is bijective onto the set of all boost transformations. Directly from (8.16) one sees that the map $\mathbb{R}^3 \ni \mathbf{v} \to e^{\mathbf{v}\cdot\mathbf{K}} \in \mathbb{R}^{16}$ is a smooth embedding and defines a submanifold of $\mathbb{R}^{16}$. One way to verify this is to identify the three components of $\mathbf{v} = (\sinh\chi)\mathbf{n}$ with the coordinates x^{ij}, when $i = 1$ and $j = 2, 3, 4$, in the space $\mathbb{R}^{16}$ of real 4×4 matrices, and prove that the hypotheses of the theorem of regular values are valid for the equations $f^{ij} := x^{ij} - x^{ij}(x^{12}, x^{13}, x^{14}) = 0$, which determine the other 12 components of the boost $e^{\mathbf{v}\cdot\mathbf{K}} \in \mathbb{R}^{16}$, referring to its explicit expression as given in (8.16). In particular the Jacobian submatrix with elements $\{\partial f^{rs}/\partial x^{ij}\}_{r,s=1,\ldots,16,i=1,j=2,3,4}$ has rank 3 and thus $(x^{12}, x^{13}, x^{14}) = \mathbf{v}$ are (global) admissible coordinates on the space of boosts, and the map $\mathbb{R}^3 \ni \mathbf{v} \to e^{\mathbf{v}\cdot\mathbf{K}} \in \mathbb{R}^{16}$, restricted to its image in the codomain, defines a diffeomorphism from $\mathbb{R}^3$ onto the space of boosts. Thus in particular, the set of boosts is homeomorphic to $\mathbb{R}^3$ when endowed with the topology induced by $\mathbb{R}^{16}$. Since the topology of $SO(1,3)\uparrow$ is that induced by $\mathbb{R}^{16}$ itself, we conclude that the set of boosts is homeomorphic to $\mathbb{R}^3$ when endowed with the topology induced by $SO(1,3)\uparrow$.

□

Recalling the kinematic discussion in Sect. 7.3.3, the meaning of the parameters χ and $\mathbf{m}$ in boosts is clear from the following last proposition.

Proposition 8.12 *Let $\mathscr{R}$ and $\mathscr{R}'$ be Minkowskian reference systems with associated co-moving Minkowskian coordinate frames x^0, x^1, x^2, x^3 and x'^0, x'^1, x'^2, x'^3 respectively and suppose that (7.15) holds with*

$$\Lambda = e^{\chi\mathbf{m}\cdot\mathbf{K}}\,.$$

Let γ represent the world line of a material point at rest with respect to $\mathscr{R}$, that is, γ admits the parametrization $x^i(\xi) = x^i_0$ constant for $i = 1, 2, 3$, $x^0 = x^0(\xi)$. The velocity of γ with respect to $\mathscr{R}'$ does not depend on x^i_0 and is constant in $\mathscr{R}'$-time, so that, indicating it by $\mathbf{v}^{(\mathscr{R}')}_{\mathscr{R}}$, it holds

$$\mathbf{v}^{(\mathscr{R}')}_{\mathscr{R}} = (\tanh\chi)\mathbf{m}\,.$$

Proof The proof follows immediately from (8.15) and Definition 7.19. □

Remarks 8.13 Boosts along $\mathbf{n} := \mathbf{e}_3$ are known in the literature as "special Lorentz transformations" along z (an analogous name is used replacing $\mathbf{e}_3$ by $\mathbf{e}_1$ and $\mathbf{e}_2$). ■

8.2.4 Decomposition Theorem for $SO(1,3)\uparrow$

In the previous subsection we have determined two classes of transformations in $SO(1,3)\uparrow$: spatial proper rotations and boosts. It is natural to wonder whether these transformations encompass, up to products of some of them, the whole group $SO(1,3)\uparrow$. The answer is positive and is stated in a decomposition theorem for $SO(1,3)\uparrow$ that we now state. The proof of the theorem will be given in a specialized next section using a nonstandard approach based on the so-called polar decomposition theorem.

Theorem 8.14 *Take $\Lambda \in SO(1,3)\uparrow$, the following holds.*

(a) *There is exactly one boost P and exactly one spatial proper rotation U (so that $P = e^{\chi \mathbf{m}\cdot\mathbf{K}}$, $U = e^{\theta \mathbf{n}\cdot\mathbf{S}}$ for some $\chi, \theta \in \mathbb{R}$ and $\mathbf{m}, \mathbf{n} \in \mathbb{S}^2$) such that*

$$\Lambda = UP\,.$$

(b) *There is exactly one boost P' and exactly one spatial proper rotation U' (so that $P' = e^{\chi' \mathbf{m}'\cdot\mathbf{K}}$, $U' = e^{\theta' \mathbf{n}'\cdot\mathbf{S}}$ for some $\chi', \theta' \in \mathbb{R}$ and $\mathbf{m}', \mathbf{n}' \in \mathbb{S}^2$) such that*

$$\Lambda = P'U'\,.$$

(c) *It holds $U' = U$ and $P' = UPU^t$.*

8.2.5 Proof of the Decomposition Theorem by the Polar Decomposition of $SO(1,3)\uparrow$

$\Lambda \in SO(1,3)\uparrow$ can be seen as a linear operator in the finite-dimensional real vector space $\mathbb{R}^4$. Therefore one may consider the polar decomposition $\Lambda = UP = P'U$ given in Theorem 6.10. U is now an orthogonal operator of $O(4)$ and P, P' are symmetric positive operators. A priori those decompositions could be physically meaningless because U and P, P' might not belong to $SO(1,3)\uparrow$: the notions of symmetry, positivity, and orthogonal group $O(4)$ are referred to the positive scalar product of $\mathbb{R}^4$ instead of the indefinite Lorentz scalar product. Nevertheless, we shall show that the polar decompositions of $\Lambda \in SO(1,3)\uparrow$ are in fact physically meaningful. Indeed, they coincide with the standard decompositions of Λ into spatial-rotation and boost parts as in Theorem 8.14.

Let us focus attention on the real vector space $V = \mathbb{R}^4$ endowed with the usual positive scalar product. In that case $\mathscr{L}(V|V) = M(4, \mathbb{R})$, orthogonal operators are the matrices of $O(4)$, and the adjoint $A^\dagger$ of $A \in \mathscr{L}(V|V)$ (see Chap. 6) coincides with the transposed matrix A^t; therefore symmetric operators are symmetric matrices. We have the following theorem which proves Theorem 8.14 as an immediate consequence.

Theorem 8.15 *If* $UP = P'U = \Lambda$ *(with* $P' = UPU^t$*) are polar decompositions of* $\Lambda \in SO(1,3)\uparrow$,

(a) $P, P', U \in SO(1,3)\uparrow$, *more precisely* P, P' *are boosts and* U *is a spatial proper rotation;*
(b) there are no other decompositions of Λ *as a product of a Lorentz boost and a spatial proper rotation different from the two polar decompositions above.*

Proof If $P \in M(4, \mathbb{R})$ we exploit the representation:

$$P = \left[\begin{array}{c|c} g & B^t \\ \hline C & A \end{array}\right], \tag{8.17}$$

where $g \in \mathbb{R}$, $B, C \in \mathbb{R}^3$, and $A \in M(3, \mathbb{R})$.

(a) We start by showing that $P, U \in O(1,3)$. As $P = P^t$, $\Lambda^t\eta\Lambda = \eta$ entails $PU^t\eta UP = \eta$. As $U^t = U^{-1}$ and $\eta^{-1} = \eta$, the obtained identity is equivalent to $P^{-1}U^t\eta UP^{-1} = \eta$, which, together with $PU^t\eta UP = \eta$, implies $P\eta P = P^{-1}\eta P^{-1}$, namely $\eta P^2\eta = P^{-2}$, where we have used $\eta = \eta^{-1}$ once again. Both sides are symmetric (notice that $\eta = \eta^t$) and positive by construction; by Theorem 6.9 they admit unique square roots, which must coincide. The square root of P^{-2} is P^{-1}, while the square root of $\eta P^2\eta$ is $\eta P\eta$, since $\eta P\eta$ is symmetric positive and $\eta P\eta\eta P\eta = \eta PP\eta = \eta P^2\eta$. We conclude that $P^{-1} = \eta P\eta$ and thus $\eta = P\eta P$ because $\eta = \eta^{-1}$. Since $P = P^t$ we have found that $P \in O(1,3)$ and thus $U = \Lambda P^{-1} \in O(1,3)$. Let us prove that $P, U \in SO(1,3)\uparrow$. $\eta = P^t\eta P$ entails $\det P = \pm 1$, on the other hand $P = P^t$ is positive and thus $\det P \geq 0$ and ${P^0}_0 \geq 0$. As a consequence, $\det P = 1$ and ${P^0}_0 \geq 0$. We have found that $P \in SO(1,3)\uparrow$. Let us determine the form of P using (8.17). $P = P^t$, $P \geq 0$, and $P\eta P = \eta$ give rise to the following equations: $C = B$, $0 < g = \sqrt{1 + B^2}$, $AB = gB$, $A = A^t$, $A \geq 0$, and $A^2 = I + BB^t$. Since $I + BB^t$ is positive, the solution of the last equation $A = \sqrt{A^2} = I + BB^t/(1+g) \geq 0$ is unique by Theorem 6.9. We have found that a matrix $P \in O(1,3)$ with $P \geq 0$, $P = P^t$ must have the form

$$P = \left[\begin{array}{c|c} \cosh\chi & (\sinh\chi)\mathbf{n}^t \\ \hline (\sinh\chi)\mathbf{n} & I - (1 - \cosh\chi)\mathbf{n}\mathbf{n}^t \end{array}\right] = e^{\chi \mathbf{n}\cdot\mathbf{K}}, \tag{8.18}$$

where we have used the parametrization $B = (\sinh\chi)\mathbf{n}$, $\mathbf{n}$ being any versor in $\mathbb{R}^3$ and $\chi \in \mathbb{R}$. By (c) in Proposition 8.11 we have found that P is a boost. (The same proof applies to P'.)

Let us pass to consider U. Since $\Lambda, P \in SO(1,3)\uparrow$, from $\Lambda P^{-1} = U$, we conclude that $U \in SO(1,3)\uparrow$. $U\eta = \eta(U^t)^{-1}$ (i.e. $U \in O(1,3)$) and $U^t = U^{-1}$ (i.e. $U \in O(4)$) entail that $U\eta = \eta U$ and thus the eigenspaces of η, E_λ (with eigenvalue λ), are invariant under the action of U. In those spaces U acts as an element of $O(\dim(E_\lambda))$ and the whole matrix U has a block-diagonal form. $E_{\lambda=-1}$ is generated by $\mathbf{e}_0$, and thus U reduces to $\pm I$ therein. The sign must be $+$ because of the requirement $U^0{}_0 > 0$. The eigenspace $E_{\lambda=1}$ is generated by $\mathbf{e}_1, \mathbf{e}_2, \mathbf{e}_3$, and therein U reduces to an element $R \in O(3)$. Actually the requirement $\det U = 1$ (together with $U^0{}_0 = 1$) implies that $R \in SO(3)$ and thus one has that $U = \Omega_R$, which is a spatial proper rotation.

(b) Suppose that $\Omega B = \Lambda \in SO(1,3)\uparrow$ where B is a pure boost and Ω is a spatial proper rotation. $\Omega \in O(4)$ by construction, and on the other hand, from (d) of Proposition 8.11, $B^t = B > 0$. Thus $\Omega B = \Lambda$ is a polar decomposition of Λ. Uniqueness of polar decomposition (Theorem 6.10) implies that $\Omega = U$ and $B = P$. The other case is analogous.

□

We emphasize here a topological consequence of this theorem with the following proposition.

Proposition 8.16 *$SO(1,3)\uparrow$ is homeomorphic to $SO(3) \times \mathbb{R}^3$, as a consequence it is connected, path-connected, and $\pi_1(SO(1,3)\uparrow) = \mathbb{Z}_2$, so that it is not simply connected.*

Proof Using the fact that $SO(1,3)\uparrow$ is bijectively identified with the product $SO(3) \times \mathbb{R}^3$, where the $\mathbb{R}^3$ factor is the space of boosts as in (e) of Proposition 8.11, it follows that $SO(1,3)\uparrow$ is homeomorphic to the topological space $SO(3) \times \mathbb{R}^3$. Indeed, the map $\mathbb{R}^{16} \ni \Lambda \mapsto (U, P) \in \mathbb{R}^{16} \times \mathbb{R}^{16}$ is continuous since the functions used to construct the factors U and P of the polar decomposition of a nonsingular matrix are continuous in the $\mathbb{R}^{16}$ topology; on the other hand the multiplication of matrices $\mathbb{R}^{16} \times \mathbb{R}^{16} \ni (U, P) \mapsto UP \in \mathbb{R}^{16}$ is continuous in the same topology; and finally the topologies used in the restricted spaces of the matrices here employed, i.e. $\Lambda \in SO(1,3)\uparrow$, $U \in SO(3)$ and P in the space of boosts, are actually those induced by $\mathbb{R}^{16}$. As soon as both factors of $SO(3) \times \mathbb{R}^3$ are connected, $SO(1,3)\uparrow$ is connected too. It is also path-connected, it being a differentiable manifold and thus admitting a topological basis made of (smoothly) path-connected open sets. Moreover, the fundamental group of $SO(1,3)\uparrow$ is, as it happens for product manifolds [2], the product of the fundamental groups of the factors. As $\pi_1(SO(3)) = \mathbb{Z}_2$ by (d) in Proposition 8.9, whereas $\pi_1\left(\mathbb{R}^3\right) = \{1\}$, one finds again $\pi_1(SO(1,3)\uparrow) = \mathbb{Z}_2$. □

References

1. Kirillov, A.: Éléments de la Théorie de représentations. Mir, Mosca (1974); Najmark, M.A., Stern, A.I.: Theory of Group Representations. Mir, Moscow (1984)
2. Sernesi, E.: Geometria I and II. Bollati Boringhieri, Torino (1990)

Chapter 9
$SL(2, \mathbb{C})$ and $SO(1, 3)\uparrow$

This chapter explains the fundamental relation between the complex matrix group $SL(2, \mathbb{C})$ and the proper orthochronous Lorentz group $SO(1, 3)\uparrow$. This relation is one of the main algebraic reasons why spinors enter relativistic physics: the Lorentz group admits a natural double covering by $SL(2, \mathbb{C})$, and many representations become simpler when lifted to this covering group.

The chapter begins with elementary properties of $SL(2, \mathbb{C})$, $SU(2)$, their Lie algebras, and the Pauli matrices. It then uses Hermitian matrices and the determinant form to construct the covering homomorphism from $SL(2, \mathbb{C})$ onto $SO(1, 3)\uparrow$. The result prepares the ground for the following chapter, where spinor spaces and spinorial tensors are introduced as the natural representation spaces associated with this group-theoretic structure.

9.1 Elementary Properties of $SL(2, \mathbb{C})$ and $SU(2)$

$SL(2, \mathbb{C})$ denotes the Lie subgroup of $GL(2, \mathbb{C})$ made of all the 2×2 complex matrices with unit determinant [1, 2, 4, 5]. $SL(2, \mathbb{C})$ can be viewed as a real Lie group when referring to real coordinates. Recall that $U(2)$ is the group of complex 2×2 unitary matrices. $SU(2) := U(2) \cap SL(2, \mathbb{C})$. Evidently, both are (Lie) subgroups of $GL(2, \mathbb{C})$.

V. Moretti, *Geometric Methods in Mathematical Physics I*, La Matematica per il 3+2 184, https://doi.org/10.1007/978-3-032-31512-0_9

9.1.1 Almost All on Pauli Matrices

The (real) Lie algebra of $SL(2,\mathbb{C})$, $sl(2,\mathbb{C})$, and that of $SU(2)$, $su(2)$, are 6-dimensional and 3-dimensional, respectively. Performing an analysis similar to the one about the Lorentz group presented in the previous chapter, one sees that

$$sl(2,\mathbb{C}) = \{M \in M(2,\mathbb{C}) \mid trM = 0\}\,,$$
$$su(2) = \{M \in M(2,\mathbb{C}) \mid M^\dagger = -M\,,\ \ trM = 0\}\,. \tag{9.1}$$

Above † denotes the Hermitian conjugate: $M^\dagger := \overline{M}^t$, and the bar indicates complex conjugation. By direct inspection one verifies that $sl(2,\mathbb{C})$ admits a basis made of the following six matrices

$$-i\sigma_1/2,\ -i\sigma_2/2,\ -i\sigma_3/2,\ \sigma_1/2,\ \sigma_2/2,\ \sigma_3/2\,, \tag{9.2}$$

and the first three define a basis of $su(2)$ as well. We have introduced the well-known *Pauli matrices*

$$\sigma_1 = \begin{bmatrix} 0 & 1 \\ 1 & 0 \end{bmatrix},\quad \sigma_2 = \begin{bmatrix} 0 & -i \\ i & 0 \end{bmatrix},\quad \sigma_3 = \begin{bmatrix} 1 & 0 \\ 0 & -1 \end{bmatrix}. \tag{9.3}$$

By direct inspection one sees that these matrices satisfy the following identities:

$$tr\sigma_i = 0\,,\ \ \sigma_i = (\sigma_i)^\dagger\,,\ \ \sigma_i\sigma_j = \delta_{ij}I + i\sum_{k=1}^{3}\epsilon_{ijk}\sigma_k\,,\quad \text{for } i,j = 1,2,3, \tag{9.4}$$

where, as usual, ϵ_{ijk} is the completely antisymmetric Ricci indicator with $\epsilon_{123} = 1$.

Remarks 9.1 As a very important consequence of the definition of the Pauli matrices, one immediately obtains the following commutation relations for the generators of $sl(2,\mathbb{C})$:

$$\left[-i\frac{\sigma_i}{2},\frac{\sigma_j}{2}\right] = \sum_{k=1}^{3}\epsilon_{ijk}\frac{\sigma_k}{2}\,,\quad \left[-i\frac{\sigma_i}{2},-i\frac{\sigma_j}{2}\right] = \sum_{k=1}^{3}\epsilon_{ijk}\left(-i\frac{\sigma_k}{2}\right),$$
$$\left[\frac{\sigma_i}{2},\frac{\sigma_j}{2}\right] = -\sum_{k=1}^{3}\epsilon_{ijk}\left(-i\frac{\sigma_k}{2}\right). \tag{9.5}$$

It is then evident that the unique real vector-space isomorphism from $sl(2,\mathbb{C})$ to the Lie algebra of $SO(1,3)\uparrow$, namely the Lie algebra of its identity component, that identifies $-i\sigma_i/2$ with S_i and $\sigma_i/2$ with K_i, $i = 1,2,3$, preserves the Lie commutator in view of (8.9). In other words $sl(2,\mathbb{C})$ and $so(1,3)$ are isomorphic as real Lie algebras. The analogous result arises by comparison of the Lie algebra of $su(2)$ and that of $so(3)$. As is known [1] this implies that there exists a local Lie-group isomorphism from a neighborhood of the identity of $SL(2, \mathbb{C})$ to a similar neighborhood of $SO(1, 3)\uparrow$, and from a neighborhood of the identity of $SU(2)$ to a similar neighborhood of $SO(3)$, respectively, whose differential maps reduce to the found Lie-algebra isomorphisms. In the following we shall study those Lie-group isomorphisms proving that, actually, they extend to surjective global Lie-group homomorphisms, which are the starting point for the theory of (relativistic) spinors. ■

For the applications to relativity, it is convenient to define some other "Pauli matrices". First of all, define

$$\sigma_0 := I \,, \tag{9.6}$$

so that σ_μ makes sense for $\mu = 0, 1, 2, 3$. In the following, as usual, $\eta_{\mu\nu}$ denotes the components of the Minkowskian metric tensor in canonical form (i.e. the elements of the matrix $diag(-1, 1, 1, 1)$), and $\eta^{\mu\nu}$ are the components of the inverse Minkowskian metric tensor. Extending the procedure of raising indices, it is customary to define the **Pauli matrices with raised indices**:

$$\sigma^\mu := \eta^{\mu\nu}\sigma_\nu \,. \tag{9.7}$$

Above $\nu, \mu := 0, 1, 2, 3$ and the convention of summation over repeated indices is assumed. Finally, essentially for technical reasons, it is also customary to define the **primed Pauli matrices**

$$\sigma'_0 := I \,, \ \sigma'_i := -\sigma_i \,, \quad i = 1, 2, 3 \,, \tag{9.8}$$

and those with raised indices

$$\sigma'^\mu := \eta^{\mu\nu}\sigma'_\nu \,. \tag{9.9}$$

It is worth noticing that $\{\sigma_\mu\}_{\mu=0,1,2,3}$ is a vector basis of the real space $\mathcal{H}(2, \mathbb{C})$ of Hermitian complex 2×2 matrices and $\{\sigma_\mu, i\sigma_\mu\}_{\mu=0,1,2,3}$ is a basis of the algebra of complex 2×2 matrices $M(2, \mathbb{C})$ viewed as a real vector space.

Exercises 9.2

1. Prove that $\sigma_0, \sigma_1, \sigma_2, \sigma_3$ is a basis of the *real* vector space of the 2×2 complex Hermitian matrices.
2. Prove that $\sigma_i\sigma_j + \sigma_j\sigma_i = 2\delta_{ij}I$ and $\sigma'_i\sigma'_j + \sigma'_j\sigma'_i = 2\delta_{ij}I$, $i = 1, 2, 3$.
3. Prove that:

$$tr(\sigma'_\mu\sigma_\nu) = -\eta_{\mu\nu} \,, \ tr(\sigma'^\mu\sigma^\nu) = -\eta^{\mu\nu} \,, \ tr(\sigma'^\mu\sigma_\nu) = -\delta^\mu_\nu \,. \tag{9.10}$$

and

$$tr(\sigma_\mu \sigma'_\nu) = -\eta_{\mu\nu}\,, \quad tr(\sigma^\mu \sigma'^\nu) = -\eta^{\mu\nu}\,, \quad tr(\sigma^\mu \sigma'_\nu) = -\delta^\mu_\nu\,. \tag{9.11}$$

(*Hint*. Use the result in Exercise 2 and the cyclic property of the trace.)

4. Defining $\epsilon := i\sigma_2$, prove that:

$$\epsilon\sigma^t_\mu \epsilon = \epsilon\overline{\sigma_\mu}\epsilon = -\sigma'_\mu\,. \tag{9.12}$$

5. Prove that $\{\sigma_\mu\}_{\mu=0,1,2,3}$ is a vector basis of $\mathcal{H}(2,\mathbb{C})$ and that $\{\sigma_\mu, i\sigma_\mu\}_{\mu=0,1,2,3}$ is a basis of the algebra $M(2,\mathbb{C})$ viewed as a real vector space.

9.1.2 Properties of Exponentials of Pauli Matrices and Consequences for $SL(2,\mathbb{C})$ and $SU(2)$

We have a technical, but very important, result stated in the following proposition.

Proposition 9.3 *The following results are valid.*

(a) If $\boldsymbol{\sigma} := (\sigma_1, \sigma_2, \sigma_3)$ *and the* $so(3)$ *generators* T_j *are defined as in Proposition 8.8,*

$$e^{-i\theta\mathbf{m}\cdot\boldsymbol{\sigma}/2}\,\mathbf{n}\cdot\sigma\, e^{i\theta\mathbf{m}\cdot\boldsymbol{\sigma}/2} = \left(e^{\theta\mathbf{m}\cdot\mathbf{T}}\mathbf{n}\right)\cdot\boldsymbol{\sigma}\,, \quad \textit{for all } \mathbf{n},\mathbf{m}\in\mathbb{S}^2, \theta\in\mathbb{R}. \tag{9.13}$$

(b) The following pair of decompositions are valid:

$$e^{\chi\mathbf{n}\cdot\boldsymbol{\sigma}/2} = \cosh\frac{\chi}{2}\, I + \sinh\frac{\chi}{2}\,\mathbf{n}\cdot\boldsymbol{\sigma}\,, \quad \textit{for all } \mathbf{n}\in\mathbb{S}^2, \chi\in\mathbb{R}, \tag{9.14}$$

$$e^{-i\theta\mathbf{m}\cdot\boldsymbol{\sigma}/2} = \cos\frac{\theta}{2}\, I - i\sin\frac{\theta}{2}\,\mathbf{m}\cdot\boldsymbol{\sigma}\,, \quad \textit{for all } \mathbf{m}\in\mathbb{S}^2, \theta\in\mathbb{R}. \tag{9.15}$$

(c) A matrix $A\in M(2,\mathbb{C})$ *has the form* $e^{-i\theta\mathbf{n}\cdot\boldsymbol{\sigma}/2}$, *for some* $\theta\in\mathbb{R}$, $\mathbf{n}\in\mathbb{S}^2$, *if and only if* $A\in SU(2)$. *The Lie group* $SU(2)$ *is homeomorphic to* $\mathbb{S}^3$.

(d) A matrix $A\in M(2,\mathbb{C})$ *has the form* $e^{\chi\mathbf{n}\cdot\boldsymbol{\sigma}/2}$, *for some* $\chi\in\mathbb{R}$, $\mathbf{n}\in\mathbb{S}^2$, *if and only if A is a Hermitian positive element of* $SL(2,\mathbb{C})$. *The set of all Hermitian positive elements of* $SL(2,\mathbb{C})$, *equipped with the topology induced by* $SL(2,\mathbb{C})$, *is homeomorphic to* $\mathbb{R}^3$ *and is not a subgroup of* $SL(2,\mathbb{C})$.

Proof (a) and (b) Fix $\mathbf{m}\in\mathbb{S}^2$ and consider the smooth functions

$$f_i(\theta) = e^{-i\theta\mathbf{m}\cdot\boldsymbol{\sigma}/2}\sigma_i e^{i\theta\mathbf{m}\cdot\boldsymbol{\sigma}/2}\,.$$

By direct inspection, making use of the first set of commutation rules in (9.5), one obtains

$$\frac{df_i}{d\theta} = \sum_{k,p=1}^{3} m^k \epsilon_{kip} f_p(\theta) \, .$$

Recalling that $(T_i)_{rs} = -\epsilon_{irs}$ and considering the other set of smooth functions

$$g_i(\theta) = \left(e^{\theta \mathbf{m}\cdot\mathbf{T}} \mathbf{e}_i\right) \cdot \boldsymbol{\sigma} \, ,$$

one verifies that, again,

$$\frac{dg_i}{d\theta} = \sum_{k,p=1}^{3} m^k \epsilon_{kip} g_p(\theta) \, .$$

Since both sets of smooth functions satisfy the same system of first-order differential equations in normal form, with the same initial condition $f_i(0) = g_i(0) = \sigma_i$, we conclude that $f_i(\theta) = g_i(\theta)$ for all $\theta \in \mathbb{R}$ by uniqueness. This result yields (9.13) immediately. The established identity implies, by exponentiation,

$$e^{-i\theta \mathbf{m}\cdot\boldsymbol{\sigma}/2} e^{s\mathbf{n}\cdot\boldsymbol{\sigma}} e^{i\theta \mathbf{m}\cdot\boldsymbol{\sigma}/2} = \exp\{s \left(e^{\theta \mathbf{m}\cdot\mathbf{T}} \mathbf{n}\right) \cdot \boldsymbol{\sigma}\}$$

for either $s \in \mathbb{R}$ or $s \in i\mathbb{R}$. Using these improved results, and rotating $\mathbf{n}$ into $\mathbf{e}_3$ with a suitable element $e^{-i\theta \mathbf{m}\cdot\boldsymbol{\sigma}/2}$ of $SU(2)$, one realizes immediately that (9.14) and (9.15) are equivalent to

$$e^{\chi\sigma_3/2} = \cosh\frac{\chi}{2}\, I + \sinh\frac{\chi}{2}\, \sigma_3 \, , \quad \text{for all } \chi \in \mathbb{R},$$

$$e^{-i\theta\sigma_3/2} = \cos\frac{\theta}{2}\, I - i \sin\frac{\theta}{2} \sigma_3 \, , \quad \text{for all } \theta \in \mathbb{R}.$$

In turn, these identities can be proved by direct inspection, using the diagonal explicit form of σ_3.

(c) If $A := e^{-i\theta \mathbf{n}\cdot\boldsymbol{\sigma}/2}$, then, making use of the well-known identity $\det(e^F) = e^{trF}$, we have

$$\det A = e^{tr(-i\theta \mathbf{n}\cdot\boldsymbol{\sigma}/2)} = e^{-i\theta \mathbf{n}\cdot tr(\boldsymbol{\sigma})/2} = e^0 = 1,$$

so that $A \in SL(2, \mathbb{C})$. We also have

$$A^\dagger = (e^{-i\theta \mathbf{n}\cdot\boldsymbol{\sigma}/2})^\dagger = e^{(-i\theta \mathbf{n}\cdot\boldsymbol{\sigma}/2)^\dagger} = e^{i\theta \mathbf{n}\cdot\boldsymbol{\sigma}/2} = A^{-1} \, .$$

This proves that every matrix of the form $e^{-i\theta \mathbf{n}\cdot\boldsymbol{\sigma}/2}$ belongs to $SU(2)$. Now we prove the converse fact. Consider a matrix $V \in SU(2)$. By the spectral theorem, it can be decomposed as $V = e^{-i\lambda}P_- + e^{i\lambda}P_+$, where $P_\pm$ are orthogonal projectors onto one-dimensional subspaces and $\lambda \in \mathbb{R}$. Consider the Hermitian operator $H := -\lambda P_- + \lambda P_+$. By construction $e^{iH} = V$. Moreover $tr\, H = 0$, hence every such Hermitian traceless matrix can be written as $H = -\theta \mathbf{n}\cdot\boldsymbol{\sigma}/2$ for some $\theta \in \mathbb{R}$ and $\mathbf{n} \in \mathbb{S}^2$. Therefore

$$V = e^{iH} = e^{-i\theta \mathbf{n}\cdot\boldsymbol{\sigma}/2} .$$

We have found that $V = e^{-i\theta \mathbf{n}\cdot\boldsymbol{\sigma}/2}$ if and only if $V \in SU(2)$ as requested.

Consider now the generic $V \in SU(2)$. In view of (b), it can be written as

$$V(\mathbf{n}, \theta) = \cos\frac{\theta}{2}\, I - i \sin\frac{\theta}{2}\, \mathbf{n}\cdot\boldsymbol{\sigma} , \quad \text{for any } \mathbf{n} \in \mathbb{S}^2, \theta \in \mathbb{R}.$$

Making explicit the form of V, with the help of the explicit expression of the Pauli matrices, one finds that the assignment of such a $V(\mathbf{n}, \theta)$ is equivalent to the assignment of a point on the surface in $\mathbb{R}^4$:

$$X^0 := \cos\frac{\theta}{2},\; X^1 = n^1 \sin\frac{\theta}{2},\; X^2 = n^2 \sin\frac{\theta}{2},\; X^3 = n^3 \sin\frac{\theta}{2}, \quad \text{for any } \mathbf{n} \in \mathbb{S}^2, \theta \in \mathbb{R}.$$

This surface is nothing but the 3-sphere $\mathbb{S}^3$,

$$(X^0)^2 + (X^1)^2 + (X^2)^2 + (X^3)^2 = 1 .$$

This one-to-one correspondence between matrices $V(\mathbf{n}, \theta)$ and the points of the 3-sphere $\mathbb{S}^3$ can be proved to be a homeomorphism between $\mathbb{S}^3$ and $SU(2)$; the idea of the proof is the same as that given for the space of boosts in the proof of (e) in Proposition 8.11.

(d) With the same procedure as in case (c), one finds that every matrix $A := e^{\chi \mathbf{n}\cdot\boldsymbol{\sigma}/2}$, with $\chi \in \mathbb{R}$, belongs to $SL(2, \mathbb{C})$ and fulfills $A = A^\dagger$. Let us prove that it is positive, too. Positivity means that $(x|Ax) \geq 0$ for every $x \in \mathbb{C}^2$, where $(\cdot|\cdot)$ is the standard Hermitian scalar product in $\mathbb{C}^2$. Notice that, if V is unitary (and thus preserves the scalar product), $(x|Ax) \geq 0$ for every $x \in \mathbb{C}^2$ if and only if $(x|V^\dagger A V x) \geq 0$ for every $x \in \mathbb{C}^2$. Using (a), one can fix the matrix $V = e^{-i\theta \mathbf{m}\cdot\boldsymbol{\sigma}/2}$ in such a way that $V^\dagger \mathbf{n}\cdot\boldsymbol{\sigma} V = \sigma_3$. Therefore $V^\dagger e^{\chi \mathbf{n}\cdot\boldsymbol{\sigma}/2} V = e^{\chi\sigma_3/2}$. Since σ_3 is diagonal, $e^{\chi\sigma_3/2}$ is diagonal as well and positivity of $e^{\chi\sigma_3/2}$ can be checked by direct inspection. We have established that every matrix of the form $e^{\chi \mathbf{n}\cdot\boldsymbol{\sigma}/2}$ is a positive Hermitian element of $SL(2, \mathbb{C})$.

Now we prove the converse fact. Consider a Hermitian positive matrix $P \in SL(2, \mathbb{C})$. By the spectral theorem for Hermitian operators it decomposes as $P = \lambda_1 P_1 + \lambda_2 P_2$, where $P_{1,2}$ are orthogonal projectors onto one-dimensional subspaces and $\lambda_{1,2} > 0$ by positivity, with $\lambda_1\lambda_2 = 1$ because $\det P = 1$. Consider the

Hermitian operator $H := \ln \lambda_1 P_1 + \ln \lambda_2 P_2$. By construction $e^H = P$. Every Hermitian 2×2 matrix H can be written as $H = t^0 I + \sum_{i=1}^3 t^i \sigma_i$ for some reals t^0, t^1, t^2, t^3, that is $H = t^0 I + \chi \mathbf{n} \cdot \boldsymbol{\sigma}/2$ for some $\chi \in \mathbb{R}$ and $\mathbf{n} \in \mathbb{S}^2$. We have found that $P = e^{t^0} e^{\chi \mathbf{n} \cdot \sigma/2}$. The requirement $\det P = 1$ imposes $t^0 = 0$, so that we have found that $P = e^{\chi \mathbf{n} \cdot \sigma/2}$ if and only if $P \in SL(2, \mathbb{C})$ is Hermitian and positive, as requested.

Consider now the generic positive Hermitian element $P \in SL(2, \mathbb{C})$. In view of (b), it can be written as

$$P(\mathbf{n}, \chi) = \cosh \frac{\chi}{2} I + \sinh \frac{\chi}{2} \mathbf{n} \cdot \boldsymbol{\sigma} , \quad \text{for any } \mathbf{n} \in \mathbb{S}^2, \chi \in \mathbb{R}.$$

Making explicit the form of P, with the help of the explicit expression of the Pauli matrices, one finds that the assignment of such a P is equivalent to the assignment of a point on the surface in $\mathbb{R}^4$:

$$X^0 := \cosh \frac{\chi}{2} , \; X^1 = n^1 \sinh \frac{\chi}{2} , \; X^2 = n^2 \sinh \frac{\chi}{2} ,$$
$$X^3 = n^3 \sinh \frac{\chi}{2} , \quad \text{for any } \mathbf{n} \in \mathbb{S}^2, \chi \in \mathbb{R} .$$

This surface is nothing but the hyperboloid

$$-(X^0)^2 + (X^1)^2 + (X^2)^2 + (X^3)^2 = -1$$

which is trivially homeomorphic to $\mathbb{R}^3$. This one-to-one correspondence between matrices $P(\mathbf{n}, \chi)$ and the points of the hyperboloid can be proved to be a homeomorphism between $\mathbb{R}^3$ and the set of all Hermitian positive elements of $SL(2, \mathbb{C})$ equipped with the topology induced by $SL(2, \mathbb{C})$, the idea of the proof being the same as that given for the space of boosts in the proof of (e) in Proposition 8.11. □

9.1.3 Polar Decomposition Theorem and Topology of $SL(2, \mathbb{C})$

We are now in a position to state and prove a decomposition theorem for $SL(2, \mathbb{C})$ which will play an important role in the following section.

Theorem 9.4 *Take $L \in SL(2, \mathbb{C})$. The following holds.*

(a) *There is exactly one positive Hermitian matrix $H \in SL(2, \mathbb{C})$ (so that $H = e^{\chi \mathbf{n} \cdot \sigma/2}$ for some $\chi \in \mathbb{R}$ and $\mathbf{n} \in \mathbb{S}^2$) and exactly one unitary matrix $V \in SU(2)$ (so that $V = e^{-i\theta \mathbf{m} \cdot \sigma/2}$ for some $\theta \in \mathbb{R}$ and $\mathbf{m} \in \mathbb{S}^2$) such that*

$$L = VH .$$

(b) *There is exactly one positive Hermitian matrix $H' \in SL(2,\mathbb{C})$ (so that $H' = e^{\chi'\mathbf{n}'\cdot\sigma/2}$ for some $\chi' \in \mathbb{R}$ and $\mathbf{n}' \in \mathbb{S}^2$) and exactly one unitary matrix $V' \in SU(2)$ (so that $V' = e^{-i\theta'\mathbf{m}'\cdot\sigma/2}$ for some $\theta' \in \mathbb{R}$ and $\mathbf{m}' \in \mathbb{S}^2$) such that*

$$L = H'V' .$$

(c) *It holds $V' = V$ and $H' = VHV^\dagger$.*

Proof This is nothing but the specialization of the polar decomposition theorem (Theorem 6.10) applied to complex 2×2 matrices. The explicit expressions for H, V, H', V' arise from (b), (c), and (d) in Proposition 9.3. □

We want to emphasize here a topological consequence of this theorem. Using the fact that $SL(2,\mathbb{C})$ is bijectively identified with the product $SU(2) \times \mathbb{R}^3$, i.e. $\mathbb{S}^3 \times \mathbb{R}^3$, where the $\mathbb{R}^3$ factor is the topological space of the Hermitian positive elements of $SL(2,\mathbb{C})$ as in (d) of Proposition 9.3, it follows (the proof is the same as that in Proposition 8.16) that $SL(2,\mathbb{C})$ is homeomorphic to the topological space $\mathbb{S}^3 \times \mathbb{R}^3$. Since the factors are connected, $SL(2,\mathbb{C})$ is connected too. It is also path-connected, it being a differentiable manifold and thus admitting a topological base made of smoothly path-connected open sets. Moreover, the fundamental group of $SL(2,\mathbb{C})$ is, as happens for product manifolds [3], the product of the fundamental groups of the factors. As $\pi_1\left(\mathbb{S}^3\right) = \pi_1\left(\mathbb{R}^3\right) = \{1\}$, one finds $\pi_1\left(SL(2,\mathbb{C})\right) = \{1\}$. We have proved the following proposition.

Proposition 9.5 *$SL(2,\mathbb{C})$ is homeomorphic to $\mathbb{S}^3 \times \mathbb{R}^3$, as a consequence it is connected, path-connected and $\pi_1\left(SL(2,\mathbb{C})\right) = \{1\}$, so that it is simply connected.*

Remarks 9.6 We can draw further conclusions from the proposition above, making use of general theorems on Lie groups [1]. Since $SL(2,\mathbb{C})$ is simply connected, it is the only simply connected Lie group (up to Lie-group isomorphisms) which admits $sl(2,\mathbb{C})$ as Lie algebra. As a consequence, it has to coincide with the universal covering of $SO(1,3)\uparrow$ (since $so(1,3)$ is isomorphic to $sl(2,\mathbb{C})$, as noticed at the beginning of this section, and since $SO(1,3)\uparrow$ is not simply connected due to Proposition 8.16). Thus there must be a Lie-group homomorphism from $SL(2,\mathbb{C})$ onto $SO(1,3)\uparrow$ which is a local Lie-group isomorphism about the units and whose differential computed in the tangent space at the group unit coincides with the Lie-algebra isomorphism which identifies $sl(2,\mathbb{C})$ and $so(1,3)$. ■

9.2 The Interplay of $SL(2,\mathbb{C})$ and $SO(1,3)\uparrow$

In this subsection we establish the main results enabling one to introduce the notion of *spinor* in relation with vectors. This result concerns the covering Lie-group homomorphism from $SL(2,\mathbb{C})$ onto $SO(1,3)\uparrow$ already mentioned in Comments 9.6.

9.2.1 Construction of the Covering Homomorphism $\Pi : SL(2,\mathbb{C}) \to SO(1,3)\uparrow$

First of all we notice that the real vector space $\mathcal{H}(2,\mathbb{C})$ of 2×2 complex Hermitian matrices admits a basis made of the extended Pauli matrices σ_μ with $\mu = 0, 1, 2, 3$. So, if t^μ denotes the μ-th component of $t \in \mathbb{R}^4$, referred to the canonical basis, there is a linear bijective map

$$\mathbb{R}^4 \ni t \mapsto H(t) := t^\mu \sigma_\mu \in \mathcal{H}(2,\mathbb{C}) . \tag{9.16}$$

Above, we have adopted the convention of summation over repeated indices, and that convention will always be assumed henceforth in reference to Greek indices. As a second step we notice the following remarkable identity, which arises by direct inspection from the given definitions,

$$\det H(t) = -\eta_{\mu\nu} t^\mu t^\nu , \quad \text{for every } t \in \mathbb{R}^4. \tag{9.17}$$

Moreover, from Exercise 9.2, one has

$$tr(\sigma'_\mu \sigma_\nu) = -\eta_{\mu\nu} , \quad tr(\sigma'^\mu \sigma^\nu) = -\eta^{\mu\nu} , \quad tr(\sigma'^\mu \sigma_\nu) = -\delta^\mu_\nu . \tag{9.18}$$

As a further step, we observe that, if $H(t) \in \mathcal{H}(2,\mathbb{C})$, then $LH(t)L^\dagger \in \mathcal{H}(2,\mathbb{C})$ because

$$(LH(t)L^\dagger)^\dagger = (L^\dagger)^\dagger H(t)^\dagger L^\dagger = LH(t)L^\dagger ,$$

and thus $LH(t)L^\dagger$ has the form $H(t')$ for some $t' \in \mathbb{R}^4$. However, if $L \in SL(2,\mathbb{C})$, it also holds $L^\dagger \in SL(2,\mathbb{C})$ (because $\det L^\dagger = \overline{\det L^t} = \overline{\det L} = 1$), and thus the determinant of $H(t)$ coincides with that of $H(t') = LH(t)L^\dagger$ by Binet's rule:

$$\det H(t') = \det(LH(t)L^\dagger) = (\det L)(\det H(t)) \det L^\dagger = 1 \cdot (\det H(t)) \cdot 1 = \det H(t).$$

Now (9.17) implies that $LH(t)L^\dagger = H(t')$ with $\eta_{\mu\nu} t^\mu t^\nu = \eta_{\mu\nu} t'^\mu t'^\nu$. This result has an immediate fundamental consequence.

Proposition 9.7 *There is a group homomorphism* $\Pi : SL(2,\mathbb{C}) \ni L \mapsto \Lambda_L \in O(1,3)$ *which is uniquely determined by*

$$LH(t)L^\dagger = H(\Lambda_L t) , \quad \textit{for all } t \in \mathbb{R}^4. \tag{9.19}$$

Proof As noticed above, $LH(t)L^\dagger = H(t')$. Since the correspondence (9.16) is bijective, for a fixed $L \in SL(2,\mathbb{C})$ a map $f_L : \mathbb{R}^4 \ni t \mapsto t'$ is well defined. So we can write $LH(t)L^\dagger = H(f_L(t))$. On the other hand f_L is linear due to the linearity and bijectivity of $t \mapsto H(t)$. Indeed, one has $LH(at + bs)L^\dagger =$

$aLH(t)L^\dagger + bLH(s)L^\dagger$, i.e. $H(f_L(at+bs)) = aH(f_L(t)) + bH(f_L(s)) = H(af_L(t)+bf_L(s))$, so that $f_L(at+bs) = af_L(t)+bf_L(s)$, because $t \mapsto H(t)$ is invertible. It remains to prove that $f_L \in O(1,3)$. We know that, as discussed above, $\eta_{\mu\nu}t^\mu t^\nu = \eta_{\mu\nu}(f_Lt)^\mu(f_Lt)^\nu$ for every $t \in \mathbb{R}^4$, namely

$$\mathbf{g}(t,t) = \mathbf{g}(f_Lt, f_Lt) , \quad \text{for every } t \in \mathbb{R}^4. \tag{9.20}$$

However, as a general fact, if $s, t \in \mathbb{R}^4$, symmetry of the pseudo-scalar product $\mathbf{g}$ entails:

$$\mathbf{g}(s,t) = \frac{1}{4}\left(\mathbf{g}(s+t,s+t) - \mathbf{g}(s-t,s-t)\right) ,$$

so that (9.20) implies

$$\mathbf{g}(s,t) = \mathbf{g}(f_Ls, f_Lt) , \quad \text{for every } s,t \in \mathbb{R}^4.$$

Let us indicate f_L by Λ_L and prove that $\Pi : L \mapsto \Lambda_L$ is a group homomorphism. Trivially, $\Lambda_I = I$ because $IH(t)I^\dagger = H(It)$ for every $t \in \mathbb{R}^4$. Moreover

$$\begin{aligned} H(\Lambda_{LL'}t) &= (LL')H(t)(LL')^\dagger = L(L'H(t)L'^\dagger)L^\dagger = L(H(\Lambda_{L'}t))L^\dagger \\ &= H(\Lambda_L\Lambda_{L'}t) , \quad \text{for all } t \in \mathbb{R}^4, \end{aligned}$$

and thus $\Lambda_{LL'} = \Lambda_L\Lambda_{L'}$. The proof is concluded. □

9.2.2 *Properties of* Π

The obtained result can be strengthened in several ways. First of all, we prove that Π is a Lie-group homomorphism.

Proposition 9.8 *The group homomorphism* $\Pi : SL(2,\mathbb{C}) \ni L \mapsto \Lambda_L \in O(1,3)$ *determined by (9.19) enjoys the following properties.*

(a) It is a Lie-group homomorphism.
(b) The following explicit formula holds:

$$(\Lambda_L)^\nu{}_\mu = -tr\left(\sigma'^\nu L\sigma_\mu L^\dagger\right) , \quad \textit{for every } L \in SL(2,\mathbb{C}). \tag{9.21}$$

(c) The kernel of Π *includes* $\{\pm I\}$ *since*

$$\Pi(L) = \Pi(L') \quad \textit{if } L' = -L. \tag{9.22}$$

Proof (a) and (b) Since the differentiable structures of $SL(2, \mathbb{C})$ and $O(1, 3)$ are those induced by $\mathbb{R}^{16}$ and the operations of taking the trace, taking the Hermitian conjugate, and multiplying matrices are trivially differentiable, (9.21) implies that the group homomorphism Π is differentiable and thus is a Lie-group homomorphism. Let us prove (9.21). Choosing $t = \delta^\alpha_\mu e_\alpha$, where $\{e_\lambda\}_{\lambda=0,1,2,3}$ is the canonical basis of $\mathbb{R}^4$, (9.19) produces:

$$L\sigma_\mu L^\dagger = Lt^\alpha \sigma_\alpha L^\dagger = (\Lambda_L)^\gamma{}_\beta\, t^\beta \sigma_\gamma = (\Lambda_L)^\gamma{}_\beta \delta^\beta_\mu \sigma_\gamma = (\Lambda_L)^\gamma{}_\mu\, \sigma_\gamma\,.$$

As a consequence

$$\sigma'^\nu L\sigma_\mu L^\dagger = (\Lambda_L)^\gamma{}_\mu\, \sigma'^\nu \sigma_\gamma\,.$$

To conclude, it is enough to apply the identity $tr(\sigma'^\nu \sigma_\gamma) = -\delta^\nu_\gamma$ (see (9.10) in Exercises 9.2).
(c) Using the very definition of Π, it is evident that $\Pi(L) = \Pi(-L)$. □

Now we study the interplay of Π with exponentials $\exp\{a\mathbf{n}\cdot\sigma\}$ as those appearing in the decomposition theorem 9.4.

Proposition 9.9 *The Lie-group homomorphism $\Pi : SL(2, \mathbb{C}) \ni L \mapsto \Lambda_L \in O(1, 3)$ determined by (9.19) enjoys the following features, using the notation of Sect. 8.2.2.*

(a) For every $\theta \in \mathbb{R}$ and $\mathbf{n} \in \mathbb{S}^2$, one has

$$\Pi\left(e^{-i\theta\mathbf{n}\cdot\boldsymbol{\sigma}/2}\right) = e^{\theta\mathbf{n}\cdot\mathbf{S}}\,, \tag{9.23}$$

so that $\Pi(SU(2)) = SO(3)$. Moreover, if $V, V' \in SU(2)$, $\Pi(V) = \Pi(V')$ if and only if $V = \pm V'$.

(b) For every $\chi \in \mathbb{R}$ and $\mathbf{n} \in \mathbb{S}^2$, one has

$$\Pi\left(e^{\chi\mathbf{n}\cdot\boldsymbol{\sigma}/2}\right) = e^{\chi\mathbf{n}\cdot\mathbf{K}}\,, \tag{9.24}$$

so that Π maps the set of 2×2 complex Hermitian positive elements of $SL(2, \mathbb{C})$ onto the set of Lorentz boosts. Moreover, this map is injective.

Proof

(a) The identity (9.13) implies immediately that, with obvious notation,

$$e^{-i\theta\mathbf{n}\cdot\boldsymbol{\sigma}/2} t^\mu \sigma_\mu \left(e^{-i\theta\mathbf{n}\cdot\boldsymbol{\sigma}/2}\right)^\dagger = e^{-i\theta\mathbf{n}\cdot\boldsymbol{\sigma}/2}(t^0 I + \mathbf{t}\cdot\boldsymbol{\sigma})e^{i\theta\mathbf{n}\cdot\boldsymbol{\sigma}/2} = t^0 I + \left(e^{\theta\mathbf{n}\cdot\mathbf{T}}\mathbf{t}\right)\cdot\boldsymbol{\sigma}$$
$$= \left(e^{\theta\mathbf{n}\cdot\mathbf{S}} t\right)^\mu \sigma_\mu\,.$$

This result, in view of the definition of Π, and since t is arbitrary, implies (9.23). Recalling that every element of $SU(2)$ has the form $e^{-i\theta\mathbf{n}\cdot\boldsymbol{\sigma}/2}$, we have also obtained that $\Pi(SU(2)) = SO(3)$. To conclude, we have to prove that $\Pi(V) = \Pi(V')$ implies $V' = \pm V$ when $V, V' \in SU(2)$. To this end, we notice that, from the definition of Π, $\Pi(V) = \Pi(V')$ is equivalent to $V_1\sigma_i V_1^\dagger = \sigma_i$ for $i = 1, 2, 3$, where $V_1 := V'V^\dagger$. Since $V_1^\dagger = V_1^{-1}$ because $V_1 \in SU(2)$, it also holds $V_1\sigma_i = \sigma_i V_1$ for $i = 1, 2, 3$. Thus V_1 commutes with all complex linear combinations of I and the Pauli matrices. But these combinations amount to all elements of $M(2,\mathbb{C})$. As a consequence it must be $V_1 = \lambda I$ for some $\lambda \in \mathbb{C}$. The requirements $\det V_1 = 1$ and $V_1^\dagger = V_1^{-1}$ imply $\lambda = \pm 1$, so that $V = \pm V'$.

(b) The identity to be proved is now

$$e^{\chi\mathbf{n}\cdot\boldsymbol{\sigma}/2} t^\mu \sigma_\mu \left(e^{\chi\mathbf{n}\cdot\boldsymbol{\sigma}/2}\right)^\dagger = \left(e^{\chi\mathbf{n}\cdot\mathbf{K}} t\right)^\mu \sigma_\mu \,,$$

that is, since $\sigma_\mu^\dagger = \sigma_\mu$,

$$e^{\chi\mathbf{n}\cdot\boldsymbol{\sigma}/2} t^\mu \sigma_\mu e^{\chi\mathbf{n}\cdot\boldsymbol{\sigma}/2} = \left(e^{\chi\mathbf{n}\cdot\mathbf{K}} t\right)^\mu \sigma_\mu \,.$$

Fixing $\mathbf{m}$ and θ suitably, and then making use of (9.13), this is equivalent to

$$e^{\chi\sigma_3/2} e^{-i\theta\mathbf{m}\cdot\boldsymbol{\sigma}/2} t^\mu \sigma_\mu e^{i\theta\mathbf{m}\cdot\boldsymbol{\sigma}/2} e^{\chi\sigma_3/2} = \left(e^{\chi\mathbf{n}\cdot\mathbf{K}} t\right)^\mu \sigma_\mu \,.$$

In turn, employing (a) in this proposition, and (b) of Proposition 8.11, this turns out to be equivalent to:

$$e^{\chi\sigma_3/2} s^\mu \sigma_\mu e^{\chi\sigma_3/2} = \left(e^{\chi K_3} s\right)^\mu \sigma_\mu \,, \tag{9.25}$$

where $s = e^{\theta\mathbf{m}\cdot\mathbf{S}} t$ is generic. (9.25) can be proved by direct inspection, expanding the exponentials via (9.14), using the last formula in (9.4), and (8.16). Since all boosts in the Lorentz group have the form $e^{\chi\mathbf{n}\cdot\mathbf{K}}$, it remains to prove that Π restricted to the Hermitian positive matrices of $SL(2,\mathbb{C})$ is injective. To this end suppose that $\Pi(e^{\chi\mathbf{n}\cdot\boldsymbol{\sigma}/2}) = \Pi(e^{\chi'\mathbf{n}'\cdot\boldsymbol{\sigma}/2})$. This is equivalent to, for every $t \in \mathbb{R}^4$:

$$e^{\chi\mathbf{n}\cdot\boldsymbol{\sigma}/2} t^\mu \sigma_\mu e^{\chi\mathbf{n}\cdot\boldsymbol{\sigma}/2} = e^{\chi'\mathbf{n}'\cdot\boldsymbol{\sigma}/2} t^\mu \sigma_\mu e^{\chi'\mathbf{n}'\cdot\boldsymbol{\sigma}/2} \,.$$

In particular, taking $\mu = 0$ and using $\sigma_0 = I$, these identities produce

$$e^{\chi\mathbf{n}\cdot\boldsymbol{\sigma}} = e^{\chi'\mathbf{n}'\cdot\boldsymbol{\sigma}} \,.$$

By direct inspection, making use of (9.14), one finds that this is possible only if

$$\cosh\chi\,\sigma_0+\sinh\chi\sum_{j=1}^{3}n^j\sigma_j=\cosh\chi'\,\sigma_0+\sinh\chi'\sum_{j=1}^{3}n'^j\sigma_j\,,$$

and thus $\chi\mathbf{n}=\chi'\mathbf{n}'$. Therefore $\chi\mathbf{n}\cdot\boldsymbol{\sigma}=\chi'\mathbf{n}'\cdot\boldsymbol{\sigma}$, and hence $e^{\chi\mathbf{n}\cdot\sigma/2}=e^{\chi'\mathbf{n}'\cdot\sigma/2}$. We have found that $\Pi(e^{\chi\mathbf{n}\cdot\sigma/2})=\Pi(e^{\chi'\mathbf{n}'\cdot\sigma/2})$ implies $e^{\chi\mathbf{n}\cdot\sigma/2}=e^{\chi'\mathbf{n}'\cdot\sigma/2}$, as wanted.

□

We are now in a position to state and prove the conclusive theorem.

Theorem 9.10 *The Lie-group homomorphism* $\Pi : SL(2,\mathbb{C})\ni L\mapsto\Lambda_L\in O(1,3)$ *determined by (9.19) satisfies the following facts.*

(a) $\Pi(SL(2,\mathbb{C}))=SO(1,3)\!\uparrow$.
(b) $\Pi(SU(2))=SO(3)$.
(c) The kernel of Π *is* $\{\pm I\}$*, that is* $\Pi(L')=\Pi(L'')$ *if and only if* $L'=\pm L''$.
(d) Π *defines a local Lie-group isomorphism (about the unit) between* $SL(2,\mathbb{C})$ *and* $SO(1,3)\!\uparrow$*, so that* $d\Pi|_I : sl(2,\mathbb{C})\to so(1,3)$ *is a Lie-algebra isomorphism.*
(e) $\Pi\restriction_{SU(2)}$ *defines a local Lie-group isomorphism (about the unit) between* $SU(2)$ *and* $SO(3)$*, so that* $d\Pi\restriction_{SU(2)}|_I : su(2)\to so(3)$ *is a Lie-algebra isomorphism.*
(f) If $L=HV$ *is the polar decomposition of* $L\in SL(2,\mathbb{C})$*, with* V *unitary and* H *positive Hermitian, then* $\Pi(H)\Pi(V)$ *is the polar decomposition of* $\Pi(L)\in SO(1,3)\!\uparrow$*, where* $\Pi(V)$ *is a spatial rotation and* $\Pi(H)$ *is a boost.*

Proof We start from (f). The proof of this statement is almost evident: if $L\in SL(2,\mathbb{C})$, the matrices of its polar decomposition in $SL(2,\mathbb{C})$ have the form $H=e^{\chi\mathbf{n}\cdot\sigma/2}$ and $V=e^{-i\theta\mathbf{m}\cdot\sigma/2}$ due to Theorem 9.4. Thus, the action of Π produces, in view of Proposition 9.9, $\Pi(V)=e^{\theta\mathbf{m}\cdot\mathbf{S}}$ and $\Pi(H)=e^{\chi\mathbf{n}\cdot\mathbf{K}}$, so that $\Pi(L)=\Pi(H)\Pi(V)=e^{\chi\mathbf{n}\cdot\mathbf{K}}e^{\theta\mathbf{m}\cdot\mathbf{S}}$. The latter is trivially a polar decomposition in $SO(1,3)\!\uparrow$ of the product of these factors and thus the *unique* one (Theorems 8.14 and 8.15). But the product of these factors is $\Pi(L)$ by construction. So $\Pi(H)\Pi(V)$ is the polar decomposition of $\Pi(L)$ in $SO(1,3)\!\uparrow$, as wanted.

(a) Let $\Lambda\in SO(1,3)\!\uparrow$, so that $\Lambda=e^{\theta\mathbf{m}\cdot\mathbf{S}}e^{\chi\mathbf{n}\cdot\mathbf{K}}$ by polar decomposition. Due to (f), we have immediately: $\Pi(e^{-i\theta\mathbf{m}\cdot\sigma/2}e^{\chi\mathbf{n}\cdot\sigma/2})=\Lambda$, and thus $\Pi(SL(2,\mathbb{C}))\supset SO(1,3)\!\uparrow$. On the other hand, if $L\in SL(2,\mathbb{C})$, as established in the proof of (f), $\Pi(L)=\Pi(H)\Pi(V)=e^{\chi\mathbf{n}\cdot\mathbf{K}}e^{\theta\mathbf{m}\cdot\mathbf{S}}$ so that $\Pi(L)\in SO(1,3)\!\uparrow$, because it is the product of two elements of that group. Therefore $\Pi(SL(2,\mathbb{C}))\subset SO(1,3)\!\uparrow$ and thus $\Pi(SL(2,\mathbb{C}))=SO(1,3)\!\uparrow$.

(b) It follows from (a) in Proposition 9.9 and (a), (b) of Proposition 8.9.

(c) Suppose that $\Pi(L')=\Pi(L'')$. This is equivalent to say that $\Pi(L)=I$, where $L=L'L''^{-1}$. By polar decomposition $L=HV$ and, by (f), $\Pi(H)\Pi(V)$ must be the polar decomposition of $I\in SO(1,3)\!\uparrow$. A polar decomposition of I is trivially obtained as the product of $I\in SO(3)$ and the trivial boost I. By uniqueness, this is the only polar decomposition of I. Hence $\Pi(H) = I$ and $\Pi(V) = I$. In view of Proposition 9.9, $H = I$ and $V = \pm I$, so that $L = \pm I$ and $L' = \pm L''$.

(d) and (e). By the general theory of Lie groups, given a basis $e_1, \ldots, e_n$ in the tangent space at the unit element (i.e. the Lie algebra of the group), the set of parameters $(t_1, \ldots, t_n)$ of the one-parameter subgroups $t_k \mapsto \exp\{t_k e_k\}$ generated by the elements of the basis, for a sufficiently small range of the parameters about 0, i.e. $|t_k| < \delta$, defines a coordinate patch, compatible with the differentiable structure, in a neighborhood of the unit element and centered at that point, i.e. the unit corresponds to $(0, \ldots, 0)$. Fix the basis $-i\sigma_1/2, -i\sigma_2/2, -i\sigma_3/2, \sigma_1/2, \sigma_2/2, \sigma_3/2$ in $sl(2,\mathbb{C})$, associated with the coordinate system $(s^1, s^2, s^3, t^1, t^2, t^3)$ about $I \in SL(2,\mathbb{C})$, and the basis $S_1, S_2, S_3, K_1, K_2, K_3$ in $so(1,3)$, associated with the coordinate system $(x^1, x^2, x^3, y^1, y^2, y^3)$ about $I \in SO(1,3)\uparrow$. By Proposition 9.9, $\Pi(e^{-is_k\sigma_k/2}) = e^{s_k S_k}$ and $\Pi(e^{t_j\sigma_j/2}) = e^{t_j K_j}$, so that, in the said coordinates, the action of Π is nothing but

$$\Pi(s^1, s^2, s^3, t^1, t^2, t^3) = (s^1, s^2, s^3, t^1, t^2, t^3)\,,$$

that is, the identity map. In other words, Π is a diffeomorphism, and thus a Lie-group isomorphism in the constructed coordinate patches about the units of the two groups.

By construction the differential $d\Pi|_I : sl(2,\mathbb{C}) \to so(1,3)$ maps each element of the first basis $-i\sigma_1/2$, $-i\sigma_2/2$, $-i\sigma_3/2$, $\sigma_1/2$, $\sigma_2/2$, $\sigma_3/2$ into the corresponding element of the other basis $S_1, S_2, S_3, K_1, K_2, K_3$. As noticed earlier in Comment 9.1, this map is also an isomorphism of Lie algebras since it preserves the commutation rules. The case concerning $SU(2)$ and $SO(3)$ has the same proof, restricting to the bases $-i\sigma_1/2$, $-i\sigma_2/2$, $-i\sigma_3/2$ and S_1, S_2, S_3. □

References

1. Kirillov, A.: Éléments de la Théorie de représentations. Mir, Mosca (1974); Najmark, M.A., Stern, A.I.: Theory of Group Representations. Mir, Moscow (1984)
2. Ruhl, W.: The Lorentz Group and Harmonic Analysis. W. A. Benjamin, New York (1970)
3. Sernesi, E.: Geometria I and II. Bollati Boringhieri, Torino (1990)
4. Streater, R.F., Wightman, A.S.: PCT, Spin and Statistics, and All That with revised preface and corrections. Princeton University Press, Princeton (2000)
5. Wightman, A.S.: L'invariance dans la Mécanique quantique relativiste. In: De Witt, C. (ed.) Relations de dispersion et Particules élémentaires. Hermann, Paris (1960)

Chapter 10
Introduction to Spinors

This chapter gives an elementary introduction to spinors from the point of view developed in the preceding chapters. Spinors are presented not as mysterious new objects, but as vectors in a two-dimensional complex space equipped with a distinguished antisymmetric tensor, the metric spinor. This structure permits one to raise and lower spinor indices and to build spinorial tensors in close analogy with the tensor calculus associated with metric vector spaces.

The chapter then explains how four-vectors in Minkowski spacetime can be encoded by spinorial objects and how the action of $SL(2, \mathbb{C})$ reproduces the proper orthochronous Lorentz transformations. The final part introduces Dirac spinors, the Dirac adjoint, the action of four-vectors on spinors, and the Dirac equation. The goal is to provide the minimal algebraic background needed to understand the role of spinors in relativistic quantum theory without developing the full general theory of Clifford algebras or spin geometry.

10.1 The Space of Weyl Spinors

In this section we use again the definition of conjugate space $\overline{W}$ and conjugate dual space $\overline{W^*}$ (which is naturally isomorphic to $\overline{W}^*$), as defined in Definition 2.1, making use of their properties as stated in Theorem 2.13. In general we shall use various definitions, notions, and results contained in Chaps. 2 and 3. However, we will confine ourselves to the case of a two-dimensional vector space W over the complex field $\mathbb{C}$. Another ingredient will be a preferred non-degenerate antisymmetric tensor $\epsilon \in W^* \otimes W^*$, called the *metric spinor*. Non-degenerate means, as usual, that, viewing ϵ as a linear map $\epsilon : W \to W^*$, it turns out to be bijective. The metric spinor will be used to raise and lower indices, similarly to the metric tensor.

V. Moretti, *Geometric Methods in Mathematical Physics I*, La Matematica per il 3+2 184, https://doi.org/10.1007/978-3-032-31512-0_10

Definition 10.1 A (Weyl) **spinor space** is a two-dimensional vector space W over the complex field $\mathbb{C}$, equipped with a preferred non-degenerate antisymmetric tensor $\epsilon \in W^* \otimes W^*$, called the **metric spinor**. The elements of W are called **spinors** (or, equivalently, **Weyl spinors**), those of $\overline{W}$ are called **conjugate spinors**, those of W^* are called **dual spinors**, and those of $\overline{W^*}$ are called **dual conjugate spinors**. The elements belonging to tensor products of $W, W^*, \overline{W}, \overline{W^*}$ are generically called **spinorial tensors**. ■

Notation 10.2

(a) Referring to a basis in W and the associated bases in W^*, $\overline{W}$, and $\overline{W^*}$, the components of spinors, i.e. vectors in W, are denoted by ξ^A (where ξ may be replaced with another Greek letter). The components of dual spinors, that is vectors of W^*, are denoted by ξ_A. The components of conjugate spinors, that is vectors of $\overline{W}$, are denoted by $\xi^{A'}$. The components of dual conjugate spinors, that is vectors of $\overline{W^*}$, are denoted by $\xi_{A'}$. Notation for spinorial tensors is similar.

(b) We sometimes make use of the *abstract index notation*, so, for instance, $\Xi^A{}_{B'}$ denotes a spinorial tensor in $W \otimes \overline{W^*}$.

(c) Sometimes it will be more convenient to use an intrinsic notation. In that case a spinor or a spinorial tensor will be indicated with a Greek letter without indices, e.g. $\xi \in W$ or $\Xi \in W \otimes W^*$.

(d) If Ξ is a spinorial tensor, and $\mathcal{B} := \{e_A\}_{A=1,2} \subset W$ is a basis, ${}_{\mathcal{B}}\Xi$ denotes the *matrix* whose elements are the components of Ξ with respect to $\mathcal{B}$, or with respect to the relevant basis canonically associated with $\mathcal{B}$ in the tensor space containing Ξ. For instance, if $\Xi \in W \otimes \overline{W}$, so that $\{e_A \otimes \overline{e}_{B'}\}_{A,B'=1,2} \subset W \otimes \overline{W}$ is the basis in $W \otimes \overline{W}$ canonically associated with $\mathcal{B}$, ${}_{\mathcal{B}}\Xi$ is the matrix of elements $\Xi^{AB'}$, where $\Xi = \Xi^{AB'} e_A \otimes \overline{e}_{B'}$. ■

10.1.1 *The Metric Spinor to Lower and Raise Indices*

Let us come to the use of the metric spinor ϵ. Exactly as in the case of the metric tensor (see Sect. 5.2), since ϵ is non-degenerate, the components of

$$\epsilon = \epsilon_{AB} e^{*A} \otimes e^{*B} , \quad \epsilon_{AB} = \epsilon(e_A, e_B) = -\epsilon_{BA} ,$$

give rise to a nonsingular antisymmetric matrix, referring to a fixed basis of W, $\{e_A\}_{A=1,2}$, and to the associated canonical bases in $W^* \otimes W^*$. The matrix whose components are ϵ^{AB}, and which satisfies

$$\epsilon_{AB}\, \epsilon^{BC} = -\delta_A^C , \tag{10.1}$$

defines a second spinorial tensor, the **inverse metric spinor**:

$$\epsilon^{AB} e_A \otimes e_B = -\epsilon^{BA} e_A \otimes e_B \ .$$

The definition turns out to be independent of the chosen basis, exactly as for the metric tensor. Notice the sign $-$ in (10.1). Following the same procedure as for the metric tensor, one sees that the metric spinor and the inverse metric spinor define a natural isomorphism and its inverse, respectively, from W to W^* and from W^* to W. In components, or in the abstract-index notation, the isomorphism corresponds to the procedure of lowering indices, and its inverse corresponds to the procedure of raising indices. These are defined, respectively, as

$$W \ni \xi^A \mapsto \xi_A := \xi^B \epsilon_{BA} \in W^* \quad \text{and} \quad W^* \ni \eta_A \mapsto \eta^A := \epsilon^{AB} \eta_B \in W \ . \tag{10.2}$$

Notice that these two procedures work by summing over *different* indices. *The order cannot be interchanged here*, because ϵ_{AB} and ϵ^{AB} are antisymmetric, differently from the corresponding metric tensors. With the given choices, it results

$$\xi^A = \epsilon^{AC}(\xi^B \epsilon_{BC}) \ .$$

The procedures extend to spinorial tensors constructed with several factors W and W^*, exactly as for the metric tensor. In particular one finds, in view of (10.1) again:

$$\epsilon_{AB} = \epsilon^{CD} \epsilon_{CA} \epsilon_{DB} \quad \text{and} \quad \epsilon^{AB} = \epsilon_{CD} \epsilon^{CA} \epsilon^{DB} \ .$$

10.1.2 The Metric Spinor in the Conjugate Spaces of Spinors

The space $\overline{W}$ is naturally anti-isomorphic to W, and the same happens for W^* and $\overline{W^*}$, as established in Theorem 2.13. As a consequence a preferred metric spinor is defined on $\overline{W}$, induced by ϵ. If $F : W \to \overline{W}$ is the natural anti-isomorphism described in Theorem 2.13,

$$\overline{\epsilon}(\xi', \eta') := \overline{\epsilon(F^{-1}(\xi'), F^{-1}(\eta'))} \ , \quad \text{for all } \xi', \eta' \in \overline{W}, \tag{10.3}$$

defines an antisymmetric bilinear map from $\overline{W} \times \overline{W}$ to $\mathbb{C}$. The corresponding tensor of $\overline{W^*} \otimes \overline{W^*}$, is indicated by $\epsilon_{A'B'}$ or, indifferently, by $\overline{\epsilon}_{A'B'}$ in the abstract-index notation. Fixing a basis $\{e_A\}_{A=1,2}$ and working in the canonically associated basis of $\overline{W^*} \otimes \overline{W^*}$, in view of the last statement of Theorem 2.13, one finds

$$\overline{\epsilon}(\overline{e}_A, \overline{e}_B) = \epsilon_{AB} \ ,$$

that is, the components of the spinorial tensor $\overline{\epsilon} \in \overline{W^*} \otimes \overline{W^*}$ are the same as those of the metric spinor. As a consequence $\overline{\epsilon}$ is non-degenerate and thus it defines a metric spinor on $\overline{W}$. There is, as a consequence, an inverse metric spinor $\overline{\epsilon}^{A'B'}$ associated with $\overline{\epsilon}_{A'B'}$, and the procedure of raising and lowering indices can be performed in the conjugate spaces as well.

Finally, one can consider spinorial tensors defined in tensor spaces with four types of factors: W, W^*, $\overline{W}$, and $\overline{W^*}$, and the procedure of raising and lowering indices can be performed in those spaces employing the relevant metric spinors.

Exercises 10.3 Prove that the inverse metric spinor $\overline{\epsilon}^{A'B'}$ defined as above may be obtained equivalently from ϵ^{AB} and the natural anti-isomorphism $G : W^* \to \overline{W^*}$ (Theorem 2.13), following the analogue of the procedure used to define $\overline{\epsilon}_{A'B'}$ from ϵ_{AB} and the natural anti-isomorphism $F : W \to \overline{W}$, as done in (10.3).

10.1.3 Orthonormal Bases and the Role of $SL(2, \mathbb{C})$

Consider the spinor space W with metric spinor ϵ. The components of the non-degenerate antisymmetric metric spinor ϵ, for a fixed basis $\{e_A\}_{A=1,2} \subset W$, are

$$\epsilon_{11} = \epsilon_{22} = 0\,, \quad \text{and} \quad \epsilon_{12} = -\epsilon_{21} = \epsilon(e_1, e_2)\,.$$

Notice that $\epsilon(e_1, e_2) \neq 0$, otherwise ϵ would be degenerate. Therefore, we can rescale e_1 and/or e_2 in order to achieve

$$\epsilon_{11} = \epsilon_{22} = 0\,, \quad \text{and} \quad \epsilon_{12} = -\epsilon_{21} = 1\,.$$

As we shall see later, these bases are important, especially in relation with standard tensors in Minkowski spacetime, so we state a formal definition.

Definition 10.4 Referring to a spinor space W with metric spinor ϵ, a basis $\{e_A\}_{A=1,2} \subset W$ is said to be (ϵ-)**orthonormal** if the components of ϵ, referred to the associated canonical basis $\{e^{*A} \otimes e^{*B}\}_{A,B=1,2} \subset W^* \otimes W^*$, read

$$\epsilon_{11} = \epsilon_{22} = 0\,, \quad \text{and} \quad \epsilon_{12} = -\epsilon_{21} = 1\,. \tag{10.4}$$

■

If $\{e_A\}_{A=1,2} \subset W$ is ϵ-orthonormal, the components of ϵ^{AB} with respect to the canonically associated basis in $W \otimes W$ are again:

$$\epsilon^{11} = \epsilon^{22} = 0\,, \quad \text{and} \quad \epsilon^{12} = -\epsilon^{21} = 1\,, \tag{10.5}$$

This happens thanks to the sign $-$ in the right-hand side of (10.1). Conversely, if, referring to some basis $\{e_A\}_{A=1,2} \subset W$ and to the canonically associated ones, one finds that (10.5) is satisfied, then one is sure that $\{e_A\}_{A=1,2} \subset W$ is ϵ-orthonormal, because it arises from (10.1) that (10.5) implies (10.4). We conclude that (10.4) and (10.5) are equivalent.

This result has an important implication concerning the role of $SL(2,\mathbb{C})$ in spinor theory. It plays the same role as the Lorentz group plays in vector theory, in relation to pseudo-orthonormal frames. If $\{e_A\}_{A=1,2} \subset W$ is ϵ-orthonormal, so that (10.5) is true, consider another basis, $\{\tilde{e}_A\}_{A=1,2} \subset W$, with $e_A = L^B{}_A \tilde{e}_B$ for some matrix $L \in GL(2,\mathbb{C})$, whose components are the coefficients $L^B{}_A$. As soon as (10.4) and (10.5) are equivalent, $\{\tilde{e}_A\}_{A=1,2}$ is orthonormal if and only if the coefficients

$$\tilde{\epsilon}^{AB} = L^A{}_C L^B{}_D \epsilon^{CD} \tag{10.6}$$

satisfy the four requirements

$$\tilde{\epsilon}^{11} = 0\,, \quad \tilde{\epsilon}^{22} = 0\,, \quad \tilde{\epsilon}^{12} = -\tilde{\epsilon}^{21}\,, \quad \tilde{\epsilon}^{12} = 1\,.$$

Taking (10.5) into account in (10.6), the only nontrivial condition among the requirements written above is the last one, which is equivalent to saying that

$$1 = L^1{}_1 L^2{}_2 - L^1{}_2 L^2{}_1\,.$$

We recognize the determinant of L on the right-hand side. In other words, if $\{e_A\}_{A=1,2} \subset W$ is ϵ-orthonormal, another basis $\{\tilde{e}_A\}_{A=1,2} \subset W$, with $e_A = L^B{}_A \tilde{e}_B$ for some matrix $L \in GL(2,\mathbb{C})$ (whose components are the coefficients $L^B{}_A$), is ϵ-orthonormal if and only if

$$\det L = 1\,.$$

Let us summarize the obtained results in a proposition that includes some further immediate consequences.

Proposition 10.5 *Consider the spinor space W with metric spinor ϵ. Let $\{e_A\}_{A=1,2} \subset W$ be a basis. The following facts hold.*

(a) Referring to the bases canonically associated with $\{e_A\}_{A=1,2} \subset W$, the following conditions are equivalent:

(i) $\{e_A\}_{A=1,2}$ is ϵ-orthonormal;
(ii) $\epsilon^{11} = \epsilon^{22} = 0$, $\epsilon^{12} = -\epsilon^{21} = 1$;
(iii) $\overline{\epsilon}_{1'1'} = \overline{\epsilon}_{2'2'} = 0$, $\overline{\epsilon}_{1'2'} = -\overline{\epsilon}_{2'1'} = 1$;
(iv) $\overline{\epsilon}^{1'1'} = \overline{\epsilon}^{2'2'} = 0$, $\overline{\epsilon}^{1'2'} = -\overline{\epsilon}^{2'1'} = 1$.

(b) Assume that $\{e_A\}_{A=1,2}$ is ϵ-orthonormal and let $\{\tilde{e}_A\}_{A=1,2} \subset W$ be another basis, with

$$e_A = L^B{}_A \tilde{e}_B\,, \quad \textit{for some } L \in GL(2,\mathbb{C}) \textit{ with components } L^B{}_A.$$

Then $\{\tilde{e}_A\}_{A=1,2}$ is ϵ-orthonormal if and only if $L \in SL(2,\mathbb{C})$.

10.2 Four-Vectors Constructed with Spinors in Special Relativity

In this section we show how it is possible to build Minkowskian four-vectors starting from spinors. This is the starting point for constructing, in theoretical physics, the differential equations describing the motion of relativistic particles with spin.

First of all we focus on the Hermitian spinorial tensors in the space $W \otimes \overline{W}$. To this end fix a basis $\mathcal{B} := \{e_A\}_{A=1,2} \subset W$ and consider the basis $\{e_A \otimes \overline{e}_{B'}\}_{A,B'=1,2}$, canonically associated with the given one, in the space $W \otimes \overline{W}$. An element $\Xi^\dagger \in W \otimes \overline{W}$ is the **Hermitian conjugate** of a given $\Xi \in W \otimes \overline{W}$ **with respect to** $\mathcal{B}$ if the matrix of the components of $\Xi^\dagger$, referred to the basis in $W \otimes \overline{W}$ canonically associated with $\mathcal{B}$, is the adjoint (i.e. Hermitian conjugate) of that of Ξ. In other words, if $\Xi = \Xi^{AB'} e_A \otimes \overline{e}_{B'}$ and $\Xi^\dagger = \Xi^{\dagger AB'} e_A \otimes \overline{e}_{B'}$, it has to be

$$\Xi^{\dagger AB'} := \overline{\Xi^{BA'}} .$$

Equivalently

$${}_{\mathcal{B}}\Xi^\dagger := \overline{{}_{\mathcal{B}}\Xi^t} .$$

Proposition 10.6 *Given two bases $\mathcal{B} := \{e_A\}_{A=1,2} \subset W$ and $\widetilde{\mathcal{B}} := \{\widetilde{e}_A\}_{A=1,2} \subset W$, $\Xi^\dagger \in W \otimes \overline{W}$ is the Hermitian conjugate of $\Xi \in W \otimes \overline{W}$ with respect to $\mathcal{B}$ if and only if it is the Hermitian conjugate of Ξ with respect to $\widetilde{\mathcal{B}}$. Thus, the notion of Hermitian conjugate is intrinsic.*

Proof Assume that $\Xi = \Xi^{AB'} e_A \otimes \overline{e}_{B'} = \widetilde{\Xi}^{AB'} \widetilde{e}_A \otimes \overline{\widetilde{e}}_{B'}$. If $e_B = L^A{}_B \widetilde{e}_A$ and L denotes the matrix whose elements are the coefficients $L^B{}_A$, then, in matrix notation, one has

$${}_{\widetilde{\mathcal{B}}}\Xi = L\, {}_{\mathcal{B}}\Xi\, L^\dagger .$$

Taking the Hermitian conjugate we achieve:

$${}_{\widetilde{\mathcal{B}}}\Xi^\dagger = (L\, {}_{\mathcal{B}}\Xi\, L^\dagger)^\dagger = L\, {}_{\mathcal{B}}\Xi^\dagger\, L^\dagger .$$

This is exactly the transformation law required for $\Xi^\dagger$ to be the Hermitian conjugate of Ξ with respect to $\widetilde{\mathcal{B}}$ as well. □

Definition 10.7 An element $\Xi \in W \otimes \overline{W}$ is said to be **real** if $\Xi = \Xi^\dagger$. $(W \otimes \overline{W})_{\mathbb{R}}$ denotes the space of real elements of $W \otimes \overline{W}$. ■

The space $(W \otimes \overline{W})_{\mathbb{R}}$ is a real four-dimensional vector space. Indeed, we know (see the previous chapter) that the matrices σ^μ, $\mu = 0, 1, 2, 3$, defined as in (9.7), form a basis of the real space of complex 2×2 Hermitian matrices $\mathcal{H}(2, \mathbb{C})$. As a consequence, if $\mathcal{B} = \{e_A\}_{A=1,2}$ is an ϵ-orthonormal basis of W, $\Xi \in (W \otimes \overline{W})_{\mathbb{R}}$ if and only if there are four real numbers ${}_{\mathcal{B}}t^0, {}_{\mathcal{B}}t^1, {}_{\mathcal{B}}t^2, {}_{\mathcal{B}}t^3$, bijectively determined by the matrix ${}_{\mathcal{B}}\Xi$ of the components of Ξ referred to the basis $\{e_A \otimes \overline{e}_{B'}\}_{A,B'=1,2}$, such that

$$ {}_{\mathcal{B}}\Xi = {}_{\mathcal{B}}t^\mu \sigma_\mu \,. \tag{10.7}$$

The inverse relation can be obtained making use of the matrices σ'^μ (9.9) and the elementary result presented in Exercise 3 of 9.2:

$$ {}_{\mathcal{B}}t^\mu = -tr({}_{\mathcal{B}}\Xi\, \sigma'^\mu) \,. \tag{10.8}$$

To proceed, the idea is to interpret the four real numbers ${}_{\mathcal{B}}t^\mu$ as the components of a four-vector in Minkowski spacetime $\mathbb{M}^4$, with respect to some $\mathbf{g}$-orthonormal frame $\{f_\mu\}_{\mu=0,1,2,3}$ of the space of translations T^4 ($\mathbf{g}$ is the metric with signature $(-1, +1, +1, +1)$).

The remarkable fact is the following. The found assignment of a vector t^μ in correspondence with a real spinorial tensor $\Xi \in W \otimes \overline{W}$ could seem to depend on the fixed basis $\mathcal{B} \subset W$. Actually this is not the case, in view of the covering homomorphism $\Pi : SL(2, \mathbb{C}) \to SO(1, 3)\!\uparrow$ as discussed in the previous chapter. The following theorem clarifies the relationship between real spinorial tensors in $(W \otimes \overline{W})_\mathbb{R}$ and four-vectors.

Theorem 10.8 *Consider a spinor space W, with metric spinor ϵ, and Minkowski spacetime $\mathbb{M}^4$ with metric $\mathbf{g}$ of signature $(-1, +1, +1, +1)$ and, as usual, define $\eta := diag(-1, 1, 1, 1)$. Finally, let $\Pi : SL(2, \mathbb{C}) \ni L \mapsto \Lambda_L \in SO(1, 3)\!\uparrow$ be the covering homomorphism discussed in Theorem 9.10.*

Fix an ϵ-orthonormal basis $\mathcal{B} := \{e_A\}_{A=1,2}$ in W and a $\mathbf{g}$-orthonormal basis $\mathcal{P} := \{f_\mu\}_{\mu=0,1,2,3} \subset T^4$. The following holds.

(a) There is a real-vector-space isomorphism

$$h_{\mathcal{B},\mathcal{P}} : (W \otimes \overline{W})_\mathbb{R} \ni \Xi \mapsto t_\Xi \in T^4$$

between the space of real spinorial tensors of $W \otimes \overline{W}$ and T^4, which is defined by (10.8), with inverse given by (10.7), where ${}_{\mathcal{B}}t^\mu$ are the components of t_Ξ referred to the basis $\{f_\mu\}_{\mu=0,1,2,3}$.

(b) The definition of $h_{\mathcal{B},\mathcal{P}}$ is independent of the fixed bases, modulo the action of Π. In other words, if $\widetilde{\mathcal{B}} = \{\widetilde{e}_A\}_{A=1,2}$ is another ϵ-orthonormal basis of W such that $e_B = L^A{}_B \widetilde{e}_A$, where the coefficients $L^A{}_B$ define $L \in SL(2, \mathbb{C})$, and $\widetilde{\mathcal{P}} := \{\widetilde{f}_\nu\}_{\nu=0,1,2,3} \subset T^4$ is the $\mathbf{g}$-orthonormal basis with $f_\nu = (\Lambda_L)^\mu{}_\nu \widetilde{f}_\mu$, then

$$h_{\mathcal{B},\mathcal{P}} = h_{\widetilde{\mathcal{B}},\widetilde{\mathcal{P}}} \,.$$

(c) *With the given notations, for every pair* $\Xi, \Sigma \in (W \otimes \overline{W})_{\mathbb{R}}$, *it holds*

$$\Xi^{AB'}\Sigma^{CD'}\epsilon_{AC}\epsilon_{B'D'} = -\eta_{\mu\nu}\, t^{\mu}_{\Xi}\, t^{\nu}_{\Sigma} \,. \tag{10.9}$$

So that, in particular, the spinorial tensor

$$\Gamma := -\epsilon \otimes \overline{\epsilon} \quad \in W^* \otimes W^* \otimes \overline{W^*} \otimes \overline{W^*} \,, \tag{10.10}$$

defines a Lorentzian metric on $(W \otimes \overline{W})_{\mathbb{R}}$ *with signature* $(-1, +1, +1, +1)$.

Proof

(a) This was proved immediately before the statement of the theorem.
(b) Adopting the notation of (10.8) and (10.7), the thesis is equivalent to

$$L\,_{\mathcal{B}}\Xi\, L^{\dagger} = (\Lambda_L)^{\mu}{}_{\nu}\, {}_{\mathcal{B}}t^{\nu}\sigma_{\mu} \,.$$

This is nothing but the result proved in Proposition 9.7.
(c) With the given definitions and making use of Exercise 4 in 9.2, we have:

$$\begin{aligned}
\Xi^{AB'}\Sigma^{CD'}\epsilon_{AC}\epsilon_{B'D'} &= t^{\mu}_{\Xi}t^{\nu}_{\Sigma}\sigma_{\mu}^{\ AB'}\sigma_{\nu}^{\ CD'}\epsilon_{AC}\epsilon_{B'D'} = t^{\mu}_{\Xi}t^{\nu}_{\Sigma}tr\left(\epsilon^{t}\sigma^{t}_{\mu}\epsilon\sigma_{\nu}\right)\\
&= -t^{\mu}_{\Xi}t^{\nu}_{\Sigma}tr\left(\epsilon\sigma^{t}_{\mu}\epsilon\sigma_{\nu}\right)\\
&= t^{\mu}_{\Xi}t^{\nu}_{\Sigma}tr(\sigma'_{\mu}\sigma_{\nu}) = -\eta_{\mu\nu}t^{\mu}_{\Xi}t^{\nu}_{\Sigma}\,,
\end{aligned}$$

where, in the last passage, we employed Exercise 3 in 9.2.

□

10.3 Dirac Spinors

In this final section we introduce *Dirac spinors*, constructed out of Weyl spinors. The point is that a Dirac spinor is obtained by juxtaposing two Weyl spinors of opposite *chirality*. This construction is natural in the formalism developed in the previous section and leads in a straightforward way to the Dirac equation. We keep throughout the convention

$$\eta = \mathrm{diag}(-1, +1, +1, +1) \,. \tag{10.11}$$

10.3.1 The Dirac Spinor Space

Let W be a two-dimensional complex spinor space equipped with the metric spinor ϵ_{AB}. We denote by $\overline{W^*}$ the space of dual conjugate spinors.

Definition 10.9 The **Dirac spinor space** associated with W is the complex vector space

$$D := W \oplus \overline{W^*} . \tag{10.12}$$

An element of D is called a **Dirac spinor**. If

$$\Psi \in D , \tag{10.13}$$

we shall write, in components,

$$\Psi = \phi^A \oplus \psi_{A'} , \tag{10.14}$$

where $\phi^A \in W$ and $\psi_{A'} \in \overline{W^*}$ are respectively called the **left-handed Weyl spinor** and the **right-handed Weyl spinor** associated with Ψ. ■

Remarks 10.10 The word *chirality* refers to the decomposition

$$D = W \oplus \overline{W^*} . \tag{10.15}$$

More precisely, the elements of the subspace W are said to have **left chirality**, whereas the elements of the subspace $\overline{W^*}$ are said to have **right chirality**. Equivalently, one introduces the **chirality operator**

$$\Gamma : D \to D , \qquad \Gamma(\phi^A \oplus \psi_{A'}) := -\phi^A \oplus \psi_{A'} . \tag{10.16}$$

Then $\Gamma^2 = I_D$, the eigenspace of eigenvalue -1 is W, and the eigenspace of eigenvalue $+1$ is $\overline{W^*}$. Therefore

$$\Pi_L = \frac{1}{2}(I_D - \Gamma) , \qquad \Pi_R = \frac{1}{2}(I_D + \Gamma) , \tag{10.17}$$

are the projectors onto the left-handed and right-handed Weyl sectors, respectively. In many texts the operator Γ is denoted by γ^5. With different sign conventions for the metric and the gamma matrices, the signs of the eigenvalues may be interchanged, but the decomposition into the two chiral sectors is the same. Dealing with quantum fields described in terms of spinor fields, for *massless fields* chirality coincides with *helicity*, whereas for *massive fields* the two notions are different. ■

Therefore a Dirac spinor is nothing but an ordered pair of Weyl spinors of opposite chirality. The decomposition

$$D = W \oplus \overline{W^*} \tag{10.18}$$

is the decomposition of the Dirac spinor into its two Weyl components. The corresponding projection operators are

$$\Pi_L(\phi^A \oplus \psi_{A'}) := \phi^A \oplus 0\,, \quad \Pi_R(\phi^A \oplus \psi_{A'}) := 0 \oplus \psi_{A'}\,, \tag{10.19}$$

so that

$$\Psi = \Pi_L \Psi + \Pi_R \Psi\,. \tag{10.20}$$

Remarks 10.11 The importance of Dirac spinors in relativistic theories is due to the fact that spacetime vectors act on them in a way that mixes the two Weyl components. More precisely, a Dirac spinor is the direct sum of a left-handed and a right-handed Weyl spinor, whereas the matrices γ^μ associated with four-vectors are off-diagonal in the Weyl representation. Therefore, when a four-vector acts on a Dirac spinor, the left-handed component is sent into the right-handed sector and the right-handed component is sent into the left-handed one. This is the reason why Dirac spinors provide the natural framework for relativistic equations involving mass. Indeed, the mass term in the *Dirac equation* (see below) couples the two chiral parts, so that they do not satisfy two separate equations. Only in the *massless* case (see below) do the two Weyl sectors decouple, giving rise to two independent chiral equations. ■

10.3.2 The Action of Four-Vectors on Dirac Spinors

Let $t \in T^4$ be a four-vector of Minkowski spacetime. As proved in the previous section, once an ϵ-orthonormal basis in W and a **g**-orthonormal basis in T^4 are fixed compatibly, t can be represented by a real spinorial tensor

$$t^{AA'} \in W \otimes \overline{W}\,. \tag{10.21}$$

Lowering indices with ϵ_{AB} and $\overline{\epsilon}_{A'B'}$, we also define

$$t_{AA'} := \epsilon_{AB}\overline{\epsilon}_{A'B'}t^{BB'}\,. \tag{10.22}$$

Definition 10.12 For every four-vector $t \in T^4$, the corresponding **Clifford action** on D is the linear map

$$\gamma(t) : D \to D \tag{10.23}$$

defined by

$$\gamma(t)(\phi^A \oplus \psi_{A'}) := t^{AA'}\psi_{A'} \oplus t_{AA'}\phi^A \,. \tag{10.24}$$

If $\{f_\mu\}_{\mu=0,1,2,3}$ is a fixed **g**-orthonormal basis of T^4, we write

$$\gamma_\mu := \gamma(f_\mu)\,, \quad \mu = 0, 1, 2, 3\,. \tag{10.25}$$

■

Referring to an ϵ-orthonormal basis $\mathcal{B} := \{e_A\}_{A=1,2}$ in W and to the basis of D canonically associated with it, the matrices of the operators γ_μ read

$$_{\mathcal{B}}\gamma_\mu = \left[\begin{array}{c|c} 0 & \sigma_\mu \\ \hline \sigma'_\mu & 0 \end{array}\right], \quad \mu = 0, 1, 2, 3\,, \tag{10.26}$$

where σ_μ and σ'_μ are the matrices introduced in the previous chapter. Consequently,

$$_{\mathcal{B}}\gamma^\mu := \eta^{\mu\nu}\, {}_{\mathcal{B}}\gamma_\nu = \left[\begin{array}{c|c} 0 & \sigma^\mu \\ \hline \sigma'^\mu & 0 \end{array}\right]. \tag{10.27}$$

Proposition 10.13 *The matrices γ_μ satisfy the* **Clifford relations**

$$\gamma_\mu\gamma_\nu + \gamma_\nu\gamma_\mu = -2\eta_{\mu\nu}I_D\,. \tag{10.28}$$

Equivalently, for every $t, s \in T^4$,

$$\gamma(t)\gamma(s) + \gamma(s)\gamma(t) = -2\mathbf{g}(t, s)I_D\,. \tag{10.29}$$

Proof Referring to the matrix representation (10.26), one has

$$_{\mathcal{B}}\gamma_\mu\, {}_{\mathcal{B}}\gamma_\nu + {}_{\mathcal{B}}\gamma_\nu\, {}_{\mathcal{B}}\gamma_\mu = \left[\begin{array}{c|c} \sigma_\mu\sigma'_\nu + \sigma_\nu\sigma'_\mu & 0 \\ \hline \sigma'_\mu\sigma_\nu + \sigma'_\nu\sigma_\mu & 0 \end{array}\right] + \left[\begin{array}{c|c} 0 & 0 \\ \hline 0 & 0 \end{array}\right], \tag{10.30}$$

that is

$$_{\mathcal{B}}\gamma_\mu\, {}_{\mathcal{B}}\gamma_\nu + {}_{\mathcal{B}}\gamma_\nu\, {}_{\mathcal{B}}\gamma_\mu = \left[\begin{array}{c|c} \sigma_\mu\sigma'_\nu + \sigma_\nu\sigma'_\mu & 0 \\ \hline 0 & \sigma'_\mu\sigma_\nu + \sigma'_\nu\sigma_\mu \end{array}\right]. \tag{10.31}$$

By the elementary identities proved for the Pauli matrices in the previous chapter, the two diagonal blocks are both equal to $-2\eta_{\mu\nu}I_2$, whence

$$_{\mathcal{B}}\gamma_\mu\, {}_{\mathcal{B}}\gamma_\nu + {}_{\mathcal{B}}\gamma_\nu\, {}_{\mathcal{B}}\gamma_\mu = -2\eta_{\mu\nu}I_4\,. \tag{10.32}$$

This is (10.28). The intrinsic version (10.29) is just the basis-independent reformulation of the same identity. □

Remarks 10.14 The sign in the right-hand side of (10.28) is due to our convention $\eta = \mathrm{diag}(-1, +1, +1, +1)$ and to the choice of γ_μ in (10.26). If one wishes to adopt the alternative convention $\widetilde{\gamma}_\mu\widetilde{\gamma}_\nu + \widetilde{\gamma}_\nu\widetilde{\gamma}_\mu = 2\eta_{\mu\nu}I$, it is enough to replace γ_μ by $i\gamma_\mu$. Thus the issue is purely conventional. ■

10.3.3 Dual Dirac Spinors and Dirac Adjoint

Since

$$D = W \oplus \overline{W^*}\,, \tag{10.33}$$

the dual space is naturally

$$D^* = W^* \oplus \overline{W}\,. \tag{10.34}$$

Its elements are called **dual Dirac spinors** or **Dirac cospinors**.

Definition 10.15 If

$$\Psi = \phi^A \oplus \psi_{A'} \in D\,, \tag{10.35}$$

its **Dirac adjoint** is the element

$$\overline{\Psi}_D := \overline{\psi}_A \oplus \overline{\phi}^{A'} \in D^*\,. \tag{10.36}$$

■

Therefore, if

$$\Phi = \alpha^A \oplus \beta_{A'} \in D\,, \tag{10.37}$$

the natural pairing between the Dirac adjoint of Ψ and Φ reads

$$\overline{\Psi}_D(\Phi) = \overline{\psi}_A\alpha^A + \overline{\phi}^{A'}\beta_{A'}\,. \tag{10.38}$$

In matrix notation, if

$$_{\mathcal{B}}\Psi = \begin{pmatrix} \phi^1 \\ \phi^2 \\ \psi_{1'} \\ \psi_{2'} \end{pmatrix}, \tag{10.39}$$

then

$$_{\mathcal{B}}\overline{\Psi}_D = \left(\overline{\psi}_1, \overline{\psi}_2, \overline{\phi}^{1'}, \overline{\phi}^{2'}\right). \tag{10.40}$$

10.3.4 The Dirac Equation

Let $\Psi : \mathbb{M}^4 \to D$ be a $C^1(\mathbb{M}; D)$ Dirac spinor field on Minkowski spacetime. Referring to a Minkowskian coordinate system x^0, x^1, x^2, x^3, we use the shorthand

$$\partial_\mu := \frac{\partial}{\partial x^\mu}. \tag{10.41}$$

Elementary particles are described by *field equations*. Two important equations are the *Dirac* and the *Klein-Gordon* ones. The former in particular describes electrons and positrons when the mass parameter m is strictly positive and has the appropriate value.

Definition 10.16 If $m \geq 0$ is a real constant, the equation

$$i\gamma^\mu \partial_\mu \Psi - m\Psi = 0 \tag{10.42}$$

is called the **Dirac equation** for a *free particle* of **mass** m. The equation and the particle are said to be **massless** if $m = 0$. ■

Writing

$$\Psi = \phi^A \oplus \psi_{A'}, \tag{10.43}$$

and using (10.27), Eq. (10.42) is equivalent to the pair of equations

$$i\sigma^\mu \partial_\mu \psi - m\phi = 0, \qquad i\sigma'^\mu \partial_\mu \phi - m\psi = 0. \tag{10.44}$$

Thus the Dirac equation is a system of two coupled **Weyl equations**. If $m = 0$, the coupling disappears and one obtains the two independent massless Weyl equations

$$i\sigma^\mu \partial_\mu \psi = 0, \qquad i\sigma'^\mu \partial_\mu \phi = 0. \tag{10.45}$$

Proposition 10.17 *If Ψ is a solution of the Dirac equation (10.42), then Ψ also satisfies the* **Klein-Gordon equation**

$$\left(\eta^{\mu\nu}\partial_\mu\partial_\nu - m^2\right)\Psi = 0\,. \tag{10.46}$$

Proof Apply the operator $i\gamma^\nu\partial_\nu + m$ to both sides of (10.42). One gets

$$0 = \left(i\gamma^\nu\partial_\nu + m\right)\left(i\gamma^\mu\partial_\mu - m\right)\Psi = -\gamma^\nu\gamma^\mu\partial_\nu\partial_\mu\Psi - m^2\Psi\,. \tag{10.47}$$

Since partial derivatives commute, only the symmetric part of $\gamma^\nu\gamma^\mu$ contributes, and thus, by (10.28),

$$-\gamma^\nu\gamma^\mu\partial_\nu\partial_\mu\Psi = -\frac{1}{2}\left(\gamma^\nu\gamma^\mu + \gamma^\mu\gamma^\nu\right)\partial_\nu\partial_\mu\Psi = \eta^{\mu\nu}\partial_\mu\partial_\nu\Psi\,. \tag{10.48}$$

Therefore

$$\left(\eta^{\mu\nu}\partial_\mu\partial_\nu - m^2\right)\Psi = 0\,, \tag{10.49}$$

as wanted. □

Remarks 10.18 The previous proposition shows that every component of a Dirac spinor describes a relativistic degree of freedom with mass m. However the Dirac equation is stronger than the Klein-Gordon equation, because it also keeps track of the spinorial structure of the field. In particular, in the massless case the two chiral components ϕ^A and $\psi_{A'}$ evolve independently, whereas in the massive case they are coupled exactly through the mass term. ■

Appendix A
Elements of Matrix Lie Group Theory

This chapter reviews the matrix Lie group theory needed in the discussion of rotations, Lorentz transformations, and their representations. The emphasis is on concrete matrix groups rather than on the most general theory of Lie groups, because the applications in this volume concern groups such as $GL(n, \mathbb{K})$, $O(n)$, $SO(n)$, $SL(2, \mathbb{C})$, and the Lorentz group.

The chapter introduces one-parameter subgroups, the exponential map, Lie algebras of matrix groups, and the special role of orthogonal groups. It also proves representation results for $O(3)$ and $SO(3)$ that are useful for understanding rotations in physical space. In this sense the chapter functions as a technical appendix, but it is written so that the reader can consult it before or during the chapters on Special Relativity and Lorentz-group structure.

A.1 Review of Lie Groups

Definition A.1 (Lie Group) A **Lie group** is a differentiable manifold (of class C^∞) G endowed with differentiable maps

$$\phi : G \ni g \mapsto g^{-1} \in G ,$$

and

$$\psi : G \times G \ni (g, h) \mapsto gh \in G ,$$

where $G \times G$ is endowed with the *product differentiable structure*, such that (G, ψ, ϕ) is a group in which ψ is the group multiplication and ϕ is the map assigning the inverse to each element of the group. ■

V. Moretti, *Geometric Methods in Mathematical Physics I*, La Matematica per il 3+2 184, https://doi.org/10.1007/978-3-032-31512-0

Remarks A.2 The differentiability requirement may be weakened by considering groups, called *locally Euclidean topological group*, which are merely C^0 manifolds (Hausdorff, 2nd countable, connected topological manifolds) with group operations continuous with respect to the topology of the manifold. Conversely, in some textbooks the differentiability requirement is strengthened to analyticity: one requires that the manifold be an analytic manifold (C^ω) and that the group operations be analytic functions. If one adopts the definition of Lie group given above, the Lie groups with C^ω structure just introduced are called *analytic* Lie groups. In fact, these apparently different definitions are *all equivalent*, due to a famous theorem of Gleason, Montgomery, and Zippin from 1952. That theorem proves that every locally Euclidean topological group always admits a compatible analytic differentiable structure with respect to which the group operations are analytic, and such a structure is uniquely determined.

Consider an element $g \in G$. It defines a differentiable map $P_g : G \ni h \mapsto P_g(h) = gh$. Therefore, if $P_{g*} := dP_g$, that map must be viewed as

$$P_{g*} : T_h G \to T_{gh} G \,.$$

Moreover, from the properties of the differential, one has $P_{fg*} = P_{f*}P_{g*}$. Fixing $A \in T_1 G$, we may then consider the first-order differential equation

$$\frac{df}{dt} = P_{f(t)*} A \,,$$

with initial condition $f(0) = 1$, the neutral element of G. The solutions of the Cauchy problem so obtained are functions with values in the group G itself, defined on open intervals $I \subset \mathbb{R}$ containing 0.

By the standard existence and uniqueness theorems for the Cauchy problem, the *maximal solution* $f_A : I_A \to G$, which includes all local solutions as restrictions, exists and is unique. Consider now, for fixed $t' \in \mathbb{R}$, the function $t \mapsto H(t) := f_A(t')f_A(t)$. By definition of P, one has

$$H(t) = f_A(t')f_A(t) = P_{f_A(t')}(f_A(t)) \,,$$

and therefore, differentiating with respect to t,

$$\frac{dH(t)}{dt} = \frac{d(f_A(t')f_A(t))}{dt} = P_{f_A(t')*}\frac{df_A(t)}{dt} = P_{f_A(t')*}P_{f_A(t)*}A = P_{H(t)*}A \,.$$

On the other hand the function $t \mapsto H'(t) := f_A(t'+t)$ satisfies the same differential equation:

$$\frac{dH'(t)}{dt} = \frac{df_A(t'+t)}{dt} = \frac{df_A(t'+t)}{d(t'+t)} = P_{f_A(t'+t)*}A = P_{H'(t)*}A \,,$$

and it satisfies the common initial condition $H(0) = H'(0) = f_A(t')$. By the uniqueness theorem we conclude that $H(t) = H'(t)$, that is, for every $A \in T_1G$ one has the **one-parameter subgroup** property

$$f_A(t')f_A(t) = f_A(t'+t)(= f_A(t)f_A(t')\ )$$

whenever $t, t', t'+t$ belong to the domain I_A of the maximal solution. The maximal solution is certainly *complete* (that is, $I_A = \mathbb{R}$) if G is compact, by the theory of differential equations on manifolds. In fact, one can prove that in the present case, where the manifold is a Lie group, f_A is always complete [3]. However, we shall not need to use this general fact below, and we shall prove it independently of the general theory in the case of matrix Lie groups, the only ones we shall use in the physical applications. ■

Definition A.3 (One-Parameter Subgroup) Let G be a Lie group. For every vector $A \in T_1G$, the maximal solution of the equation

$$\frac{df}{dt} = P_{f(t)*}A\ ,$$

with initial condition $f_A(0) = 1$, denoted by

$$I_A \ni t \mapsto \exp(tA)\ ,$$

is called the **one-parameter subgroup generated by** A. ■

Remarks A.4 Observe in particular that $(\exp(tA))^{-1} = \exp(-tA)$ whenever both t and $-t$ belong to the domain I_A of the function. ■

Consider now fixed $A \in T_1G$ and the family of maps parametrized by t and defined in a neighborhood of 0:

$$F_{t,A} : G \ni g \mapsto \exp(tA)\ g\ \exp(-tA)\ .$$

Since $F_{t,A}(1) = 1$, the differential $dF_{t,A}|_1$ defines a map from T_1G into the tangent space at the point $\exp(tA)\ 1\ \exp(-tA)$, namely again T_1G.

Definition A.5 (Adjoint and Commutator) Let G be a Lie group and define, for $A \in T_1G$,

$$F_{t,A} : G \ni g \mapsto \exp(tA)\ g\ \exp(-tA)\ .$$

(a) The differential $dF_{t,A}|_1$, denoted by

$$Ad\ F_{t,A} : T_1G \to T_1G\ ,$$

is called the **adjoint** of $F_{t,A}$.

(b) The **commutator** is the map from $T_1G \times T_1G$ into T_1G defined by

$$[A, B] := \frac{d}{dt}|_{t=0}(Ad\ F_{t,A})\ B\ .$$

for every pair $A, B \in T_1G$. ■

Proposition A.6 *Referring to the definitions above, the following properties of the commutator are easily verified: left linearity:*

$$[aA + bB, C] = a[A, C] + b[B, C]\ ,$$

antisymmetry:

$$[A, B] = -[B, A]\ ,$$

(which implies that linearity also holds on the right), and the Jacobi identity:

$$[A, [B, C]] + [B, [C, A]] + [C, [A, B]] = 0\ ,$$

where $A, B, C \in T_1G$ and $a, b \in \mathbb{R}$ are arbitrary. ■

Proof See [3]. □

Remarks A.7 Notice that left linearity and antisymmetry immediately imply right linearity, so that the commutator is bilinear. ■

Definition A.8 (Lie Algebras) The algebraic structure given by a vector space V together with an operation, called commutator, $\{\ ,\ \} : V \times V \to V$, which enjoys left linearity, antisymmetry, and the Jacobi identity, is called a **Lie algebra**. Moreover:

(a) given two Lie algebras $(V, \{\ ,\ \})$ and $(V', \{\ ,\ \}')$, a vector-space isomorphism $\phi : V \to V'$ is called a **Lie algebra isomorphism** if it preserves the Lie-algebra structure, namely if it also satisfies $\{\phi(A), \phi(B)\}' = \phi(\{A, B\})$ for every $A, B \in V$;

(b) if G is a Lie group, the Lie algebra given by the tangent space T_1G together with the commutator $[\ ,\ ]$ of Definition A.5 is called the **Lie algebra of the group** G. ■

A famous theorem due to Lie proves the following.

Theorem A.9 (Lie's Theorem) *If two Lie groups G and G' have isomorphic Lie algebras, then there exists a diffeomorphism from a neighborhood of the identity of G to a neighborhood of the identity of G' which is also a group isomorphism: such groups are said to be* **locally isomorphic**.

Moreover, for every Lie algebra there exists a simply connected Lie group admitting that Lie algebra as the Lie algebra of the group, and such a group is determined uniquely up to Lie-group isomorphisms, that is, diffeomorphisms which are also group isomorphisms. ■

Proof See [3]. □

Definition A.10 (Lie Subgroup) If G is a Lie group and $G' \subset G$ is an embedded submanifold of G which is still a group with respect to the restrictions to G' of the group operations of G, then G' naturally acquires a Lie-group structure induced by that of G and is called a **Lie subgroup** of G. ■

Remarks A.11

(1) It is immediate to prove that the Lie algebra of G' turns out to be a **Lie subalgebra** of that of G, in the sense that $T_1 G'$ is a vector subspace of $T_1 G$ and the commutator on $T_1 G'$ is the restriction of the commutator on $T_1 G$ to $T_1 G'$.

(2) There exists a very powerful theorem (see, for instance, [1]) asserting that every closed subgroup K of an analytic Lie group G is an analytic Lie subgroup of G. The proof requires advanced techniques. In fact, the theorem is valid in general for Lie groups with our definition: the analyticity assumption may be omitted because of the theorem of Gleason, Montgomery, and Zippin, which ensures that every Lie group is also an analytic Lie group with respect to a uniquely determined analytic differentiable structure. ■

In view of the last remark we can state the following fundamental theorem.

Theorem A.12 *If $G' \subset G$ is a closed subgroup of the Lie group G, then it is also a Lie subgroup of G.* ■

Exercises A.13

1. If G is a Lie group and $A_1, \ldots, A_n$ is a basis of its Lie algebra, since the commutator is bilinear on the algebra and takes values in the algebra, it must be represented by a tensor

$$C \in T_1^* G \otimes T_1^* G \otimes T_1 G\,.$$

In components,

$$[A_i, A_j] = C_{ij}{}^k A_k\,.$$

The components of C are called the **structure constants** of the group.
Show that if two Lie groups have the same structure constants with respect to two bases in their respective Lie algebras, then they are *locally isomorphic* in the sense of Lie's theorem.

Hint. Prove that the linear map identifying the two bases is a Lie-algebra isomorphism.

2. Let G be a Lie group and let $A_1, \ldots, A_n$ be a basis of its Lie algebra. Consider the map defined on a sufficiently small neighborhood of the origin of $\mathbb{R}^n$

$$F : (x^1, \ldots, x^n) \mapsto e^{\sum_{k=1}^n x^k A_k} .$$

Show that this map (1) is well defined, namely that the sufficiently small neighborhood above really exists, and that (2) defines a coordinate system for the differentiable structure of G in a neighborhood of the identity of the group. This coordinate system is called a *coordinate system of the first kind.*

Hint. From the standard theorems on the dependence on parameters of the solutions of first-order differential equations on manifolds, it follows that the stated function is locally well defined. Moreover, if $e_1, \ldots, e_n$ is the canonical basis of $\mathbb{R}^n$, then $dF_0 : e_i \mapsto A_i$ for every $i = 1, \ldots, n$, so that $dF_0 : \mathbb{R}^n \to T_1G$ is injective (and thus surjective, since the two spaces have the same dimension). The implicit-function theorem immediately implies that F is a local diffeomorphism between $\mathbb{R}^n$ and G with values in a neighborhood of $1 \in G$.
3. Let G be a Lie group and let $A_1, \ldots, A_n$ be a basis of its Lie algebra. Consider the map defined on a sufficiently small neighborhood of the origin of $\mathbb{R}^n$

$$H : (x^1, \ldots, x^n) \mapsto e^{x^1 A_1} \cdots e^{x^n A_n} .$$

Show that this map (1) is well defined, namely that the sufficiently small neighborhood above really exists, and that (2) defines a coordinate system for the differentiable structure of G in a neighborhood of the identity of the group. This coordinate system is called a *coordinate system of the second kind.*

Hint. Each one-parameter subgroup $x^i \mapsto e^{x^i A_i}$ is defined for x^i in a neighborhood of 0, so that there exists a rectangular neighborhood of the origin of $\mathbb{R}^n$ where the function is defined. Moreover, if $e_1, \ldots, e_n$ is the canonical basis of $\mathbb{R}^n$, one easily proves that $dH_0 : e_i \mapsto A_i$ for every $i = 1, \ldots, n$, so that $dH_0 : \mathbb{R}^n \to T_1G$ is injective (and thus surjective, since the two spaces have the same dimension). The implicit-function theorem immediately implies that H is a local diffeomorphism between $\mathbb{R}^n$ and G with values in a neighborhood of $1 \in G$.

A.2 Matrix Lie Groups

Let us now specialize the theory to matrix Lie groups. It is important to stress that this is not a severe restriction, since one can prove that every compact Lie group is isomorphic to a matrix Lie group (for noncompact groups the theorem fails; a typical counterexample is the universal covering of the group $SL(2, \mathbb{R})$). In the following $GL(n, \mathbb{K})$ denotes the group of nonsingular $n \times n$ matrices (that is, with nonvanishing determinant) over the field $\mathbb{K}$, and $M(n, \mathbb{K})$ denotes the set of all $n \times n$ matrices over the field $\mathbb{K}$, where $\mathbb{K}$ is always $\mathbb{R}$ or $\mathbb{C}$.

The natural coordinates of $\mathbb{K}^{n^2}$, $x^1, \ldots, x^{n^2}$, determine a matrix in $M(n, \mathbb{K})$ if one identifies the rows of the latter with the n consecutive groups of n entries. From now on we shall follow this convention.

We may endow $M(n, \mathbb{K})$ with the topology induced by the natural norm of $\mathbb{K}^{n^2}$, which makes it a complete normed space. We may rewrite that norm as follows: if $A \in M(n, \mathbb{K})$,

$$||A|| := \sqrt{\sum_{i,j=1}^{n} |A_{ij}|^2} \,.$$

We now need some results concerning the extension of the exponential function to values in $M(n, \mathbb{C})$. These results are stated in the following three lemmas. The first one is actually more general.

Lemma A.14 *Consider the map*

$$M(n, \mathbb{K}) \ni A \mapsto ||A|| := \sqrt{\sum_{i,j=1}^{n} |A_{ij}|^2} \,.$$

with $\mathbb{K} = \mathbb{C}$ or $\mathbb{R}$. The following properties hold.

(1) *That map is a norm on $M(n, \mathbb{K})$.*
(2) *If $A, B \in M(n, \mathbb{K})$, then*

$$||AB|| \leq ||A||||B|| \,.$$

(3) *The space $M(n, \mathbb{C})$ endowed with the norm above is complete.*
(4) *If $\mathbb{K} \ni z \mapsto A(z) \in M(n, \mathbb{R})$ is an arbitrary function with values in $M(n, \mathbb{R})$, then*

$$\lim_{z \to z_0} A(z) = A_0$$

if and only if for every $i, k = 1, \ldots, n$

$$\lim_{z \to z_0} (A(z))_{ik} = (A_0)_{ik} \,.$$

■

Proof

(1) The map defined above is nothing but the norm of $\mathbb{K}^{n^2}$ written in a different form.

(2)

$$||AB||^2 = \sum_{i,k=1}^{n} \left| \sum_{j=1}^{n} A_{ij} B_{jk} \right|^2 \leq \sum_{i,k=1}^{n} \left(\sum_{j=1}^{n} |A_{ij}|^2 \right) \left(\sum_{p=1}^{n} |B_{pk}|^2 \right) ,$$

where we have applied the Cauchy–Schwarz inequality. Hence

$$||AB||^2 \leq \sum_{i,j=1}^{n} |A_{ij}|^2 \sum_{k,p=1}^{n} |B_{pk}|^2 = ||A||^2 ||B||^2 .$$

Taking square roots on both sides gives the thesis.

(3) It is a general result that every finite-dimensional normed vector space is complete with respect to the topology induced by the norm (see [2]).

(4) This is the standard property, from elementary analysis, of functions from $\mathbb{K}$ to $\mathbb{K}^p$, taking into account that the defined norm is precisely the norm of $\mathbb{K}^p$ with $p = n^2$.

□

Remarks A.15 The set of complex $n \times n$ matrices, together with the norm defined above, provides an example of a *Banach algebra*. ■

Lemma A.16 *If $A \in M(n, \mathbb{C})$ and $z \in \mathbb{C}$, the series*

$$\sum_{k=0}^{\infty} z^k \frac{A^k}{k!}$$

converges absolutely (with respect to the norm of $M(n, \mathbb{C})$) and uniformly, in the topology of $M(n, \mathbb{C})$, on every disk $|z| \leq R < +\infty$, and it may be differentiated term by term with respect to z infinitely many times. In particular:

(1) *the function*

$$\mathbb{C} \ni z \mapsto e^{zA} := \sum_{k=0}^{\infty} z^k \frac{A^k}{k!} \tag{A.1}$$

is well defined and infinitely differentiable on $\mathbb{C}$;

(2) *for every $z \in \mathbb{C}$ one has*

$$\frac{d}{dz} e^{zA} = A e^{zA} = e^{zA} A ; \tag{A.2}$$

(3) *for every pair $z, z' \in \mathbb{C}$ one has*

$$e^{zA+z'A} = e^{zA}e^{z'A} = e^{z'A}e^{zA}\,. \tag{A.3}$$

■

Proof The proof is based on the validity of the analogous theorems for series of complex-valued functions, which extend straightforwardly to series of functions from $\mathbb{C}$ with values in Banach spaces. In particular, Weierstrass' theorem ensures that if a series is dominated by a convergent series of positive numerical terms, then absolute and uniform convergence follow. In our case, if $|z| \leq R$, by Lemma A.14,

$$\left|\left|z^k\frac{A^k}{k!}\right|\right| \leq |z|^k\frac{||A||^k}{k!} \leq R^k\frac{||A||^k}{k!}\,,$$

and the series of positive terms

$$\sum_{k=0}^{\infty} R^k\frac{||A||^k}{k!}$$

converges, namely to $\exp(R||A||)$.
(1) Similarly, the series of derivatives of order p converges uniformly on $|z| \leq R$ because it is dominated by the convergent series of positive terms

$$\sum_{k=0}^{\infty} R^k||A||^p\frac{||A||^k}{k!} = ||A||^p e^{R||A||}\,.$$

This implies that for every integer p and every pair $i, j = 1, \ldots, n$ the series of complex-valued functions

$$\sum_{k=0}^{\infty}\frac{d^p}{dz^p}z^k\frac{(A^k)_{ij}}{k!}$$

converges absolutely and uniformly. Using iteratively a standard theorem of analysis, one then finds that, for every integer q and every pair $i, j = 1, \ldots, n$, the q-th derivative of $(\exp(zA))_{ij}$ exists and satisfies

$$\frac{d^q}{dz^q}(\exp(zA))_{ij} = \sum_{k=0}^{\infty}\frac{d^q}{dz^q}z^k\frac{(A^k)_{ij}}{k!}\,.$$

By part (4) of Lemma A.14, the q-th derivative of $\exp(zA)$ exists in the norm $|| \; ||$ and satisfies

$$\frac{d^q e^{zA}}{dz^q} = \sum_{k=0}^{\infty} \frac{d^q}{dz^q} z^k \frac{A^k}{k!} .$$

In particular the sum of the series may be differentiated term by term infinitely many times.

Let us pass to (2). Simply differentiating term by term, one has

$$\frac{d}{dz} e^{zA} = \sum_{k=1}^{\infty} A z^{k-1} \frac{A^{k-1}}{(k-1)!} = \sum_{k=1}^{\infty} z^{k-1} \frac{A^{k-1}}{(k-1)!} A .$$

Redefining the summation index and using Lemma A.14, from which one gets

$$||A(\exp(zA) - S_n)|| \leq ||A||||(\exp(zA) - S_n)|| \to 0$$

and

$$||(\exp(zA) - S_n)A|| \leq ||(\exp(zA) - S_n)||||A|| \to 0$$

where S_n is the n-th partial sum of the series, one finds

$$\sum_{p=0}^{\infty} A z^p \frac{A^p}{p!} = A \sum_{p=0}^{\infty} z^p \frac{A^p}{p!} = \left(\sum_{p=0}^{\infty} z^p \frac{A^p}{p!} \right) A ,$$

which is the thesis.

To conclude, the proof of (3) is identical to the one for the complex-valued exponential function, *taking into account the fact that* $(zA)(z'A) = (z'A)(zA)$, which is used in one passage. □

Remarks A.17 Property (3) can be strengthened, with the same proof, to

$$e^{A+B} = e^A e^B = e^B e^A$$

under the assumption

$$AB = BA .$$

Without that assumption, the stated property is generally *false*. ■

Lemma A.18 *If $A \in M(n, \mathbb{C})$, then for every $t \in \mathbb{C}$*

$$det\, e^{tA} = e^{t tr\, A} ,$$

in particular

$$det\, e^A = e^{tr\, A} .$$

■

Proof Consider the map

$$\mathbb{C} \ni t \mapsto \det e^{tA} .$$

We want to compute its derivative for arbitrary t. Namely,

$$\lim_{h\to 0} \frac{\det e^{(t+h)A} - \det e^{tA}}{h} = \lim_{h\to 0} \frac{\det(e^{tA}e^{hA}) - \det e^{tA}}{h} = \det e^{tA} \lim_{h\to 0} \frac{\det e^{hA} - 1}{h}$$

provided the last limit exists. One has

$$e^{hA} = I + hA + ho(h) ,$$

where $o(h) \to 0$ as $h \to 0$ in the metric topology of $\mathbb{C}^{n^2}$, so that

$$\lim_{h\to 0} \frac{\det e^{(t+h)A} - \det e^{tA}}{h} = \det e^{tA} \lim_{h\to 0} \frac{\det(I + hA + ho(h)) - 1}{h} .$$

Recall that

$$\det B = \epsilon^{i_1 \dots i_n} B_{1i_1} \cdots B_{ni_n} ,$$

whence one immediately finds

$$\det(I + hA + ho(h)) = 1 + h \sum_{i=1}^{n} A_{ii} + ho(h) .$$

Substituting above, we obtain

$$\frac{d \det e^{tA}}{dt} = \det e^{tA} \mathrm{tr} A .$$

This also proves that the considered function is infinitely differentiable. Hence the infinitely differentiable function $f_A : \mathbb{C} \ni t \mapsto \det e^{tA}$ satisfies the differential equation

$$\frac{df_A(t)}{dt} = (\mathrm{tr}\, A) f_A(t) .$$

The infinitely differentiable function $g_A : \mathbb{C} \ni t \mapsto e^{t\,\mathrm{tr}\,A}$ trivially satisfies the same differential equation

$$\frac{dg_A(t)}{dt} = (\mathrm{tr}\,A)g_A(t)\,.$$

Both functions satisfy the initial condition $f_A(0) = g_A(0) = 1$, and therefore, by uniqueness of solutions of first-order differential equations, the two functions coincide for every $t \in \mathbb{C}$, so that

$$\det e^{tA} = e^{t\,\mathrm{tr}\,A}\,.$$

This implies the thesis for $t = 1$. □

Proposition A.19 (The Lie Group $GL(n,\mathbb{R})$**)** *The group $GL(n,\mathbb{R})$ of invertible real $n \times n$ matrices is a Lie group with the differentiable structure induced by $\mathbb{R}^{n^2}$. Moreover the following properties hold.*

(1) *The Lie algebra of $GL(n,\mathbb{R})$ is given by the set of real $n \times n$ matrices, $M(n,\mathbb{R})$, and the commutator is defined by*

$$[A, B] = AB - BA\,,$$

for every pair $A, B \in M(n,\mathbb{R})$.

(2) *The one-parameter subgroups of $GL(n,\mathbb{R})$ are complete (that is, they are defined on all of $\mathbb{R}$ in the parameter) and have the form*

$$\mathbb{R} \ni t \mapsto e^{tA}$$

for every $A \in M(n,\mathbb{R})$, where the exponential is the one defined in Lemma A.16.

Proof Think of the real $n \times n$ matrices as points of $\mathbb{R}^{n^2}$ whose natural coordinates determine, when read in consecutive groups of n entries, the n rows of the matrix. To prove that $GL(n,\mathbb{R})$ is a Lie group, it is sufficient to show two things:

(1) $GL(n,\mathbb{R})$ is a differentiable manifold;
(2) multiplication and inversion are differentiable with respect to the differentiable structure induced by $\mathbb{R}^{n^2}$.

The determinant function from $\mathbb{R}^{n^2}$ to $\mathbb{R}$ is certainly differentiable, since it is a polynomial in the global coordinates of $\mathbb{R}^{n^2}$. The inverse image of $\mathbb{R} \setminus \{0\}$ under that function is clearly an open set, since the determinant is continuous. We may endow that set with a differentiable-manifold structure by restricting the global chart of $\mathbb{R}^{n^2}$ to it. This defines a differentiable structure on $GL(n,\mathbb{R})$, which is nothing but the set of points of $\mathbb{R}^{n^2}$ where the determinant does not vanish. The second property is immediate for multiplication. For inversion it is also immediate, using Cramer's rule: the components (coordinates in $\mathbb{R}^{n^2}$) of the inverse matrix of $A \in GL(n, \mathbb{R})$ are rational functions of the coordinates of A, with denominators that never vanish on $GL(n, \mathbb{R})$.

Let us pass to the second part. We prove (1) and (2) together. If $A \in M(n, \mathbb{R})$, consider the function $\mathbb{R} \ni t \mapsto e^{tA}$. By Lemma A.18 this function takes values only in $GL(n, \mathbb{R})$ because $\det e^{tA} = e^{t \operatorname{tr} A} > 0$ for every $A \in M(n, \mathbb{R})$. Using part (2) of Lemma A.16, we obtain

$$\frac{d}{dt} e^{tA} = A e^{tA} = e^{tA} A .$$

For $t = 0$ one recovers the matrix A, which therefore belongs to $T_1 GL(n, \mathbb{R})$ because $e^{0A} = I$. We have proved that $M(n, \mathbb{R}) \subset T_1 GL(n, \mathbb{R})$. Since by construction $\dim T_1 GL(n, \mathbb{R}) \leq n^2$ and $\dim M(n, \mathbb{R}) = n^2$, it follows that $M(n, \mathbb{R}) = T_1 GL(n, \mathbb{R})$. If $A \in T_1 GL(n, \mathbb{R}) = M(n, \mathbb{R})$, the equation defining the one-parameter subgroup generated by A reads simply

$$\frac{d f_A(t)}{dt} = f_A(t) A ,$$

because the group product is ordinary matrix multiplication. The function $\mathbb{R} \ni t \mapsto e^{tA}$ defined in Lemma A.16 plainly satisfies the differential equation above because of part (2) of that lemma. By uniqueness of the solution, this is the unique solution and thus the explicit form of the one-parameter subgroup generated by A. Notice that by construction the solution is complete. Applying part (b) of Definition A.5, one immediately finds that the commutator of the Lie algebra of $GL(n, \mathbb{R})$ is

$$[A, B] = \frac{d}{dt}\Big|_{t=0} \left(e^{tA} B e^{-tA}\right) = AB - BA$$

for every pair $A, B \in M(n, \mathbb{R})$. □

We may therefore give the following definition.

Definition A.20 (Real $n \times n$ Matrix Lie Group) A real $n \times n$ matrix Lie group is a Lie subgroup of $GL(n, \mathbb{R})$. ■

We have the following elementary properties of matrix Lie groups, which are immediately verified.

Proposition A.21 (Properties of Matrix Lie Groups) *Let G be a real $n \times n$ matrix Lie group. Then the following properties hold.*

(1) *G is a Lie group with the differentiable structure induced by $\mathbb{R}^{n^2}$.*

(2) *The Lie algebra of G is a Lie subalgebra of $M(n, \mathbb{R})$ endowed with the commutator*

$$[A, B] = AB - BA .$$

(3) *The one-parameter subgroups of G are complete and have the form*

$$\mathbb{R} \ni t \mapsto e^{tA}$$

for every $A \in T_1 G$, where the exponential is the one defined in Lemma A.16. ■

Proof Everything is an immediate consequence of Proposition A.19. □

A.3 The Lie Groups $O(3)$ and $SO(3)$

Let us examine a very important Lie group, the group of n-dimensional rotations $O(n)$:

$$O(n) := \{R \in M(n, \mathbb{R}) \mid RR^t = I\} .$$

We have the following theorem.

Theorem A.22 (Elementary Properties of the Lie Group $O(n)$) *Consider the set of matrices*

$$O(n) := \{R \in M(n, \mathbb{R}) \mid RR^t = I\} .$$

Then the following properties hold.

(1) *$O(n)$ is a real $n \times n$ matrix Lie group, closed under matrix transposition, and is called the* **group of n-dimensional rotations**.
(2) *The Lie algebra of $O(n)$, denoted by $o(n)$, is the vector space of real antisymmetric $n \times n$ matrices. That space has dimension $n(n-1)/2$.*
(3) *$O(n)$ is compact.* ■

Proof

(1) First of all we show that $O(n)$ is a group and is closed under transposition. Clearly $I \in O(n)$, and if $R, R' \in O(n)$ then

$$(RR')(RR')^t = RR'R'^t R^t = RIR^t = RR^t = I ,$$

so that $O(n)$ is closed under multiplication. It is sufficient to prove, in order to conclude that $O(n)$ is a group, that it is closed under inversion. The fact that $RR^t = I$ implies that $\det R \det R^t = 1$, that is, $(\det R)^2 = 1$, whence $\det R =$

± 1; in any case $\det R \neq 0$, so the inverse matrix of every $R \in O(n)$ exists and coincides with the transpose of R. If we prove that $R^{-1} \in O(n)$, we also prove closure under transposition. If $R \in O(n)$, since the right inverse coincides with the left one, one has $R^{-1}R = I$. Transposing, $R^t(R^{-1})^t = I$, and since $R^t = R^{-1}$, it follows that $R^{-1}(R^{-1})^t = I$, namely $R^{-1} \in O(n)$. Thus we have proved that $O(n)$ is a group with respect to matrix multiplication, hence it is a subgroup of $GL(n, \mathbb{R})$. To conclude point (1), since $O(n) \subset GL(n, \mathbb{R})$, it is enough to prove that $O(n)$ is an embedded submanifold of $\mathbb{R}^{n^2}$. Recall a standard theorem, the *regular-value theorem* [3]:

> *let M, N be differentiable manifolds of dimensions m, n respectively, with $m > n$, and let $f : M \to N$ be a differentiable map. If $p \in N$ is such that for every point $x \in P := f^{-1}(\{p\})$ the differential df_x is surjective (equivalently, the associated Jacobian matrix has rank n), then $P \subset M$ is an embedded submanifold of M of dimension $m - n$.*

Notice that the requirement that df_x be surjective is equivalent to requiring that

> *the kernel of df_x have dimension $m - n$.*

Consider now the definition of $O(n)$. In coordinates on $\mathbb{R}^{n^2}$,

$$\delta^{ik} - \sum_{j=1}^{n} x^{ij}x^{kj} = 0\,.$$

These equations are not all independent: if one exchanges i and k, one gets the same equation. The independent equations are, for instance, those corresponding to the upper triangular part of a symmetric $n \times n$ matrix, namely the equations above with the restriction $i \leq k$. Hence there are $n(n+1)/2$ of them. Define then a function

$$f : \mathbb{R}^{n^2} \to \mathbb{R}^{n(n+1)/2}$$

with components

$$f^{ik}(x^{11}, x^{12}, \ldots, x^{nn}) = \delta^{ik} - \sum_{j=1}^{n} x^{ij}x^{kj}$$

where $i \leq k$. The function f is clearly differentiable. Let us show that the points $R \in O(n) = f^{-1}(\{0\})$, where 0 is the zero vector of $\mathbb{R}^{n(n+1)/2}$, are such that the kernel of df_R has dimension $n^2 - n(n+1)/2 (= n(n-1)/2)$. By the regular-value theorem, this shows that $O(n)$ is an embedded submanifold of $\mathbb{R}^{n^2}$ (and hence of $GL(n, \mathbb{R})$, which contains $O(n)$ and has the differentiable structure induced by $\mathbb{R}^{n^2}$) of dimension $n(n-1)/2$. Let us start by proving that the kernel of df_I has dimension $n(n-1)/2$. To compute the dimension of $\ker df_I$, consider all differentiable curves $R = R(u)$ satisfying $R(u)R(u)^t = I$ and $R(0) = I$, and compute the dimension of the space generated by the tangent vectors at the point I to such curves. By construction that space is the kernel of df_I. From $R(u)R(u)^t = I$ one gets, if $\dot{R}$ denotes the tangent vector at $u = 0$, namely at I,

$$\dot{R} + \dot{R}^t = 0$$

that is,

$$\dot{R} = -\dot{R}^t .$$

The matrices thus obtained are real antisymmetric $n \times n$ matrices. Conversely, if A is a real antisymmetric $n \times n$ matrix, the curve $R(u) := e^{uA}$, which has A as tangent vector at I, satisfies $R(u)R(u)^t = I$; directly from the definition of the exponential,

$$e^{uA}(e^{uA})^t = e^{uA}e^{uA^t} = e^{uA}e^{-uA} = e^{(u-u)A} = I .$$

Therefore the kernel of df_I is precisely the real vector space of real antisymmetric $n \times n$ matrices. It is easy to prove that space has dimension $n(n-1)/2$.

Let us now pass to the generic point $g \in O(n)$. As above, consider all differentiable curves $R = R(u)$ satisfying $R(u)^t R(u) = I$ (which is equivalent to $R(u)R(u)^t = I$) and $R(0) = g$, and compute the dimension of the space generated by the tangent vectors at the point g to such curves. By construction that space is the kernel of df_g. From $R(u)^t R(u) = I$ one gets, if $\dot{R}$ denotes the tangent vector at $u = 0$, namely at g,

$$\dot{R}^t g + g^t \dot{R} = 0$$

that is, since $g^t = g^{-1}$,

$$g^{-1}\dot{R} = -(g^{-1}\dot{R})^t .$$

This means that $\dot{R}$ belongs to the kernel of df_g if and only if

$$\dot{R} = gA$$

where $A = g^{-1}\dot{R}$ is real antisymmetric because of $\dot{R}^t g + g^t \dot{R} = 0$. Thus A belongs to the kernel of df_I. The map generated by g is, by construction, P_{g*} and hence is an isomorphism of vector spaces, being the differential of the diffeomorphism $P_g : GL(n, \mathbb{R}) \to GL(n, \mathbb{R})$. Therefore the kernel of df_g has the same dimension as that of df_I, and this proves that we are in the hypotheses of the regular-value theorem: $O(n)$ is an embedded submanifold of $GL(n, \mathbb{R})$.

(2) In fact the proof has already been given above, since the tangent vectors at I in $O(n)$ are obtained as the tangent vectors at $u = 0$ of curves $R = R(u)$ satisfying $R(u)R(u)^t = I$ and $R(0) = I$. We have seen above that space coincides with the space of real antisymmetric $n \times n$ matrices, which has dimension $n(n-1)/2$.

(3) It is enough to prove that $O(n)$ is a closed and bounded subset of $\mathbb{R}^{n^2}$ because, in $\mathbb{R}^k$, closed and bounded sets are compact. As for closedness, it is clear that $O(n)$ contains all its accumulation points: if $A_k \in O(n)$ and $A_k \to A \in \mathbb{R}^{n^2}$ as $k \to \infty$, then clearly $A_k^t \to A^t$ and $I = A_k A_k^t \to AA^t$. Hence $A \in O(n)$. Boundedness in the norm of $\mathbb{R}^{n^2}$ is obvious. Indeed, if $R \in O(n)$,

$$||R||^2 = \sum_{i=1}^{n} \left(\sum_{j=1}^{n} R_{ij} R_{ij} \right) = \sum_{i=1}^{n} \delta_{ii} = n \, .$$

□

Theorem A.23 (Topological Structure of $O(3)$ with Respect to Connectedness) *The matrix Lie group $O(3)$ has two connected components, namely:*

the compact and connected matrix Lie group

$$SO(3) := \{R \in O(3) \mid det\ R = 1\}$$

called the **special group of three-dimensional rotations**, *and the compact set (which is not a subgroup)*

$$PSO(3) := \{PR \in O(3) \mid R \in SO(3)\}$$

where $P := -I$ is the **parity inversion**. ■

Proof Let $R \in O(3)$. Then det $R = \pm 1$ directly from the definition of $O(3)$. It is immediate to prove that the subset of matrices in $O(3)$ with determinant 1 forms a subgroup of $O(3)$, which we denote by $SO(3)$. Let us denote by $PSO(3)$ the set containing the matrices with determinant -1, namely $O(3) \setminus SO(3)$. It is clear that these two sets are disjoint and that their union is $O(3)$. Define $P := -I$; then $P \in PSO(3)$. If $R \in O(3)$, it can only happen that $R \in SO(3)$ or $R \in PSO(3)$. Since $PP = I$, it follows that $PR \in PSO(3)$ if and only if respectively $R \in SO(3)$. Consequently, $PSO(3) = \{PR \in O(3) \mid R \in SO(3)\}$.

We now show that $SO(3)$ is open, connected, and compact. (The same properties hold also for $PSO(3)$ because the action of P is a diffeomorphism. However $I \notin PSO(3)$, so the latter is not a subgroup.)

Consider $R \in SO(3)$. Since R is real and the associated characteristic polynomial has degree three, by the fundamental theorem of algebra it follows that R must have at least one real eigenvalue λ, corresponding to a real root of that polynomial (the other two are either real or complex conjugates). The matrix R may always be viewed as a unitary matrix on $\mathbb{C}^3$; in that case it is well known that (1) it is diagonalizable and (2) its eigenvalues are of the form $e^{i\alpha}$ for some $\alpha \in \mathbb{R}$. Summing up: one of the three eigenvalues, say λ_1, must be real with modulus 1, and the other two must also be either real or complex conjugates, and in any case all have modulus 1. Therefore one can only have $\lambda_1 = +1$ or $\lambda_1 = -1$. In the second case, in order to have determinant 1, that is, the product of the eigenvalues equal to 1, the other two eigenvalues must also be real; moreover, being of the form $e^{i\alpha}$, at least one of the two must be 1. In conclusion: *every* $R \in SO(3)$ *always admits the eigenvalue* 1.

Let $\mathbf{e}_1$ be a normalized eigenvector associated with the eigenvalue 1. Complete $\mathbf{e}_1$ to an orthonormal basis of $\mathbb{R}^3$: $\mathbf{e}_1, \mathbf{e}_2, \mathbf{e}_3$. The transformation from the canonical basis to the obtained one is represented by a matrix L in $O(3)$, because both bases are orthonormal. Therefore there exists $L \in O(3)$ such that

$$LRL^t = R' := \begin{bmatrix} 1 & 0 & 0 \\ e & a & b \\ f & c & d \end{bmatrix} . \tag{A.4}$$

The matrix on the right-hand side necessarily belongs to $O(3)$ because it is a product of matrices in $O(3)$. Binet's rule shows that $R \in SO(3)$ implies $R' \in SO(3)$. Imposing the condition of belonging to $SO(3)$, one immediately sees that

$$R' = \begin{bmatrix} 1 & 0 & 0 \\ 0 & a & b \\ 0 & c & d \end{bmatrix} . \tag{A.5}$$

where $a^2 + b^2 = c^2 + d^2 = 1$, $ac + bd = 0$, and $ad - bc = 1$. One concludes easily that the matrix R' must have the form

$$R_\theta := \begin{bmatrix} 1 & 0 & 0 \\ 0 & \cos\theta & -\sin\theta \\ 0 & \sin\theta & \cos\theta \end{bmatrix} . \tag{A.6}$$

where $\theta \in \mathbb{R}$. We have therefore proved that for every $R \in SO(3)$ there exists $L \in O(3)$ such that

$$R = L^t R_{\theta_R} L ,$$

for some $\theta_R \in \mathbb{R}$. If we let θ vary from 0 to θ_R, we obtain a continuous curve $\theta \mapsto L^t R_\theta L$ joining I to R. Hence $SO(3)$ is path connected. Consequently it is connected. Observe that $SO(3)$ is also an open subset (in the relative topology) of $O(3)$, being the intersection of $O(3)$ with the inverse image of $(0, +\infty)$ of the *continuous* determinant function on $GL(n, \mathbb{R})$.

As already said, since $R \mapsto PR$ is a diffeomorphism of $O(3)$, it maps open connected sets into open connected sets. Therefore, since $SO(3)$ is open and connected, $PSO(3)$ is open and connected as well. Since the two sets are connected, disjoint, and their union is $O(3)$, they are the connected components of $O(3)$.

By construction, since $SO(3)$ is a connected component of $O(3)$, it is also trivially an embedded submanifold of $O(3)$ in the induced differentiable structure; hence $SO(3)$ is a Lie subgroup of $O(3)$ and therefore a matrix Lie group. Since connected components are both open and closed, $SO(3)$ is closed in the compact set $O(3)$, hence it is compact. Since $I \in SO(3)$, it also follows that the Lie algebra of $SO(3)$ coincides with that of $O(3)$. □

Corollary *One has $so(3) = o(3)$, where $so(3) := T_1 SO(3)$.* ■

Proof $PSO(3)$ is not a subgroup of $O(3)$ because it does not contain the identity. Only the connected component $SO(3)$ contains the identity, and this immediately implies that

$$T_1 SO(3) = T_1 O(3) .$$

□

A.4 Representation Theorem for $O(3)$ and $SO(3)$

Introduce a particular basis of $so(3)$ given by the matrices

$$(S_i)_{jk} = -\epsilon_{ijk} ,$$

where ϵ is the usual Levi-Civita symbol, explicitly

$$S_1 := \begin{bmatrix} 0 & 0 & 0 \\ 0 & 0 & -1 \\ 0 & 1 & 0 \end{bmatrix} , \tag{A.7}$$

$$S_2 := \begin{bmatrix} 0 & 0 & 1 \\ 0 & 0 & 0 \\ -1 & 0 & 0 \end{bmatrix} , \tag{A.8}$$

$$S_3 := \begin{bmatrix} 0 & -1 & 0 \\ 1 & 0 & 0 \\ 0 & 0 & 0 \end{bmatrix} , \tag{A.9}$$

These matrices are antisymmetric and therefore belong to $so(3)$; moreover, it is immediate to prove that they are linearly independent, so they form a basis of $so(3)$. The structure constants assume a simple form in this basis, as is proved by direct computation:

$$[S_i, S_j] = \epsilon_{ij}{}^k S_k \,, \tag{A.10}$$

where $\epsilon_{ij}{}^k = \epsilon_{ijk}$ are the components of the usual Levi-Civita symbol.

We finally wish to prove a representation theorem for the matrices of $SO(3)$. We need some preliminary result.

Lemma A.24 *For every $R \in SO(3)$, the identity*

$$RS_i R^t = \sum_{j=1}^{3} R^t_{ij} S_j$$

holds. ■

Proof Start from the well-known relation

$$\epsilon^{ijk} R_{pi} R_{qj} R_{rk} = (\det R)\epsilon_{pqr} \,,$$

where the indices have been raised and lowered with the metric δ_{ij} and the matrix R is arbitrary. If $R \in SO(3)$, the relation above becomes, changing the signs on both sides,

$$-\epsilon^{ijk} R_{pi} R_{qj} R_{rk} = -\epsilon_{pqr}$$

that is,

$$\sum_i R_{pi} (RS_i R^t)_{qr} = (S_p)_{qr}$$

and therefore, multiplying by R_{pj} and summing over p,

$$\sum_i \delta_{ij} (RS_i R^t)_{qr} = \sum_p R_{pj} (S_p)_{qr}$$

which can be rewritten as

$$(RS_j R^t) = \sum_p R^t_{jp} (S_p) \,.$$

□

The lemma just proved allows one to conclude with a representation theorem for $SO(3)$.

Theorem A.25 (Representation of $SO(3)$) *One has $R \in SO(3)$ if and only if there exist a unit vector $n \in \mathbb{S}^2$ and a number $\theta \in \mathbb{R}$ such that*

$$R = e^{\theta n \cdot S}$$

where

$$n \cdot S := \sum_{i=1}^{3} n^i S_i$$

and the matrices $S_i \in T_1 SO(3)$ are given by

$$(S_i)_{jk} = -\epsilon_{ijk} .$$

■

Proof As proved in the proof of Theorem A.23, if $R \in SO(3)$ then there exists $L \in O(3)$ such that $R = L^t R_\theta L$, where R_θ is given in (A.6). We may always replace, if necessary, L by $PL = -L$ in order to have $L \in SO(3)$ without altering the identity above. It is easy to prove that

$$R_\theta = e^{\theta S_1} .$$

Indeed, differentiating the first member with respect to θ through (A.6), one readily finds

$$\frac{dR_\theta}{d\theta} = S_1 R_\theta .$$

On the other hand one also has

$$\frac{de^{\theta S_1}}{d\theta} = S_1 e^{\theta S_1} .$$

Therefore the two functions satisfy the same differential equation and also the same initial condition:

$$R_0 = e^{0 S_1} = I ,$$

hence they coincide. It follows that, using the notation $n \cdot S := \sum_i n^i S_i$,

$$R = L^t e^{\theta S_1} L = L^t e^{\theta \mathbf{e}_1 \cdot S} L .$$

Using the power-series expansion of the exponential, one has

$$L^t e^{\theta \mathbf{e}_1 \cdot S} L = L^t (I + \theta \mathbf{e}_1 \cdot S + \frac{1}{2!} (\theta \mathbf{e}_1 \cdot S)^2 + \cdots) L = I + \theta \mathbf{e}_1 \cdot L^t S L + \frac{1}{2!} (\theta \mathbf{e}_1 \cdot L^t S L)^2 + \cdots$$

so that

$$L^t e^{\theta \mathbf{e}_1 \cdot S} L = e^{\theta \mathbf{e}_1 \cdot L^t S L} .$$

Using Lemma A.24 with obvious notation,

$$\mathbf{e}_1 \cdot L^t S L = (L\mathbf{e}_1) \cdot S .$$

In conclusion, taking into account that n is a unit vector of $\mathbb{S}^2$ if and only if $n = A\mathbf{e}_1$ for some $A \in SO(3)$, we have proved that, if $R \in SO(3)$, then for some $\theta \in \mathbb{R}$ and $n \in \mathbb{S}^2$,

$$R = e^{\theta n \cdot S} .$$

Conversely, if we choose $n \in \mathbb{S}^2$ and $\theta \in \mathbb{R}$ and define

$$R := e^{\theta n \cdot S} ,$$

then, recalling that the matrices S_i are antisymmetric,

$$RR^t = e^{\theta n \cdot S} (e^{\theta n \cdot S})^t = e^{\theta n \cdot S} e^{\theta n \cdot S^t} = e^{\theta n \cdot S} e^{-\theta n \cdot S} = e^{(\theta - \theta) n \cdot S} = I .$$

Hence $R \in O(3)$. On the other hand

$$\det R = e^{\theta \operatorname{tr}(n \cdot S)} = e^0 = 1 ,$$

so that $R \in SO(3)$. □

We have obviously the following corollary.

Corollary *One has $R \in PSO(3)$ if and only if there exist a unit vector $n \in \mathbb{S}^2$ and a number $\theta \in \mathbb{R}$ such that*

$$R = -e^{\theta n \cdot S} .$$

■

Proof Since $PP = I$, one has $R \in PSO(3)$ if and only if $PR \in SO(3)$, whence the thesis immediately follows, since $P = -I$. □

Remarks A.26

(1) One may ask whether, for a given $R \in SO(3)$, the pair θ, n representing R through Theorem A.25 is unique. Clearly, if $\theta \in \mathbb{R}$, the answer is negative. If we restrict the range of θ to $[0, 2\pi]$, the answer is still negative because a rotation by θ around the unit vector n is the same as a rotation by $2\pi - \theta$ around the unit vector $-n$ (one must associate rotations with unit vectors according to a fixed orientation, for instance the "right-hand rule"). We may then choose the range of θ in $[0, \pi]$. In that case the only ambiguities are the description of I, corresponding to $\theta = 0$ and arbitrary $n \in \mathbb{S}^2$, and the rotations by π around n and $-n$, which are actually the same rotation.

(2) The discussion above yields a more precise description of the topology of $SO(3)$. Indeed, the points of $SO(3)$ may be parameterized by the vectors $\theta n \in \mathbb{R}^3$ with $\theta \in [0, \pi]$ and $n \in \mathbb{S}^2$. In this way the points of $SO(3)$ correspond to the points of a closed ball in $\mathbb{R}^3$ of radius π, with the rule that, for every diameter, the antipodal points *on the boundary* are identified. Those identified points correspond to the rotations by π around $\pm n$, which are, as observed above, the same rotation. This description shows that the resulting topological space is homeomorphic to the real projective space $\mathbb{R}P^3$, and in particular

$$\pi_1(SO(3)) = \mathbb{Z}_2 .$$

■

Exercises A.27

1. Consider the commutative group of translations on $\mathbb{R}^n$, namely $\mathbb{R}^n$ itself: if $t \in \mathbb{R}^n$,

$$L_t : \mathbb{R}^n \ni x \mapsto x + t \in \mathbb{R}^n .$$

Show that group is a matrix Lie group, a subgroup of $GL(n+1, \mathbb{R})$, with Lie algebra identifiable with $\mathbb{R}^n$ itself.
Hint. Consider $\mathbb{R}^n$ as the affine hyperplane in $\mathbb{R}^{n+1}$ of equation

$$\{(x^1, \ldots, x^{n+1}) \in \mathbb{R}^{n+1} \mid x^{n+1} = 1\} .$$

Then consider the class of matrices

$$\left[\begin{array}{c|c} I & t \\ \hline 0 & 1 \end{array}\right] . \tag{A.11}$$

2. Show that the Galilei group is a matrix Lie group, a subgroup of $GL(5,\mathbb{R})$.
Hint. Consider the class of matrices

$$\left[\begin{array}{c|c} \Omega_{R,V_0} & C \\ \hline 0 & 1 \end{array}\right], \tag{A.12}$$

where $C \in \mathbb{R}^4$ is the column vector with $C^t = (c, X_0)$ and

$$\Omega_{R,V_0} := \left[\begin{array}{c|c} 1 & 0 \\ \hline V_0 & R \end{array}\right], \tag{A.13}$$

where $V_0 \in \mathbb{R}^3$ and $R \in O(3)$.

3. Show that the Lie algebra of the Galilei group is isomorphic to the direct sum $so(3) \oplus \mathbb{R}^7$ as a vector space, but not as a Lie algebra (that is, the commutator does not satisfy

$$[(A,t),(A',t')] = ([A,A'],0)\,,$$

where $A, A' \in so(3)$ and $t, t' \in \mathbb{R}^7$).

Index of Symbols

This list collects the main symbols used throughout the text. The meaning of a symbol may be specialized in a local context; when this happens, the local definition in the text prevails.

General Notation and Linear Algebra

Symbol	Meaning
$\mathbb{K}$	Base field of scalars, usually $\mathbb{R}$ or $\mathbb{C}$.
$\mathbb{R}, \mathbb{C}, \mathbb{N}, \mathbb{Z}$	Real, complex, natural and integer numbers.
V, W, U, H	Vector spaces; H is often used for Hilbert spaces.
V^*	Dual vector space of V, made of linear functionals on V.
$\overline{V}$	Complex conjugate vector space associated with V.
V^{**}	Bidual vector space of V.
$\dim(V)$	Dimension of the vector space V.
$\{e_i\}_{i\in I}$	A basis of a vector space.
$\{e^{*i}\}_{i\in I}$	The dual basis of V^* associated with $\{e_i\}_{i\in I}$.
$\langle v, u^*\rangle$	Canonical pairing between a vector and a covector.
$GL(V)$	General linear group of the vector space V.
$GL(n, \mathbb{K})$	Group of invertible $n \times n$ matrices with entries in $\mathbb{K}$.
Id_V, I	Identity map on V; identity matrix when a basis is fixed.
A^t	Transposed matrix of A.
A^{-1}	Inverse of the invertible matrix or linear map A.
$\det A$	Determinant of the matrix or endomorphism A.
$\oplus$	Direct sum of vector spaces or direct sum decomposition.
$\times_{\alpha\in A} S_\alpha$	Cartesian product of the family of sets $\{S_\alpha\}_{\alpha\in A}$.

V. Moretti, *Geometric Methods in Mathematical Physics I*, La Matematica per il 3+2 184, https://doi.org/10.1007/978-3-032-31512-0

Tensor Products and Abstract Index Notation

Symbol	Meaning
$V_1 \otimes \cdots \otimes V_n$	Tensor product of the vector spaces $V_1, \ldots, V_n$.
$v_1 \otimes \cdots \otimes v_n$	Decomposable tensor determined by $v_1, \ldots, v_n$.
$V^{\otimes p}$	p-fold tensor product of V with itself.
$V^{*\otimes q}$	q-fold tensor product of V^* with itself.
$T^p_q(V)$	Space of tensors of type (p, q) over V.
$\mathscr{A}_{\mathbb{K}}(V)$	Tensor algebra generated by the vector space V.
$t^{i_1\ldots i_p}{}_{j_1\ldots j_q}$	Components of a tensor of type (p, q) in a chosen basis.
δ^i_j	Kronecker delta; components of the identity tensor.
A, B, C	Abstract indices, often collecting strings of tensorial indices.
$\sigma^{\otimes}$	Tensorial action induced by a permutation σ.
$\mathscr{P}_n$	Group of permutations of n objects.
ϵ_σ	Signature of a permutation σ.
$S^p(V)$	Space of symmetric p-tensors over V.
$\Lambda^p(V)$	Space of antisymmetric p-tensors, or p-vectors, over V.
$\Lambda^p(V^*)$	Space of p-forms over V.
$\Lambda(V)$	Exterior, or Grassmann, algebra of V.
$\mathcal{A}$	Antisymmetrization operator.
$\wedge$	Exterior product.
$\tilde{\wedge}$	Alternative normalization of the exterior product.
$\lrcorner$	Interior product, also called contraction.
$\tilde{\lrcorner}$	Interior product associated with $\tilde{\wedge}$.
$\binom{n}{p}$	Binomial coefficient, equal to $\dim \Lambda^p(V)$ when $\dim V = n$.

Scalar Products, Metric Tensors and Densities

Symbol	Meaning
$(\cdot\|\cdot)$, $\langle\cdot,\cdot\rangle$	Scalar product or pseudo-scalar product, according to context.
$\mathbf{g}$, g	Metric or pseudo-metric tensor.
g_{ij}	Covariant components of the metric tensor.
$\tilde{\mathbf{g}}$, $\tilde{g}^{ij}$	Contravariant metric tensor and inverse metric components.
g^{ij}	Components obtained by raising indices with the inverse metric.
$h : V \to V^*$	Natural isomorphism induced by a non-degenerate metric tensor.
$h^{-1} : V^* \to V$	Inverse isomorphism induced by the inverse metric tensor.
$O(m, p)$	Pseudo-orthogonal group of signature (m, p).
$SO(m, p)$	Special pseudo-orthogonal group of signature (m, p).
η	Canonical diagonal matrix of a pseudo-metric; in relativity, $\eta = \mathrm{diag}(-1, +1, +1, +1)$.
$\epsilon_{i_1 \dots i_n}$	Completely antisymmetric Ricci symbol.
$\mathcal{E}_{i_1 \dots i_n}$	Ricci pseudo-tensor associated with an orientation and a metric structure.
$\mathrm{sgn}(\det A)$	Sign of the determinant of a change-of-basis matrix.
$\|\det g\|$	Absolute value of the determinant of the metric matrix.

Groups and Representations

Symbol	Meaning
G	A group; the identity element is usually denoted by e or I.
$H \leq G$	Subgroup H of G.
$H \triangleleft G$	Normal subgroup of G.
G/H	Quotient group when H is normal in G.
$G_1 \times G_2$	Direct product of groups.
$G_1 \rtimes G_2$	Semidirect product of groups.
$D : G \to GL(V)$	Linear representation of G on V.
$D_1 \otimes D_2$	Tensor product representation.
$\ker D$	Kernel of the representation or homomorphism D.
$\mathrm{Im}\, D$	Image of the representation or homomorphism D.
$\mathfrak{G}$	Lie algebra of a Lie group G, when this notation is used.
$[X, Y]$	Lie bracket or matrix commutator.
$\exp X$	Exponential map, or matrix exponential in the matrix-group case.
$O(3)$, $SO(3)$	Orthogonal and special orthogonal groups in three dimensions.
$so(3)$	Lie algebra of $SO(3)$.

Special Relativity

Symbol	Meaning
$\mathbb{M}^4$	Four-dimensional Minkowski spacetime.
T^4	Vector space of translations of Minkowski spacetime.
$T_p\mathbb{M}^4$	Tangent vector space at the event p.
$T_p^*\mathbb{M}^4$	Cotangent vector space at the event p.
x^0, x^1, x^2, x^3	Minkowskian coordinates.
∂_{x^a}	Coordinate basis vector associated with x^a.
dx^a	Dual basis covector associated with x^a.
$\mathbf{e}_0, \mathbf{e}_1, \mathbf{e}_2, \mathbf{e}_3$	Pseudo-orthonormal basis of T^4.
c	Speed of light.
$\overline{V_p}$, V_p, ∂V_p	Closed light cone, open light cone and light cone at p.
$\gamma(\tau)$	Worldline parametrized by proper time τ.
$\dot{\gamma}(\tau)$	Four-velocity of a material point.
$P = m\dot{\gamma}$	Four-momentum of a material point of mass m.
P^0	Time component of the four-momentum in a Minkowskian frame.
$E_{\mathscr{R}} = cP^0_{\mathscr{R}}$	Mechanical energy in the reference system $\mathscr{R}$.
$K_{\mathscr{R}}$	Relativistic kinetic energy in the reference system $\mathscr{R}$.
F	Four-force.
T^{ab}	Stress-energy tensor components.
$O(1, 3)$	Lorentz group.
$SO(1, 3)$	Proper Lorentz group.
$O(1, 3)\uparrow$	Orthochronous Lorentz group.
$SO(1, 3)\uparrow$	Proper orthochronous Lorentz group.
$IO(1, 3)$	Inhomogeneous Lorentz group.
$ISO(1, 3)\uparrow$	Proper orthochronous Poincaré group.
Λ	Lorentz transformation matrix.
(Λ, T)	Poincaré transformation, with Lorentz part Λ and translation part T.
Ω_R	Lorentz transformation induced by a spatial rotation R.
K_1, K_2, K_3	Boost generators in the Lorentz Lie algebra.
S_1, S_2, S_3	Spatial rotation generators in the Lorentz Lie algebra.
$\mathbf{K}, \mathbf{S}$	Formal vectors collecting the boost and rotation generators.

$SL(2, \mathbb{C})$, Pauli Matrices and Spinors

Symbol	Meaning
$SL(2, \mathbb{C})$	Complex 2×2 matrices with determinant 1.
$SU(2)$	Unitary matrices in $SL(2, \mathbb{C})$.
$sl(2, \mathbb{C})$, $su(2)$	Lie algebras of $SL(2, \mathbb{C})$ and $SU(2)$.
$\sigma_1, \sigma_2, \sigma_3$	Pauli matrices.
$\boldsymbol{\sigma}$	Formal vector whose components are the Pauli matrices.
σ_μ, σ'_μ	Four matrices used to identify four-vectors with Hermitian 2×2 matrices.
$H(t)$	Hermitian matrix associated with the four-vector $t \in \mathbb{R}^4$.
$\Pi : SL(2, \mathbb{C}) \to SO(1, 3)\uparrow$	Covering homomorphism from $SL(2, \mathbb{C})$ to the proper orthochronous Lorentz group.
Λ_L	Lorentz transformation associated with $L \in SL(2, \mathbb{C})$ by the homomorphism Π.
W	Two-dimensional complex Weyl spinor space.
$\overline{W}$	Conjugate Weyl spinor space.
ϵ_{AB}	Metric spinor on W.
ϵ^{AB}	Inverse metric spinor used to raise spinor indices.
A, B, C	Unprimed spinor indices.
A', B', C'	Primed spinor indices referring to conjugate spinor spaces.
ξ^A, η_A	Spinor and cospinor components in abstract-index notation.
$t^{AA'}$	Spinorial representative of a four-vector.
$D = W \oplus \overline{W^*}$	Dirac spinor space.
$\Psi = \phi^A \oplus \psi_{A'}$	Decomposition of a Dirac spinor into left-handed and right-handed Weyl components.
Γ	Chirality operator on D.
Π_L, Π_R	Projectors onto the left-handed and right-handed Weyl sectors.
$\gamma(t)$	Clifford action of the four-vector t on the Dirac spinor space.
γ_μ, γ^μ	Dirac gamma matrices with lowered and raised spacetime indices.
I_D	Identity operator on the Dirac spinor space D.
$\overline{\Psi}_D$	Dirac adjoint of the Dirac spinor Ψ.
∂_μ	Coordinate partial derivative $\partial/\partial x^\mu$.
$i\gamma^\mu\partial_\mu - m$	Free Dirac operator of mass m.
$\eta^{\mu\nu}\partial_\mu\partial_\nu - m^2$	Klein-Gordon operator associated with the adopted Minkowski metric.

Bibliography

1. Kirillov, A.: Éléments de la Théorie de représentations. Mir, Mosca (1974); Najmark, M.A., Stern, A.I.: Theory of Group Representations. Mir, Moscow (1984)
2. Rudin, W.: Real and Complex Analysis. McGraw-Hill (1987)
3. Warner, F.W.: Foundations of Differentiable Manifolds and Lie Groups. Springer, New York (1983)

Index

V. Moretti, *Geometric Methods in Mathematical Physics I*, La Matematica per il 3+2 184, https://doi.org/10.1007/978-3-032-31512-0

MIX
Papier aus verantwortungsvollen Quellen
Paper from responsible sources
FSC® C105338

If you have any concerns about our products,
you can contact us on
ProductSafety@springernature.com

In case Publisher is established outside the EU,
the EU authorized representative is:
Springer Nature Customer Service Center GmbH
Europaplatz 3, 69115 Heidelberg, Germany

Printed by Libri Plureos GmbH
in Hamburg, Germany